A Quick Course in Statistical Process Control

A Quick Course in Statistical Process Control

MICK NORTON
College of Charleston

NETEFFECT SERIES

PEARSON
Prentice Hall

Upper Saddle River, New Jersey
Columbus, Ohio

Library of Congress Cataloging-in-Publication Data

Norton, Mick.
A quick course in statistical process control / Mick Norton.
p. cm.
Includes bibliographical references and index.
ISBN 0-13-093062-8
1. Quality control—Statistical methods. I. Title.
TS156.N65 2005
658.5'62—dc22

2004005801

Executive Editor: Debbie Yarnell
Production Editor: Louise N. Sette
Production Supervision: Carlisle Publishers' Services
Design Coordinator: Diane Ernsberger
Cover Designer: Kristi Holmes
Production Manager: Deidra Schwartz
Marketing Manager: Jimmy Stephens

This book was set in Goudy WtcTReg by Carlisle Communications, Ltd. It was printed and bound by R. R. Donnelley & Sons Company. The cover was printed by The Lehigh Press, Inc.

Pearson Education Ltd.
Pearson Education Singapore Pte. Ltd.
Pearson Education Canada, Ltd.
Pearson Education—Japan
Pearson Education Australia Pty. Limited
Pearson Education North Asia Ltd.
Pearson Educación de Mexico, S.A. de. C.V.
Pearson Education Malaysia Pte. Ltd.

PEARSON
Prentice Hall

10 9 8 7 6 5 4 3 2 1
ISBN 0-13-093062-8

Preface

This book is aimed at those who come in contact in any way with statistical process control (SPC) or people who simply want an introduction to the subject. The first four chapters examine why control charts work, how to update charts, how to interpret what the charts are saying, how to set up a chart, and how to assess the capability of a process being charted. Besides being of interest to those engaged in any of these specific activities, these chapters also should give decision makers some ideas about what kinds of variables they can track or control. Just enough probability is introduced, lightly, for the reader to understand how infrequently out-of-control signals may be expected to occur when a process is in control and why they will occur much more frequently when it isn't. The book introduces statistics from scratch, making no assumption that the reader has been exposed to fundamental statistical ideas or to statistical measures.

Chapter 5 is devoted to a deeper understanding of probability. The first section examines basic rules used for computing probabilities and conditional probabilities and shows how probability may be used to answer questions in business, industry, and other areas. The second section details some of the probability distributions used most often to answer questions of interest in the quality arena. Along with being able to apply these distributions in a variety of settings, the reader will gain further expertise with and a deeper understanding of control charts. The third section is devoted to the topic of reliability of a product or system. Frequently used reliability measures are introduced—for example, the reliability function, failure rate, and mean time to failure.

Using statistics and probability properly, while essential in making good process decisions, is not the whole of the quality picture. A company can optimize how it uses statistics and still fail because of its management process, how it treats customers or employees, or how it solves problems.

Chapter 6 is devoted to *thinking* quality. Topics include the management philosophies of Deming and the thinking of other quality visionaries such as Juran and Ishikawa. Additionally, there is an overview of total quality management and of the Six Sigma movement. Problem-solving tools are introduced—for example, Pareto charts, cause-and-effect diagrams, scatter diagrams, flowcharts, and methods a group can use when it can't reach consensus on choosing one option from among three or more.

This book grew out of two sources—a course I designed and taught while consulting with a manufacturing company, and a course I designed and taught for the MS Program in Mathematics at the College of Charleston, where I have "professed" mathematics and statistics for many years. This book has a definite "how to" point of view. I repeat. This is a practical book.

Portions of the input and output contained in this book are printed with the permission of Minitab Inc. MINITAB® and the MINITAB logo® are registered trademarks of Minitab Inc. The reader who has access to Excel or MINITAB will have some opportunities to use them, and even instructions for how to do so. However, no software is required to understand this book. Occasional exercises that are included for those who have such software may be safely skipped.

I wish to thank some good people. A number of years ago, Jim Lynch invited me to sit in on an SPC short course when I was on sabbatical at the University of South Carolina. Seeing how such a course could work has benefited my teaching in many ways. Several years later, I was able to spend a sabbatical year working on SPC issues with a manufacturing company. Solving mathematical and statistical problems that help a company make a product better gives a satisfaction that is hard to top and, well, is just plain fun. Gordon Jones, then-Dean of the School of Sciences and Mathematics at the College of Charleston, liked the idea of such a sabbatical and approved it. Way to go, Gordon. I have consulted with some wonderful people on projects for various companies, but particularly want to acknowledge Gregg Adams, Neal Tonks, and Ben Bruner. Also, thanks are due to Debbie Yarnell at Prentice Hall and the following reviewers for suggesting additional topics and ideas that would improve the scope of the book: J.K. Crain, Texas A&M University—Commerce; Grace Duffy, Trident Technical College; Mark Durivage, Owens Community College; Samuel H. Huang, University of Cincinnati; Young J. Kim, Mississippi State University; Kellie Knox, Southwest Wisconsin Technical College; Jooh Lee, Rowan University; George Pillinayagam, Lorain County Community College; and Carl Wargula, Gateway Community College.

My son Andrew didn't squawk much when I wanted him to quit running over pedestrians and running into other vehicles so that I could use the computer. My daughter Susan tended to come in the room whenever I happened to be typing a personal or family anecdote for a homework problem that asks the reader to identify a principle of TQM or one of Deming's

14 points that is being violated. I would ask her to read the problem and whether she remembered the outrageous incident. She would always read the problem, say she remembered, then invariably tell me that the purpose of her visit was to let me know that she was on her way out the door to do this or that. Thanks for your patience, Susan. We're glad you're you. Last but definitely not least is Libby, who has given me much encouragement toward working with industry and with this book. Additionally, when you're writing a book and the computer gets temperamental, like when you get the idea to paste all of the sections into one big file called *The Whole Enchilada* and the computer chokes, it is particularly helpful to have a wife who is a computer programmer. Really.

RMN

Contents

A Quick Course in Statistical Process Control

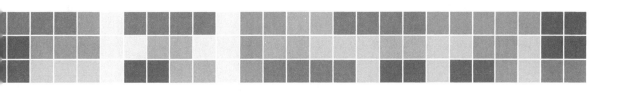

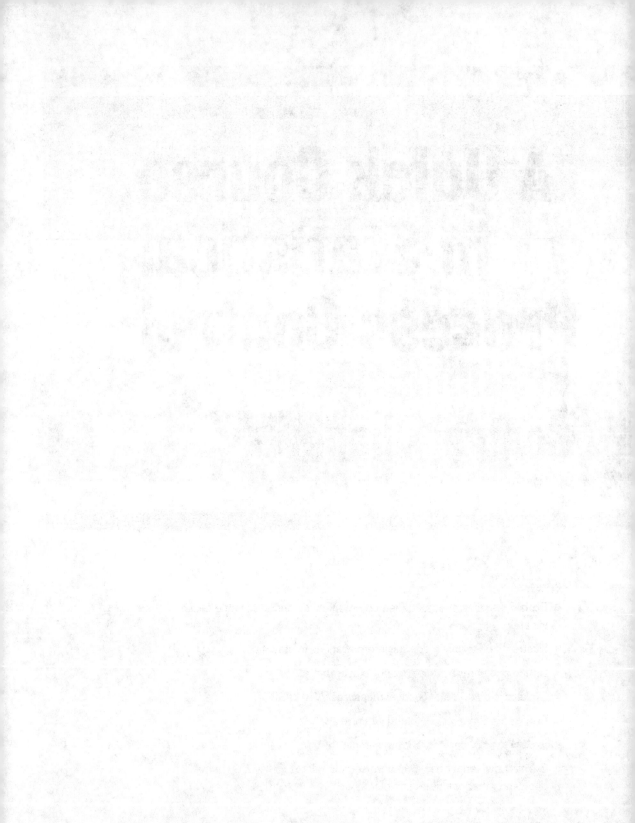

1

Statistical Preliminaries for Control Charts

OBJECTIVES

- Explain the difference between common cause and assignable cause variation
- Illustrate commonly made misinterpretations in statistics
- Define and show how to recognize overcontrol
- Explain fundamental statistical measures used in SPC
- Explain the Empirical Rule and when to apply it
- Explain the concept of random sample
- Explain how variation in a random sample relates to variation in the population that has been sampled

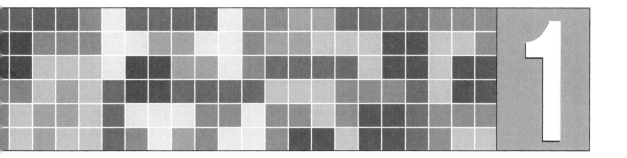

1.1 ADOPTING A STATISTICAL POINT OF VIEW

> ***Statistical Process Control (SPC):*** *The use of statistical methods to control/improve a process.*

The general goal of SPC is to produce better goods and services. People are better off when the products they use last a long time and function reliably. Also, people who are customers—and that's everyone—are consistently happier when service providers know their stuff, treat them in a friendly, professional manner, and make them willing to be long-term customers. When data can be collected to measure the quality of a manufactured product or of a service, the potential is there to use the data to tell if quality is slipping or holding steady, and whether efforts to improve quality are working. The whole philosophy of SPC is to use data to continually improve quality. One of the twentieth-century giants of statistical process control, W. Edwards Deming, once said "In God we trust. All others must bring data." Of course, having data is not the same as understanding what the data reveal. This is where statistical methods come in. The basic statistical tools of SPC make a number of things possible.

One can *monitor a process to determine if it is working to the best of its capability*. Any system will degrade without proper oversight and care. Quality may be slipping or holding steady, and the tools of SPC, such as control charts, can determine which.

Generally, quality does not improve by accident. Improving the quality of manufactured goods or of services means making changes in the system. ***Control charts*** can be used to *determine if intentional changes in the process are having the desired effect*. For example, has the new drill press reduced the proportion of drilled holes that don't meet specifications? The fact that there may be fewer ***out-of-spec*** holes produced the day after the new drill press was put to use than the day before doesn't make a statistical case that the process has improved. *Too many people want to take credit because today's numbers are better than yesterday's, or worse, blame someone because today's numbers are not as good as yesterday's.* Even the same drill press would be expected to produce a different number of out-of-spec holes from day to day. The same customer service representative will not generate exactly the same satisfaction level from one customer to the next. Of course, it is also possible that today's number is so much better than yesterday's that the difference is beyond reasonable chance. But, how big of a difference is required to be sure that the positive effect is beyond reasonable change? It is also possible that the combined weight of several days-worth of numbers, no one of which is compelling by itself, can indicate improvement beyond reasonable chance. Control charts can signal when either level of improvement has occurred.

The previous point indicates that understanding variation in numbers is important. Control charts and summary statistical measures we will discuss, such as the range and standard deviation, can be used to *detect when there are unwanted sources of variation in a process.* No two items made are perfectly alike. In any production process there is always variation from item to item, batch to batch, at different points in a continuous flow, and so on. Some amount of variation is impossible to avoid. Still, in business and industry, variation is the enemy. Whether there is *needlessly high* variation from product unit to product unit is a different question than whether data will show that the average product unit is on target. A manufacturer wants to reduce unit-to-unit variation so that customers will come to appreciate the *consistency* of the product. Monitoring variation is important so that it may be kept to a minimum. Control charts and statistical measures are tools used in this effort.

Additionally, SPC has methods for measuring how well a process is performing—that is, measuring what is called the ***capability of the process.*** It is natural to ask how capable a process was this week, this month, during a particular product run, and so on. By examining the data from different time periods one can determine if the process is more or less capable, currently, than it used to be. An actual statistical measure of process capability is called a ***process capability index.*** It is a very common practice for Company A to ask to see Company B's capability indices for a particular time period. The purpose is to help Company A decide whether it should purchase goods made during that period by Company B. The corollary is that to be competitive, Company B needs to be collecting the kind of show-and-tell data that produce control charts and capability indices. Needless to say, companies that want access to the largest markets are companies that make continuous efforts to improve process capability.

Uses of SPC, such as those just discussed, are tied to two related objectives:

- continuous improvement in product and process, and
- increasing product and process consistency by reducing variation.

Many companies prominently display posters with slogans such as "Continuous Improvement," "Zero Defects," "Get It Right the First Time," or "Work Smarter, Not Harder." But these slogans are useless if the posters are seen but don't reflect how employees think or act. Positive things happen when continuous improvement is a workforce state of mind with action consequences. This equates *continuous improvement* with *never being satisfied with how things are.*

Merely having production meet product specifications should not be viewed as good enough. If the average product measurement is off target this week, what can be done to bring the process closer to target next week? What process changes can be made to reduce variation so that the product will be more consistent? Remember: Variation is the enemy. The less variation there is, the less rework there will be, the less off-spec production there will be, and the less scrap there will be. Can standard operating procedures be changed in order to speed up the change-out of parts? To reduce downtime? To reduce the frequency of laboratory tests? To reduce the incidence of mislabeled shipments? or to reduce the drain on workers caused by morale busters? Questions like these reflect the thinking process that accompanies the tools and practices of SPC. This state of mind has the best chance of prevailing in the workplace if employees see that management has acquired it first and "walks the talk." This type of thinking gets at the "soft," or philosophical, side of SPC, which is introduced in Chapter 6.

So where does statistics fit into the picture? The answer is that understanding and controlling variation is what statistics is all about. We will use tools such as the *standard deviation* and *range* to measure the amount of variation present in a process. This will enable us to set up control charts and then use them to determine if variation is stable or changing—for the better or the worse.

Variation can be viewed as having one of two causes:

Common cause variation is variation inherent in the process. This variation is natural in a process that is working as it was designed to work.

Assignable cause variation (aka *special cause variation*) is variation attributable to a source present that makes the process not work as designed.

Even when a process is functioning at its highest capability, there will be variation. The resistance in consecutively produced insulators will be different if measured to enough decimal places, two ball bearings will have slightly different diameters, two customers who order the same thing will wait in line different amounts of time at a fast food restaurant, a clerk will occasionally misplace an order or enter the wrong number on a customer order form, and so on. The variation in a system that is working up to its design capability is *common cause variation*. Some level of variation has to be tolerated in any system—not wanted, just tolerated. Further, there should be a never-ending effort to reduce this kind of variation. But one can never expect to completely eliminate variation. Even the best of people sometime make mistakes. No system is perfect.

One of the key purposes of control charts is to signal when *assignable cause variation* may be present. In order to set up a chart that can do this, one first must measure the level of common cause variation in the process. This brings us to the central idea of why control charts work:

Knowing the level of common cause variation in a process makes it possible to define limits that, when exceeded, suggest that one should investigate whether an assignable cause is present.

How Not to Think—An Example

The following data are taken from an AP article that appeared in the Charleston, South Carolina *Post & Courier* in June of 1991 (see the article "Senate Panel Cites 'Plague' of Rural Criminal Violence"). A U.S. senator who will go unnamed proposed spending $100,000,000 to investigate a "plague" of crime in the rural United States. One of a number of faulty "statistical" justifications given in the article was the following comparison of the increase in crime in Montana, an example of a rural (that is, low population density) state, to Los Angeles, an example of a high population density region:

Murders	1989	1990	Increase
Montana	23	30	30%
Los Angeles	877	983	12%

Reprinted with permission of The Associated Press.

The idea here is that when the murder count in Montana increased by 7 from one year to the next, that amounted to a 30% increase, whereas during the same time frame, the count in Los Angeles increased by 106, just a 12% increase. What the article attempts to convey is that a 30% increase is more alarming than a 12% increase, which supports the idea that a plague of crime exists in the rural United States.

There are several problems with this reasoning. One is that one year's number is being compared to the previous year's number. Depending on the numbers, a statistician might be able to make a compelling argument about a Montana crime problem if murder counts had risen systematically over a succession of years rather than just over two years. In the absence of more than two year's worth of data, a case still could be made with just two year's worth of data, as we shall see, but it would not be made on the basis of percent increase. The problem with using percent increase is that when working with small numbers, small changes can translate into big percent increases. An increase of 7 murders amounts to a considerable percentage increase because the reference point is 23 murders. By the article's reasoning, an even stronger case could be made for a rural state that had one murder in one year followed by two the next. The additional one murder represents a whopping 100% increase. In the meantime, the additional 106 murders in Los Angeles represents only a 12% increase because the reference point is 877. The issue is whether a murder increase of seven is unusual. To decide this, we must first understand how much fluctuation is typical from one year to the next. Once the amount of typical fluctuation is assessed, it will be possible to define a limit that, when an annual murder count exceeds

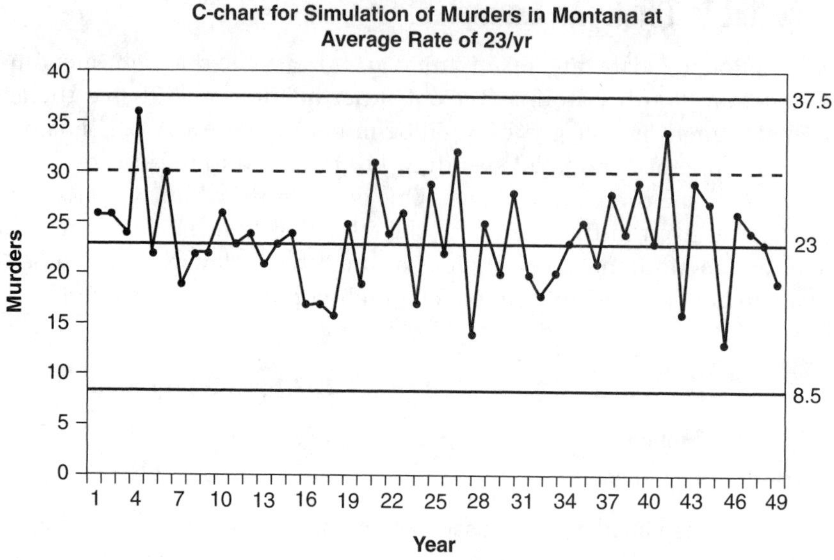

FIGURE 1.1 *C*-Chart for Simulation of Murders in Montana at Average Rate of 23 per Year.

it, suggests that one should investigate whether an assignable cause is present. We can illustrate this concept by looking at our first example of a control chart.

Figure 1.1 is an Excel chart that represents how 50 consecutive years of murder counts in Montana might graph under the assumption that the expected number of murders in any one year is 23. Many statistical computer packages can generate, that is, *simulate,* such data based on probability models. The numbers produced by these models, the number of events of a given kind that occur in each of a succession of time periods—in our case, the number of murders each year for 50 years—will closely mimic the kinds of numbers that nature produces. To generate such data, all one needs to know is the expected event rate per time period. In our case, this is the annual murder rate. Since no one was alarmed in the year when 23 murders occurred, we take the annual murder rate to be 23, and look at the simulated data to help determine if 30 or more murders in a year is unusual. The important thing here is that Figure 1.1 shows the kind of year-to-year variation expected from a system that is perfectly stable over the course of many years.

Figure 1.1 is an example of a *c*-chart, which we will meet in more detail later. At this point, we simply note that the chart has three horizontal reference lines. One has a *y*-intercept equal to the average annual murder rate, 23. As expected, some years have more than 23, some less, and there appears to be a random mix of points above and below this line.

The other two lines are used to identify any points that are unusually far from 23. The *y*-intercepts of these lines, 37.5 and 8.5, are called ***control limits.***

Later we will see how these limits are determined. For now it suffices simply to understand how the limits might be used in a manufacturing process.

The *c*-chart is commonly used to monitor, over a sequence of time periods, the number of ***nonconforming*** items produced, or items which do not ***conform*** to specifications. Some examples are:

- the number of defective bumpers produced in a day,
- the combined number of blemishes in the paint jobs of a three-car sample (the sample should be considered as representative of all the cars painted on a particular day),
- the number of accidents in a month, and
- the number of flaws in each 1,000-yard sample of copper wire insulation (the 1,000-yard sample should be considered as representative of all the wire produced on a particular 12-hour shift).

When a point occurs that does not fall between the control limits, the event is considered so unusual that the manufacturer would investigate to determine if there is an assignable cause that can explain the unusual count. If an assignable cause is found, any problem must be fixed so as not to affect future counts. If the problem is serious enough, the manufacturing process may need to be shut down, process permitting. Shutdowns are never good for the company, its profits, or its employees.

On most control charts the control limits are determined so that, by chance, only about 3 points in 1,000 would be expected to not fall between the control limits when the process is working up to its capability. As indicated, a process that is working as it was designed to will sometimes produce points falling outside the control limits. But this is rare, occurring roughly three times in one thousand, on average. If no assignable cause can be connected to such a point, it is assumed that the point is one of those "three in one thousand," and the process is allowed to continue.

Note, too, that there is a probability symmetry for the control limits. When the process is working up to its capability, only about 1.5 points in 1,000 would be expected to fall *above* the larger of these (called the ***upper control limit [UCL]***), and about 1.5 would be expected to fall *below* the other (the ***lower control limit [LCL]***).

We now return to Figure 1.1 and the question of whether 30 or more murders in a year is unusual when the expected murder rate remains a constant 23 over a span of years. A reference line, a dashed line with *y*-intercept of 30, is shown in the figure. The murder count is 30 in year 6, and exceeds 30 in four other years. There are also three other points "knocking at the door," with 29 murders in those years. In any case, in one-tenth of the years, the count is 30 or more. In statistics, an event that occurs in one-tenth of repeated experiments is not considered to be unusual. Of course, had we simulated another 50 years of murder counts, the fraction of years in which there are 30 or more

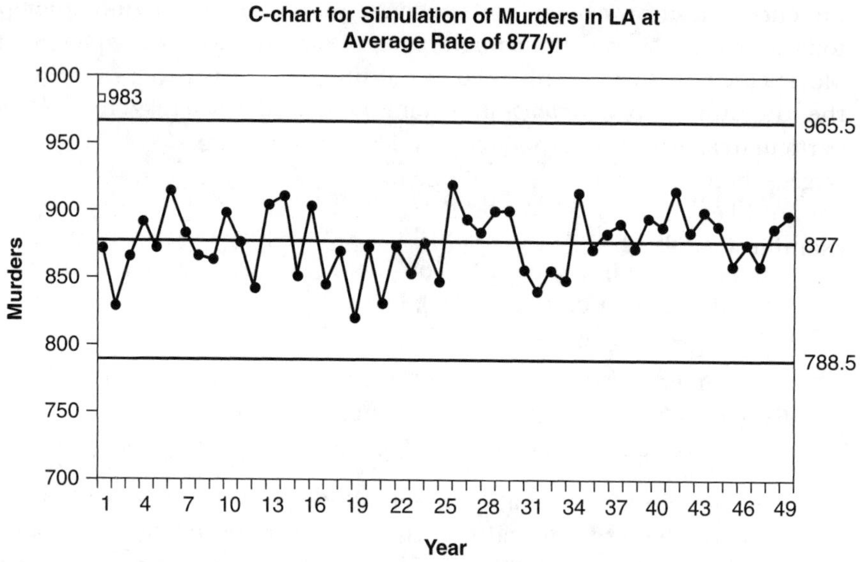

FIGURE 1.2 *C*-Chart for Simulation of Murders in Los Angeles at Average Rate of 877 per Year.

murders might be slightly different than one-tenth. But the idea here is that having 30 or more murders in a year is not a big deal for a stable process that averages 23 murders a year over the course of many years.

We now consider 50 years of simulated murders for Los Angeles, assuming a stable process that averages 877 murders a year over the course of many years. Refer to Figure 1.2.

In Figure 1.2, a small square is shown next to the vertical axis to show where 983 is located. Clearly, 983 is well above the upper control limit of 965.5. Since having a point above the upper control limit is something that would happen at a rate of 1.5 points per 1,000 in a process that is stable and averages 877 murders a year over the course of many years, an assignable cause should be sought. However, there may not be an assignable cause. Whether or not there is an assignable cause, it is the Los Angeles data that reveals a potential problem, not the Montana data.

Understanding basic probability and statistical principles protects people from being misled and from following false trails. Specific to this example, the point being made is as follows:

> Whether a change from one time period to the next is deemed significant should be based on probability, not on percentage increase.

The first control charts were introduced in the 1920s by Walter Shewhart, who worked at Bell Telephone Laboratories (see section 6.1 in Chapter 6 for a discussion of some of the pioneers in the quality movement). Because they have a simple design and are highly sensitive to the presence of assignable cause, these charts are still used by many companies today to monitor and control manufacturing and service-industry processes. More recently, other kinds of charts have been designed to be sensitive to the presence of assignable cause in different ways. For example, it may be particularly important to know if a process has wandered even slightly off target where, by the nature of the process, large deviations from the target are not likely. There are charts that cater to many situations. Later we will examine a variety of charts and how to interpret them, including Shewhart-type charts. Most control charts have three major reference lines:

- center line (target or historical mean),
- upper control limit (UCL), and
- lower control limit (LCL).

Figure 1.1 and Figure 1.2 illustrate the three lines. If the quality variable being measured has a target specification, that value would usually be the y-intercept for the center line. In this case, the control chart may be used to help keep the process centered on the target value. If there is no target value, usually the center line will be based on historical data.

Again, control limits are usually set so that only about 3 points in 1,000 would be ***expected*** to be outside the control limits when the process is ***in control*** (i.e., is working as it was designed to). For a process that is in control, only common cause variation is present. Points should appear randomly scattered on both sides of the center line. Few should be near the control limits.

A point beyond either control limit is called an ***out-of-control signal.***

- An out-of-control signal doesn't necessarily mean there is a problem.
- An out-of-control signal does suggest looking for an assignable cause—depending on severity and feasibility, the system may need to be shut down, process permitting.

Later we will be introduced to other kinds of out-of-control signals. It is possible that a succession of points, all of which are between the control limits, can display a trend of such contrast with the expected pattern of random fallout relative to the center line that an assignable cause problem is suggested. For the time being, the only out-of-control signals we consider are points that fall outside the control limits.

When an out-of-control signal occurs, there will be one of several outcomes. One clean outcome is that an assignable cause is found and can be remedied. A better outcome is that as assignable cause is discovered and is trivial. For example, a number was miscopied. In this case there would be nothing to "fix." Another outcome is that no cause might be found—this could be one of those 3 points in 1,000—and life goes on. This is usually an unsettling outcome because those who allow the process to continue will worry that the next data collected will yield another out-of-control signal, that there really is a problem, and that an assignable cause will be difficult to find.

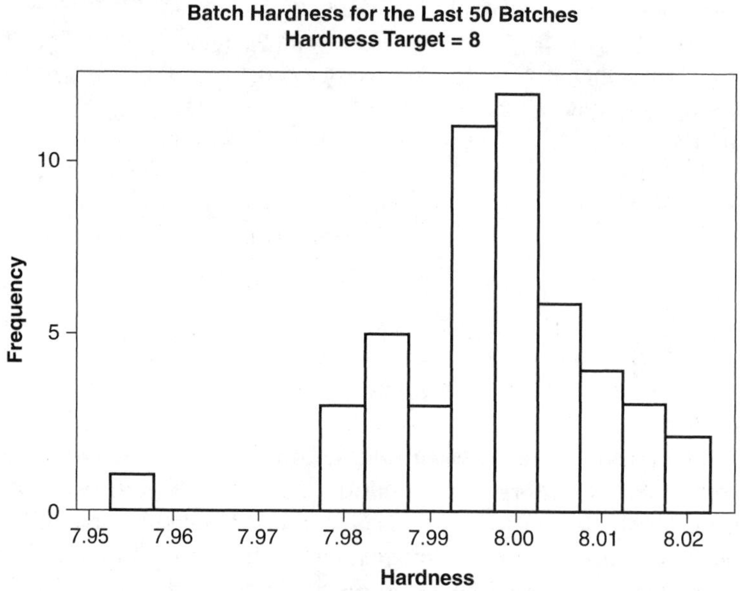

FIGURE 1.3 Batch Hardness for the Last 50 Batches (Hardness Target = 8).

Sometimes it is difficult to distinguish common cause variation from assignable cause variation. Knowing the process will help. That Fred miscopied a number can be either a common cause problem or an assignable cause problem. Everybody makes an occasional mistake. Such mistakes are inherent in the process and should be viewed as contributing to common cause variation. Fred should not be "chewed out" for such a mistake. On the other hand, Fred may be the frequent source of this particular kind of problem. The system is not supposed to work that way. Fred's miscopying should be viewed as an assignable cause, and the reason for the frequent occurrences explored. Retraining may be a solution. On the other hand, Fred may be dyslexic, in which case he should be considered for other duties, ones for which dyslexia won't pose problems.

Mound-Shaped Distributions

When the output variable is a sum of independent inputs, unless the magnitude of certain inputs dominate the others, the distribution of the output variable will be mound shaped. Figure 1.3 illustrates such a distribution for a chemical batch process. At this point in the production process, quality is measured by the hardness of the completed batch. Hardness is affected by a number of variables:

Inputs

Chemical A amount	High or Low
Chemical B amount	High or Low
Chemical C amount	High or Low
Temperature	High or Low
Operator touch	High or Low
Chemical A purity	High or Low

Chemicals A, B, and C are mixed according to a recipe. A chemical reaction follows, and the results of the reaction harden in a protected environment at a set temperature. The equipment that delivers prescribed amounts of the chemicals has inherent variation. Hence the amount of Chemical A in a given batch would be a little on the high side or a little on the low side, and virtually never the *exact* amount called for by the recipe. Independently of Chemical A, the amount of Chemical B would be slightly over- or under-dispensed. The same is true for Chemical C. While there is vigilance in maintaining the environment temperature, this temperature is affected by operators opening and closing room and observation doors and by the temperature outdoors. It follows that the environment temperature is virtually always slightly high or slightly low. The purity of Chemical A is virtually always slightly above or slightly below the average purity for this chemical, with the same being true for Chemicals B and C. Additionally, operator "touch" on the equipment affects hardness, and so on.

A simplistic way to view the hardness of a batch is the result of a pinball cascade down an array of pins (Figure 1.4).

In this scenario, a ball is dropped onto the pin directly beneath it. The pin keeps the ball from falling straight down, and the ball is equally likely to make a left turn or a right turn, according to whether the amount of Chemical A is a little high or a little low. Depending on the direction of the

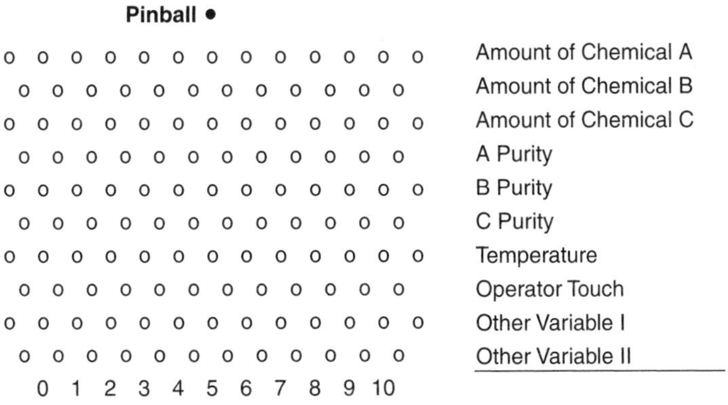

FIGURE 1.4 How Batch Hardness Variables Affect One Batch.

turn, the ball hits one of the two center pins in the second row, and moves to left or right according to whether the amount of Chemical B is a little high or a little low. The ball then hits one of the three middle pins in the third row. Note that the ball has only one way of arriving at the leftmost of the three pins, and only one way of arriving at the rightmost—but there are two ways to arrive at the middle pin. The ball then continues on its way, always turning left or right, until falling through the last row of pins, at which point the batch has been made.

The ball makes 10 turns altogether, and the number of right turns is noted beneath where the ball exits the bottom row of pins. By this simple model, let's say that the hardness of the batch is given by substituting the number of right turns in the formula

$$H = 7.95 + \frac{\text{Number of Right Turns}}{100}$$

If all the pins are removed, that is, if all the inputs are exactly the values they are supposed to be, the ball would drop straight down across the 5-exit, yielding the perfect batch, one with a hardness measure of 8.00.

The model is simplistic because it accords all of the input variables the same magnitude of impact on hardness. Nevertheless, one can reason from the model why the output distribution for a sample of 50 batches might have the mound shape that is seen in Figure 1.3. There are many paths through the array to the 5-exit. The effect of a high amount of Chemical A can be offset by the impact of a low temperature, or the effects of two variables can be offset by two other variables. The farther an exit is from the 5-exit, the smaller the number of paths there are to that exit. One extreme case would be the 0-exit. There is only one path that leads to it, the path of all left turns: LLLLLLLLLL. To have a hardness of 7.95, by chance, every input variable would need to be off-target in the direction that reduces hardness. There are 10 paths that lead to the 1-exit:

RLLLLLLLLL
LRLLLLLLLL
LLRLLLLLLL
LLLRLLLLLL
LLLLRLLLLL
LLLLLRLLLL
LLLLLLRLLL
LLLLLLLRLL
LLLLLLLLRL
LLLLLLLLLR

It follows that a ball has 10 times the chance of coming through the 1-exit as it does the 0-exit. Many introductory probability and statistics texts show that the number of paths with a given number of right turns is equal to the value of a binomial coefficient. Specific to our model we have

Number of Right Turns	0	1	2	3	4	5	6	7	8	9	10
Number of Paths with This Many Right Turns	1	10	45	120	210	252	210	120	45	10	1

Again, when the output variable is a sum of independent inputs, unless the magnitude of certain inputs dominate the others, the distribution of the output variable will be mound shaped. *Also, as long as no inputs dominate, the more inputs there are, the more similar to a "normal" distribution the output distribution will be.* This is a nonmathematically stated version of the Central Limit Theorem, one of the most important facts in statistics. The shape of a normal distribution is sometimes called a *bell curve.*

Because many variables in industry are the result of a number of independent inputs, their distributions have an approximate bell-curve shape. For this reason, many kinds of control charts are based on normal distributions. Further, how one determines the control limits and why one would expect 997 points in 1,000 to fall between the control limits when the process is in control are both tied to normal distributions.

The array of pins described in Figure 1.4 and the idea of aiming a ball at a particular pin in the top row and allowing it to fall downward through the pins describes an SPC training device called a *quincunx.* The balls are really small beads that fall into columns when they exit the array of pins. The device is covered with a clear plastic cover so the observer can see everything that happens. One can use a marker on the plastic cover to label the columns 0, 1, 2, and so on. A quincunx may be used as a training tool to illustrate a number of statistical truths.

One truth illustrated with a quincunx activity is that making process changes based on naive statistical intuition can be harmful. Out-of-control signals on control charts indicate that there are potential problems in the process. These signals are statistically designed to indicate either single, very exceptional points, or unusual trends based on a succession of points. An untrained person might be tempted to react to single measurements that differ in any way from a target value. This is called **overcontrol.** It is also called ***reacting to noise***.

STATISTICAL FACT: When only common cause variation is present, one increases variation by making process adjustments in reaction to a single undesirable data value, incident, or customer complaint.

EXAMPLE

This activity is similar in spirit to W. Edwards Deming's funnel experiment (see Deming, 1986, page 327). Fred is "making batches" on the quincunx. He knows what the target hardness value is.

Whenever a ball comes out of an exit other than the 5-exit, he decides to make an obvious compensation for the system being off target. We follow Fred through a sequence of adjustments.

A ball falls into the 6-column, one column to the right of target. Fred "compensates," dropping the next ball on the top pin in the 4-column. The ball falls into the 2-column, three columns to the left of the target 5-column. Fred adjusts by dropping the next ball on the top pin in the 7-column, three columns to the right of where the last ball was dropped. The ball falls into the 5-column, and Fred makes no adjustment for the next ball, dropping it from atop the 7-column as well. This ball ultimately lands in the 7-column, two columns to the right of target. So Fred drops the next ball from atop the 5-column. It falls into the 2-column, so Fred drops the next ball from atop the 8-column. It lands in the 11-column, so Fred drops the next ball from atop the 2-column. It lands in the 1-column, and so on.

When all balls are dropped from atop the 5-column, we anticipate finished "batches" having a mound-shaped distribution centered on or near the 5-column. That is, there will be a concentration of balls near the 5-column. However, Fred's actions of continually recentering the process will produce a more spread-out distribution of batches. The batches will have less of a concentration near the 5-column and a higher proportion far away from the 5-column. It follows that we expect a higher proportion of batches to be off-spec when Fred operates the equipment: Fred is guilty of overcontrol. Even equipment centered slightly off-target will produce a smaller percentage of bad product than when Fred handles the equipment. It is best to react to statistical signals, not noise. An extreme sentiment to this end is given by Warren Bennis:

> *"In future factories, due to automated tasks, instead of lots of workers, there will be just a man and a dog. The man is there to feed the dog, the dog is there to keep the man from touching the equipment."*

STATISTICAL FACT: In many workplaces, employees are mistakenly rewarded or blamed for common cause variation.

This illustrative quincunx activity is similar in spirit to W. Edwards Deming's red bead experiment (see Deming, 1986, page 346). The instructor tells the students that when the hardness of a batch is less than 7.98, the batch must be scrapped. A bold line should be drawn on the plastic cover separating the 2-exit from the 3-exit. Beads falling into columns to the left of the line represent the batches that must be scrapped. Five or six students are called upon to serve as shift workers. One at a time, each worker drops 20 beads (makes 20 batches). These twenty batches represent a sample of all

TABLE 1.1 Batch Results.

Worker	Number of Scrapped Batches
Mary	2
Marcel	1
Leslie	0
Diane	1
Paul	3

the batches made by the worker on his or her shift. The number of batches that must be scrapped is recorded. Table 1.1 illustrates possible results.

After the last worker is done, the instructor announces that it is time to give merit raises. Having made no bad batches, Leslie is praised, and it is commented that she clearly is careful and deliberate. With such nice results, she will receive an 8% merit raise. Marcel and Diane are encouraged and given 4% raises. Mary is told that management believes her heart is in the right place. However, she has cost the company serious money. She will receive a 2% raise and is informed that she must attend a two-day worker retraining session, beginning the following Monday at 8 A.M. The retraining session is to help her understand the importance of quality, the cost that is associated with scrapped batches, and the procedures to be followed in making batches. Finally, Paul is asked if he knows how much it costs the company when a batch is scrapped. $10,000, he is told. A sarcastic remark can be made to the effect that on top of the scrap cost, the company is paying him a salary while he is making bad batches. He is informed that he probably has a great future, just not with this company. Finally, Paul is told that his services are terminated.

The students are asked what they think of management's actions. Usually, a student will state that the actions are unfair because "It's all chance." The key here is that the process was the same for all the workers. They were rewarded or blamed for common cause variation. There was no attempt to measure the amount of common cause variation. This must be done first. Only after the level of common cause variation is measured can control limits be determined and used to indicate when the number of scrapped batches in a sample of 20 batches is unusual.

Blaming (or rewarding) people for outcomes beyond their control (i.e., for common cause variation or for variation of the system) fosters resentment in the workplace and is an example of reacting to noise. Control charts can be used to help detect unusual variation. To do this, common cause variation must be assessed first.

PROBLEMS

1. A company that makes automobile bumpers monitors daily the number of bumpers that customers want replaced due to unsatisfactory appearance. Lately the company has averaged 12 such bumpers per day. The upper control limit is 22.39 and the lower control limit is 1.608. Data for 25 days of production in March are as follows:

12	14	10	11	11	14	13	0	13	10
10	14	14	10	23	19	12	9	12	9
8	13	16	8	9					

   Do any of these days give an out-of-control signal? Draw a rough sketch of what a *c*-chart would look like with these data plotted.

2. A company that makes a cleaning solvent takes a sample every six hours and measures the viscosity in a laboratory. The target viscosity is 115 Pas, the upper control limit is 139 Pas, and the lower control limit is 91 Pas. Data for 30 days of production follow. Do any of these days give an out-of-control signal? Draw a rough sketch of what a control chart would look like with these data plotted.

112	105	123	126	100	115	129	105	120	109	106	122	113
110	103	107	110	113	117	129	117	116	107	108	111	123
130	125	113	121									

3. Identify whether the following statements are true or false.
   a. An out-of-control signal on a control chart means there is a problem that needs to be fixed.
   b. A control chart with no out-of-control signals means that assignable cause problems are not present in the system.
   c. An out-of-control signal indicates that an assignable cause problem may be present.
   d. A point outside the control limits is a signal to check records to determine an assignable cause.

4. On a continuous basis, liquids are mixing and flowing through pipes at a plant that makes a polyurethane liquid. Every 15 seconds, the company measures the viscosity of the finished product at a point in the flow just after the mixed liquids have completed a chemical reaction. These measurements are obtained by a viscometer that not only measures the viscosity of the liquid passing through it, but also adjusts the flow of one particular chemical, Chemical A, that goes into the mix at a point upstream, just prior to where the mixed liquids begin to undergo the chemical reaction. Essentially, the

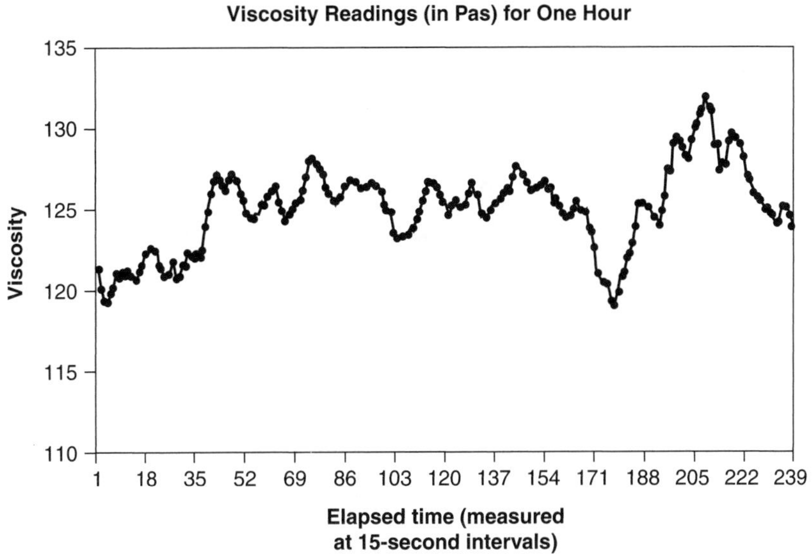

FIGURE 1.5 Viscosity Readings (in Pas) for One Hour.
Source: Reprinted from M. Norton, "Detecting Overcontrol for a Continuous Variable from a Graph Diagnostic," *Quality Engineering* 14, no. 1 (2001–2002): 9–12 by courtesy of Marcel Dekker, Inc.

viscometer determines the difference between the viscosity measurement and the target viscosity of 125 Pas, and adjusts the input flow of Chemical A to correct for 100% of this difference. It takes eight to nine minutes for the flow to travel from this input point to the viscometer.

a. What word describes a design flaw with this process? What is the probable consequence?
b. A graph of viscosity versus time, covering a one-hour period, is examined to see if unwanted consequences of adjustments show up eight to nine minutes after the adjustments are made (see Figure 1.5). Figure 1.6 shows two copies of this graph, with one copy lagging the other by 8 minutes and 45 seconds. What does Figure 1.6 reveal?
c. Figure 1.7 shows viscosities with time-period markers spaced 8 minutes and 45 seconds apart. During the last such time period shown, there is a three-mountain-pattern. How far back can the three-mountain-pattern be traced?
d. Notice in Figure 1.7 that viscosity is centered close to the target viscosity of 125 Pas. In general, how might the graph have appeared if adjustments had been based on statistical signals rather than on *every* viscosity reading?

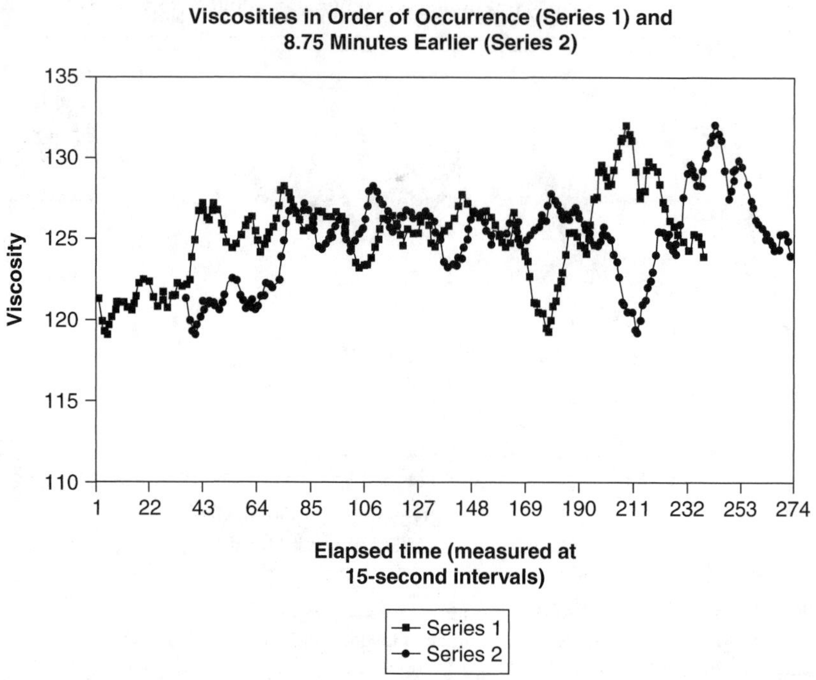

FIGURE 1.6 Viscosities in Order of Occurrence (Series 1) and 8.75 Minutes Earlier (Series 2).

Source: Reprinted from M. Norton, "Detecting Overcontrol for a Continuous Variable from a Graph Diagnostic," *Quality Engineering* 14, no. 1 (2001–2002): 9–12 by courtesy of Marcel Dekker, Inc.

1.2 STATISTICAL MEASURES NEEDED FOR BASIC CONTROL CHARTS

Before exploring some facts about normal distributions and control charts, we first must be familiar with three key statistical measures, the mean, the standard deviation, and the range.

REFERENCE EXAMPLE: An hourly sample of five ball bearings is taken from production and their diameters measured. The target diameter is 4.20 mm.

3 P.M. sample: 4.10 4.69 4.07 4.43 4.17

One statistical measure the company is very likely to focus on is the sample ***mean*** (or average). The mean is obtained by adding all of the values and dividing this total by the number of values that were added. If we let n represent the number of values that are added and let $X_1, X_2, X_3, \ldots, X_n$ represent the values to be added, then the sample mean may be expressed symbolically as:

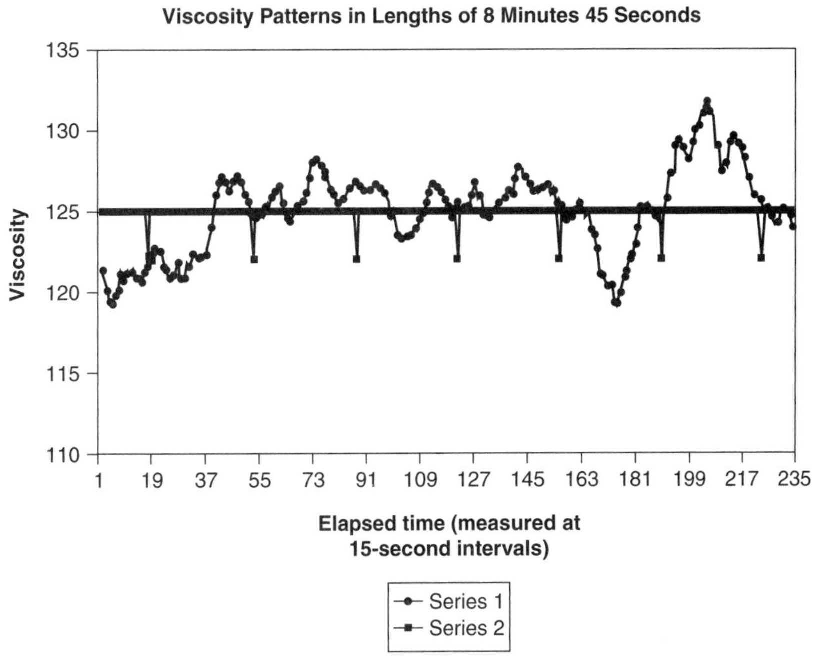

FIGURE 1.7 Viscosity Patterns in Lengths of 8 Minutes and 45 Seconds.
Source: Reprinted from M. Norton, "Detecting Overcontrol for a Continuous Variable from a Graph Diagnostic," *Quality Engineering* 14, no. 1 (2001–2002): 9–12 by courtesy of Marcel Dekker, Inc.

$$\overline{X} = \frac{\sum_{i=1}^{n} X_i}{n}$$

The capital Greek letter Σ in the formula is a mathematical symbol that means "add up the following." The "1" and "n" at the bottom and top of Σ indicate that the X values to be added are the ones that have subscripts from 1 to n. When "1" and "n" at the bottom and top of Σ are omitted, it is understood that all n values are to be added. In the case of the reference example's 3 P.M. sample, $n = 5$ and

$$\overline{X} = \frac{4.10 + 4.69 + 4.07 + 4.43 + 4.17}{5} = 4.292 \ mm$$

The units for the mean always will be the same as for the sample values (millimeters in our example). The sample **mean** specifies a single value on the number line at which the group of sample values may be viewed as being located (or centered). If there is a target value for the product, naturally, the company would like all sample means to be near the target.

Figure 1.8 shows two possible histograms—which would the company rather see? Since the target diameter is 4.2 mm, the sample averages displayed

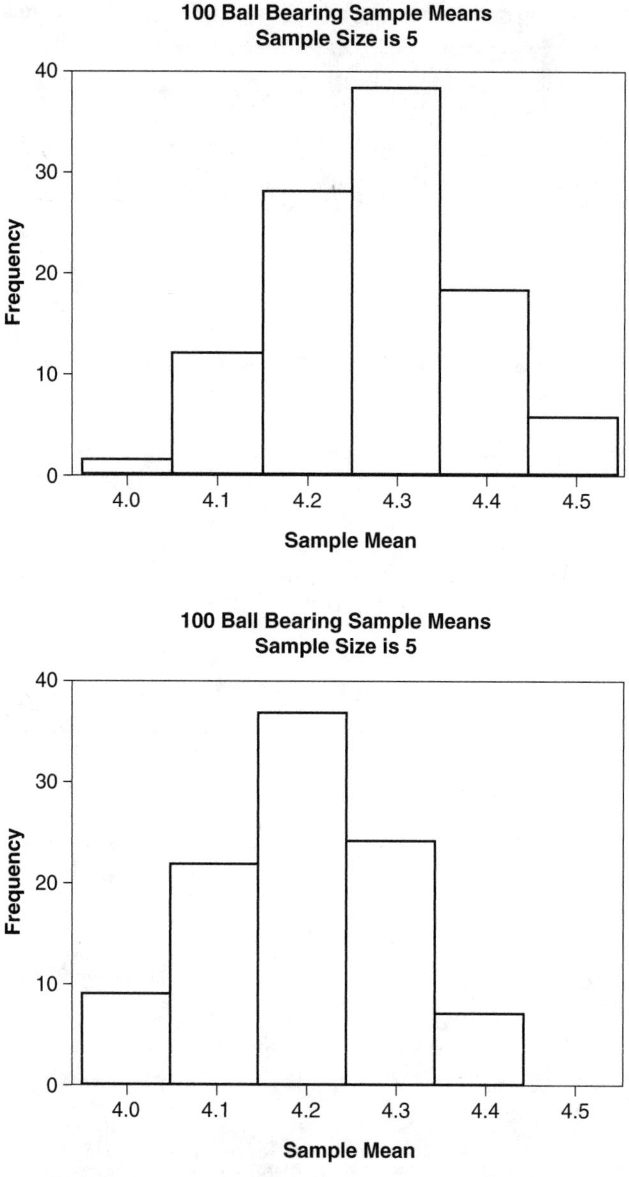

FIGURE 1.8 Which Histogram Would the Company Rather See?

in the bottom histogram of Figure 1.8 are closer to target than those displayed in the top histogram.

Some control charts are the plots of successive sample means. The two sets of sample means described by the histograms in Figure 1.8 are plotted on control charts in Figure 1.9. This is a case where one can see at a glance

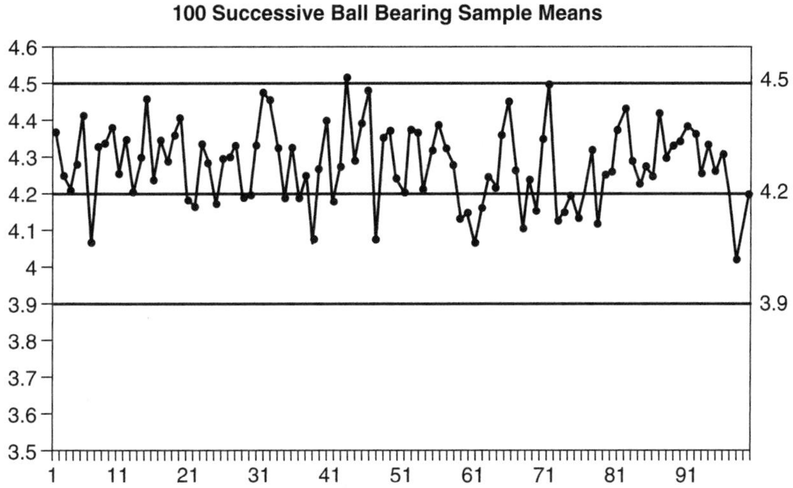

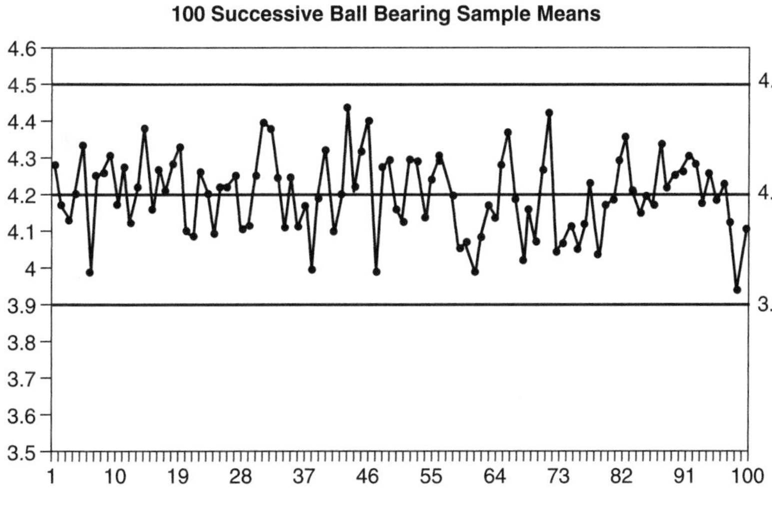

FIGURE 1.9 Control Charts for Histograms in Figure 1.8.

the imbalance of points on one side of the target. The other control chart reflects a process that is centered close to target.

A second statistical interest is the amount of variability (or dispersion) in the property being measured. The more predictable and consistent the product, the less variability there will tend to be in the values of a sample from production.

4.10	4.69	74.07	4.43	4.17
4.15	4.54	4.12	4.43	4.22

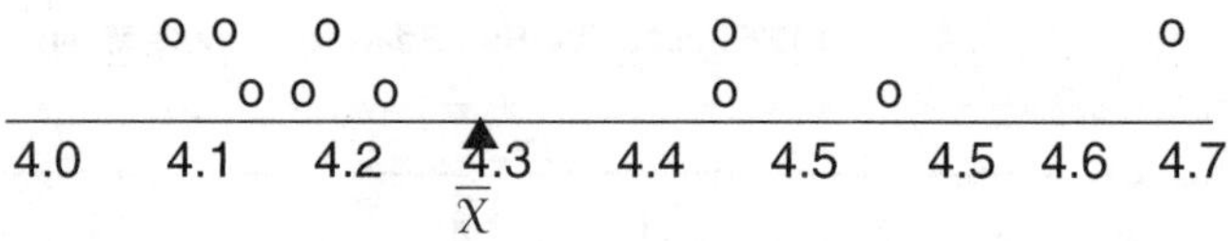

FIGURE 1.10 Ball-Bearing Diameters.

TABLE 1.2 Procedures for Computing a Sample Standard Deviation One Step at a Time.

A	B	C	D	E	F
Sample 1	$\overline{X}$	$X_i - \overline{X}$	$(X_i - \overline{X})^2$	**D sum**	***S***
4.10 mm	4.292	−.192	.036864	.27848	.264 mm
4.69 mm	4.292	.398	.158404		
4.07 mm	4.292	−.222	.049284		
4.43 mm	4.292	.138	.019044	(D sum)/4	
4.17 mm	4.292	−.122	.014884	.06962	
Sample 2					
4.15 mm	4.292	−.142	.020164	.13548	.184 mm
4.54 mm	4.292	.248	.061504		
4.12 mm	4.292	−.172	.029584		
4.43 mm	4.292	.138	.019044	(D sum)/4	
4.22 mm	4.292	−.072	.005184	.03387	

A common measure of variation is the sample ***standard deviation.*** The formula for the standard deviation of a sample is somewhat intimidating by sight:

$$S = \sqrt{\frac{\Sigma (X_i - \overline{X})^2}{n - 1}}$$

Consider the two samples of five ball bearing diameters shown in Figure 1.10.

In Figure 1.10, each sample mean is 4.292, so both groups of numbers are "centered" in the same place on a number line. But one group of numbers exhibits more variation, as we shall see.

To evaluate S, the formula shows that the sample mean is subtracted from each value in the sample, and that each such difference is squared. The symbol Σ says that all of these squares are to be added. This sum is then divided by $n - 1$, and finally, a square root is taken. Using the spreadsheet format displayed in Table 1.2, one may follow this step-by-step process one column at a time. In so doing, the sense in which the sample standard deviation measures variation will become apparent.

The entries in column C are the results of subtracting the sample mean from each sample value. The −.192 at the top of column C means that on a number line, the sample value 4.10 is .192 units to the left of the mean. The value 4.15 in Sample 2 also lies to the left of the mean, but is closer (the entry in column C is −.142). Accordingly, when these two entries in column C are squared to produce entries in column D, the column D entry for the value in Sample 1 is the larger of the two column D entries because 4.10 is the more distant from the mean. Similarly, 4.69 from Sample 1 produces a larger value in column D than its Sample 2 counterpart, 4.54, because it is the more distant of the two values from the mean. In this example, the Sample 1 values are all farther from the mean than their corresponding values in Sample 2. Therefore each of the first five entries in column D is greater than its counterpart in the last five entries. It follows that the sum of the first five entries in column D, .27848, must exceed the sum of the bottom five entries, .13548 (see column E). We then divide each of these two sums of squares by 4 (why?—refer to the formula for S and remember that $n=5$). Naturally, the bigger of the two sums from column D yields the bigger value after division by 4. Lastly, we take a square root. Likewise, the square root of the bigger positive number is greater than the square root of the smaller positive number. That is, the standard deviation of Sample 1, .264 mm, is greater than the standard deviation of Sample 2, .184 mm. The units for the standard deviation will always be the same as for the sample values (millimeters in this example).

This illustrates that the farther a sample value is from the sample mean, the greater the square of the distance from the mean will be (the column D entry), and hence the greater its contribution toward the sample standard deviation. On the other hand, a sample value close to the sample mean makes only a small contribution toward S. It follows that when the sample values are clustered near the mean, the standard deviation will be small. Note that no entry in column D can be negative (why?), and a standard deviation can never be negative. The more values there are far from the mean, the larger the standard deviation will be.

The larger the standard deviation, the greater the variation in the sample values. The smaller the standard deviation, the less variation in the sample values.

A standard deviation can never be negative, but it can be zero. The reader should mimic the steps displayed in Table 1.2 for the sample values 7, 7, 7, 7, 7, 7, 7, 7, 7, 7, 7, and 7. In fact, the only way S can be zero is if all the sample values are equal. When this happens, each sample value equals the mean, so that every entry in column D would be zero, and so would S.

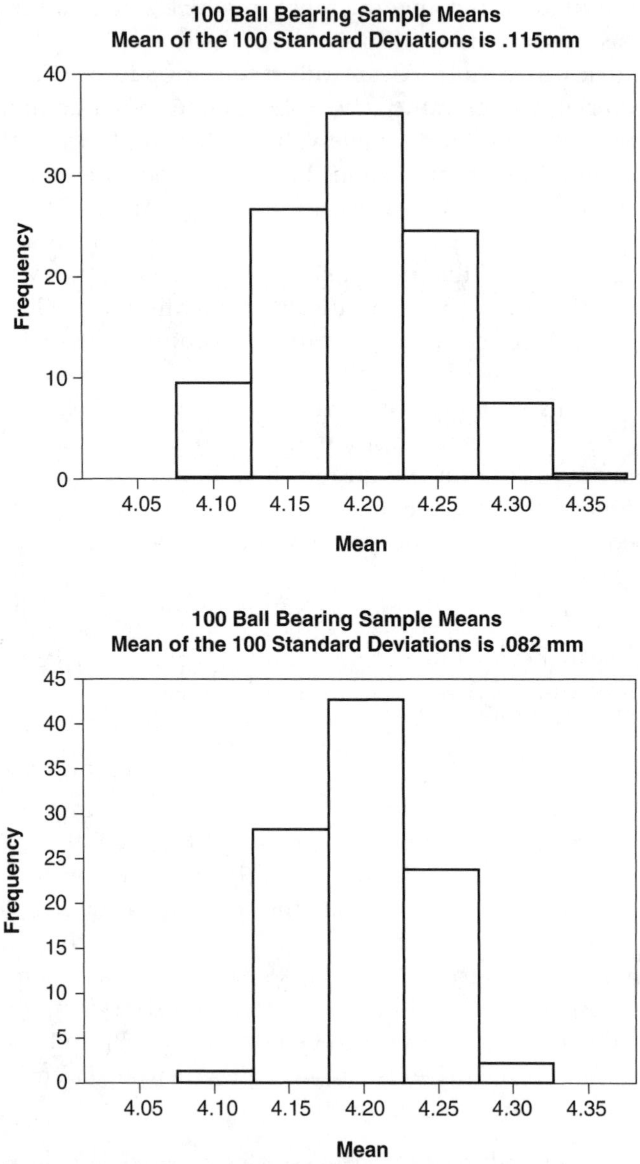

FIGURE 1.11 Which Histogram Would the Company Rather See?

Figure 1.11 shows two possible histograms. Both histograms show that the sample means are concentrated near the target diameter of 4.2 mm. However, one can confirm at a glance that the means in the bottom histogram are more concentrated near the target. The process that produced these ball bearings is making a more consistent product. If the histograms were not available, the same message would be conveyed by the fact that the average standard deviation is noticeably smaller for the 100 samples that produced the bottom histogram.

In practice, no one calculates standard deviation by hand. In many production facilities, equipment that takes measurements automatically feeds these values into spreadsheets where software packages such as MINITAB or Excel can evaluate means, standard deviations, and other summary measures. Many inexpensive calculators also do these calculations.

Using the Texas Instruments TI-83 Calculator to Obtain Summary Statistics

Turn on the calculator and hit the STAT key (near the top center of the keypad). In order to enter data, use the Up, Down, Left, or Right arrows if necessary to move the cursor to the EDIT heading and onto the menu item beneath it: "1:Edit . . . " Hit the ENTER key (lower right corner of the keypad). If there are numbers showing in list L1, clear the list by moving the cursor directly onto the "L1" by means of the Up, Down, Left, or Right arrows. Hit the CLEAR key, and move the cursor down to the top entry position in the list, which is now clear. Refer to the 3 P.M. sample of ball bearing diameters used to illustrate the sample mean. Key in 4.10 (or 4.1) and hit ENTER. Key in 4.69 and hit ENTER. Continue until all five values have been entered into L1. Hit the STAT key again, but then use the arrows to move the cursor onto CALC and then onto the top item beneath CALC: "1: 1-Var Stats." Hit ENTER. Hit ENTER again. A number of "1-Variable" statistics will have their values shown on the screen.

$$\bar{x} = 4.292$$

$$\sum x = 21.46$$

$$\sum x^2 = 92.3848$$

$$Sx = .2638560213$$

$$\sigma x = .236$$

$$n = 5$$

We see from the last equation that there were five numbers in the sample, that the sample mean is 4.292, and that the sample standard deviation (Sx) is approximately .2639. Other information, such as the sum of the sample values (21.46), is there, too.

Using Excel to Obtain Summary Statistics

Enter the sample values into the spreadsheet, say into the first five rows of column A. Refer to Table 1.3. Click on a cell where the mean is to be displayed,

TABLE 1.3 Sample Values, Their Mean and Standard Deviation.

A	B	C
4.1		4.292
4.69		0.263856
4.07		
4.43		
4.17		

for example, cell C1. Type the characters =**average(** and then drag the mouse across all of the values to be averaged. The cell entry now looks like this: =**average(A1:A5.** Then type **)** and hit the ENTER key (having the matching right parenthesis is not necessary). The sample mean 4.292 should now show in the cell. Now click on a cell where the standard deviation is to be displayed. Type =**stdev(** and then drag the mouse across all of the values in the sample. The cell entry now looks like this: =**stdev(A1:A5.** Type a right parenthesis and hit ENTER (or just hit ENTER). The standard deviation .263856 should appear in the cell.

Using MINITAB to Obtain Summary Statistics

Enter the sample values into some column in the MINITAB spreadsheet, say into the first five rows of C1 (i.e., column 1). After entering the last value, be sure to click on a cell in some other column. Select the **Stat** menu at the top of the screen and click on **Basic Statistics.** On the submenu that appears, click on **Display Descriptive Statistics.** A box will appear with three white screens in it. One screen will list which columns contain data. The other box needed is named **Variables:.** We now tell MINITAB which column has the data we want to use by either double-clicking on C1 in the box that identifies columns containing data or by typing C1 directly into the **Variables:** box. Then click **OK.** The following information will appear in the Session window on the screen:

Descriptive Statistics: C1

Variable	N	N*	Mean	SE Mean	StDev
C1	5	0	4.292	0.118	0.264

Minimum	Q1	Median	Q3	Maximum
4.070	4.085	4.170	4.560	4.690

As with the TI-83 calculator, we get the mean, standard deviation, the number of values in the sample, and other summary measures we will not need for our purposes.

There are many statistical measures we do not address, choosing to focus just on those that are commonly used in control charts. So to describe at what point on a number line a group of numbers may be viewed as being *located,* we have used only the sample mean. To describe the extent of *variation* in the group of numbers, we have used the sample standard deviation. We also need to discuss the *range,* which measures the variation in a group of numbers, but does so in a different way.

The ***range*** of a group of numbers is the largest value in the group minus the smallest value in the group.

We return to the 3 P.M.-sample of ball-bearing diameters in order to illustrate the range, which usually is denoted by the letter R:

3 P.M.-sample:	4.10	4.69	4.07	4.43	4.17

According to how the range is defined, we see that R = 4.69 mm − 4.07 mm = .62 mm. The units for the range will always be the same as for the sample values (millimeters in this example). The concept for using the spread between the largest and smallest values in a group to measure variation is that for numbers that exhibit a lot of variation, this spread should be large, while for numbers that are more concentrated, this spread should be small. *As with the standard deviation, the larger the value of the range, the more variation in the data there is understood to be.*

In the days before calculators and computers, the range was a very popular measure of variation because it was easy to understand and trivial to compute by hand. It still is the measure of variation used by many companies. However, because it uses only the two most extreme values in a group of numbers to summarize the variation within the entire group—or said another way, *ignores all the values between them*—the standard deviation does a better job of measuring variation according to statistical theory, when $n > 2$. Generally, a company will choose one of these to serve as its measure of variation. We treat both R and S since one of the two will be seen in the workplace.

We now need a theoretical construct. Consider some particular variable of interest—hardness of a batch, diameter of a ball bearing, and so on. For something concrete, let's say it is the latter.

The collection of all ball bearings that could be made by the process with the capability it possesses at the time a sample is drawn is called the ***population.***

The ball bearings in the sample obtained at 3 P.M. may be seen, held in one's hand, and measured. They are tangible. However, the population is a conceptual thing. One can only *imagine* the millions upon millions of ball bearings that *could* be made by the equipment before the equipment degrades or has some problem. Further, one has to imagine that the diameter of every one of them could be measured. Once we can so imagine these things, we can use summary data from the sample to estimate statistical features of the population.

> The population mean is the mean value of the variable of interest taken over the whole population. It usually is denoted by the Greek letter μ.

The population mean usually cannot be evaluated. The population of all ball bearings that *could* be made by the equipment has to be conceptual. On the other hand, one might try to predict, based on preelection interviews of some voters, how the population consisting of all the voters in the state will vote on the day of the election. This population is tangible. However, even though measuring everyone in the population—that is, finding out how they will vote—is possible, it is not easy or practical. Usually, any statistical fact about a *population* is regarded as unknown or unknowable.

> The population standard deviation is the standard deviation of the variable of interest taken over the whole population. It usually is denoted by the Greek letter σ.

For completeness, we make the technical observation that the formula for evaluating σ is slightly different than the formula used to evaluate the sample standard deviation, S. It is

$$\sigma = \sqrt{\frac{\Sigma (X_i - \mu)^2}{n}}$$

where n is the number of values in the population. This makes the quotient under the square root a perfect average of the squares of distances from the population values to the population mean. A natural question to ask is why the divisor in the formula for σ is n while the divisor used in computing S is $n - 1$. The answer is that a primary use of S is to estimate the unknown σ. Theory shows that using n in the divisor of S would result in a consistent underestimation of σ. In any case, there usually is little occasion to use the formula for σ because the population is almost always inaccessible.

It is very common for people to be confused by the fact that are two kinds of means and two kinds of standard deviations. It will be helpful to remember that the mean and standard deviation one routinely computes using data obtained from items produced at a particular time are the *sample* mean and the *sample* standard deviation:

$$\overline{X} = \frac{\Sigma X_i}{n} \qquad S = \sqrt{\frac{\Sigma (X_i - \overline{X})^2}{n - 1}}$$

The population mean μ and population standard deviation σ usually are unknown or unknowable. However, these unknowns are important enough

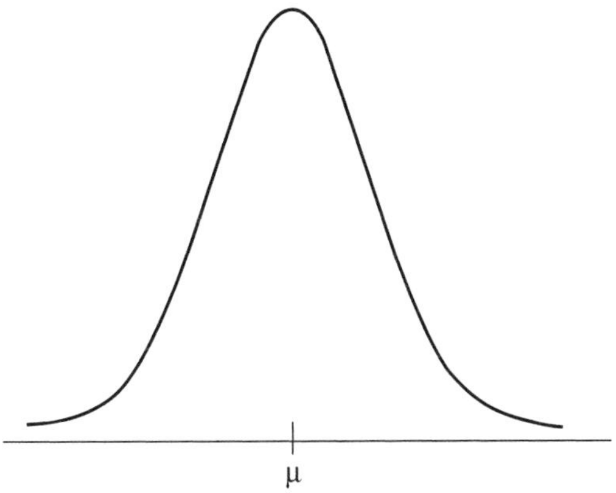

FIGURE 1.12 Distribution of Diameters of Ball Bearings.

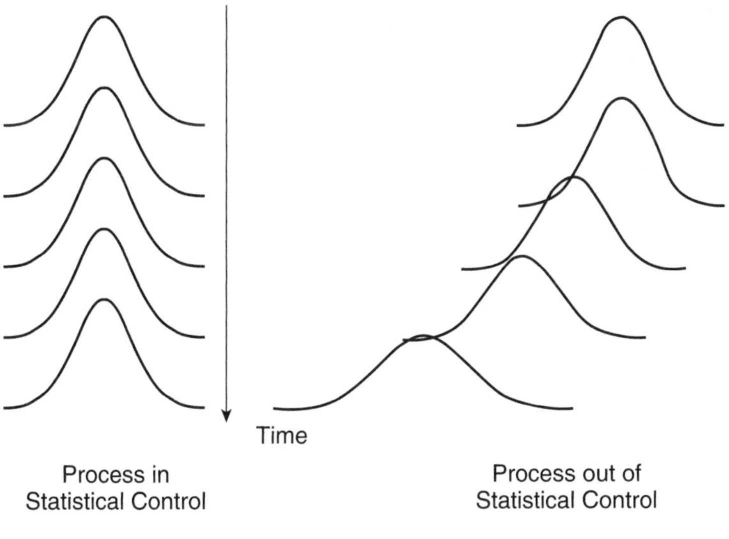

FIGURE 1.13 Population Change Over Time.

that we will want to estimate their values. This is because μ tells where the process is centered at the time the sample is taken and σ measures the current level of variation in the system. These are commonly called the *process mean* and *process standard deviation.*

The distribution of diameters for all ball bearings in the population might look like as shown in Figure 1.12.

In order to make a consistent product, one would like to have the population not change over time, except as the result of intentional improvements. Refer to Figure 1.13. The left portion of the figure depicts what the population looks like at five different times when samples are taken from production. When the population does not change over time, the process is said to be stable, or to be **in control.** The process depicted on the right side of Figure 1.13 is not in control, a situation commonly referred to as out of control. One reason is that the location of the process is drifting over time. Another is that σ is changing—incidentally, for the worse (why?). For either reason, one may conclude that there is a lack of control.

The variables one measures may be viewed as falling into one of two categories: continuous or discrete.

A variable is ***continuous*** when the measurement scale is on a continuum.

Having the measurement scale be on a continuum means that there are no gaps in the range of values that the variable could equal. One way to determine if a variable is continuous is to ask the question, "Could we measure more decimal places with a more precise measuring instrument?" If the answer is "yes," then the variable is continuous. Height, weight, diameter, and viscosity are some examples of continuous variables. It is a good bet that you were less than 3 feet long at birth and are taller than 4 feet now. In this case, there would have been a time when your height was precisely π feet, regardless of the fact that a person's height usually is measured to the nearest inch. More decimal places of height are possible with more precise measuring instruments, which makes height continuous.

A variable is ***discrete*** when possible values of the variable have gaps between them.

The number of murders in Montana in a year and the number of flaws in a bumper are examples of discrete variables. They have gaps in the measurement range—there could be three flaws in a bumper or four, but nothing between.

We shall see later that there are many different kinds of control charts. Some are used to monitor or control process location, while others do the same for process variation. Whether the variable being tracked is discrete or continuous also has an impact on what kind of chart is used.

PROBLEMS

Unless specified otherwise, problem evaluation may be done on a calculator or by a computer package such as MINITAB. Adopt the following rounding convention in the problems that follow. Let the mean and standard deviation be reported to one more decimal place than the raw data. The range, being the difference between the largest and smallest values in a data set, should carry the same number of decimal places as the raw data.

1. Find the sample mean, sample standard deviation, and range for the following weights of off-spec spools (in grams) of copper wire.

21.9	14.3	19.1	18.7	13.1
16.8	21.2	16.6	14.4	14.6

2. Find the sample mean, sample standard deviation, and range for the following distances that the centers of drilled holes were from their target centers (in millimeters).

2.86	3.31	3.78	2.64	0.50	0.97	2.46	2.93

3. Find the sample mean, sample standard deviation, and range for the number of vehicles that pass through an intersection between 5 P.M. and 6 P.M. on 20 consecutive workdays.

98	98	94	100	97
104	103	97	100	95
105	98	98	100	100
93	104	94	99	96

4. By hand, find the mean and sample standard deviation of the values 11, 3, 2, 3, and 1.
5. By hand, find the mean and sample standard deviation of the values 8, 7, 12, 7, 4, 6, and 12.
6. Six ball bearings taken off the production line at 3 A.M. have a mean diameter of 3.70 mm and a standard deviation of .00 mm. Give an example of data having these properties. If diameter were measured to more decimal places, would the standard deviation still be "zero"?
7. In a particular production process, the specification limits for bolt lengths are 15.99 cm and 16.01 cm. The four bolts in every box are inspected and any bolt with length not meeting spec limits is replaced. In a box having all four bolts meet specs, what is the maximum possible value of S? If one bolt is scrapped, what is the maximum possible value of S for the remaining three bolts?
8. Suppose a small population consists of the integer values 8, 11, 5, 10, 6, 9, and 7. Find the population mean, μ, and standard deviation, σ.
9. Identify whether the following variables are discrete or continuous variables.
   a. Temperature
   b. Bolt length
   c. The customer's cost to photocopy one page at a copy shop
   d. The number of batches that must be scrapped in a day
   e. The concentration of a particular carcinogen in a soil sample
   f. The cost of a package of postage stamps
   g. Flow rate of water through a pipe

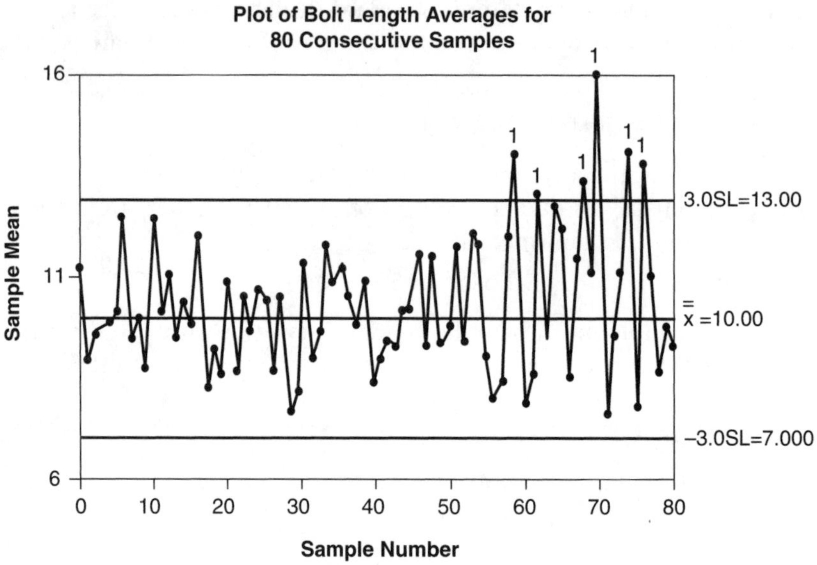

FIGURE 1.14 Plot of Bolt Length Averages for 80 Consecutive Samples.

10. The lengths (in centimeters) of five bolts from production are measured and averaged four times a day. The target length is 10.0 cm. The averages are plotted in sequential order in Figure 1.14. Does bolt length appear to be on target throughout the period? Does variation appear to be consistent throughout the period? Does the process appear to be in control?
11. The resistance (in ohms) of five electrical devices from production is measured and averaged four times a day. The target resistance is 17.0 ohms. The averages are plotted in sequential order in Figure 1.15. Does resistance appear to be on target throughout the period? Does variation appear to be consistent throughout the period? Does the process appear to be in control?

1.3 SOME FACTS ABOUT NORMAL DISTRIBUTIONS

As noted earlier, many variables in nature come close to having normal distributions. This is because many variables can be viewed as a sum of a large number of independent inputs. Often with just three or four such input variables, the output distribution will be mound shaped and closely approximate a normal distribution.

For something concrete, consider a variable such as batch hardness. Suppose that the population of hardness values has a normal distribution.

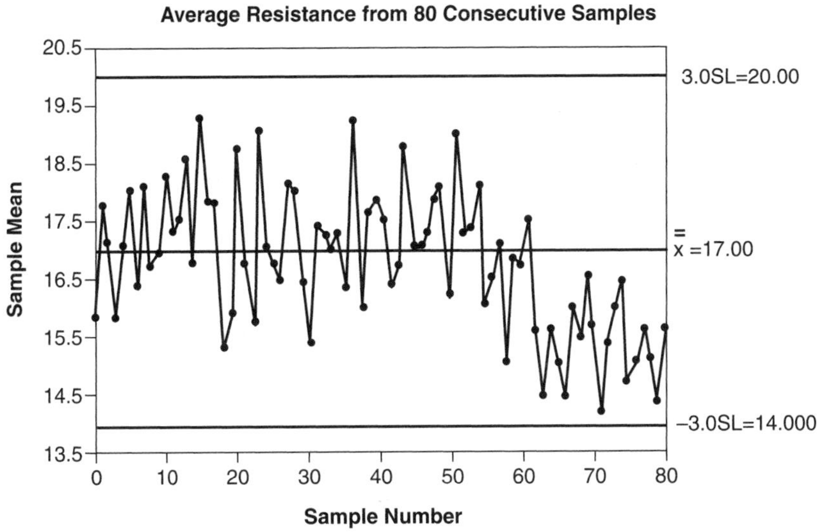

FIGURE 1.15 Average Resistance from 80 Consecutive Samples.

There is an equation that gives the bell-curve shape for the distribution. It is

$$y = \frac{e^{-\frac{(x-\mu)^2}{2\sigma^2}}}{\sigma\sqrt{2\pi}}$$

The graph of the equation is shown is Figure 1.16. The y-axis is suppressed so that we can focus on shape issues.

The equation may appear to be formidable because, besides x, it involves μ, σ, e, and π, and it involves an exponent and a square root. However, we know that π is just a constant, approximately 3.14159. Similarly, e is just another constant, approximately equal to 2.71828. Like π, e is a value that occurs often enough in higher mathematics to warrant having a symbol to identify it. The other values needed to produce the graph are values we usually would not know: μ, the population mean, and σ, the population standard deviation. But if we knew their values, we could substitute any x value into the equation, evaluate to get the y-coordinate, and plot the point.

As indicated in Figure 1.16, the highest point on the curve lies above the value σ on the x-axis. The curve is symmetric with respect to the line $x = \mu$. The value of σ determines whether the curve appears spike-like or as a gentle rolling hill. If the standard deviation is small, there is little variation in the population, the hardness measurements for most batches would be concentrated near μ, the curve will have an abrupt rise and fall near μ,

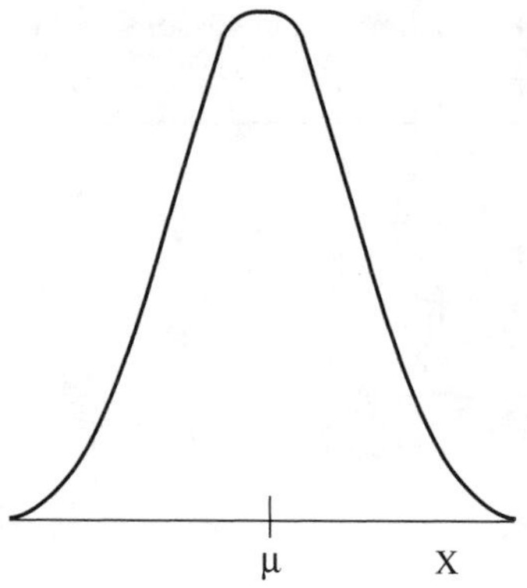

FIGURE 1.16 The Bell-Shaped Curve of a Normal Distribution.

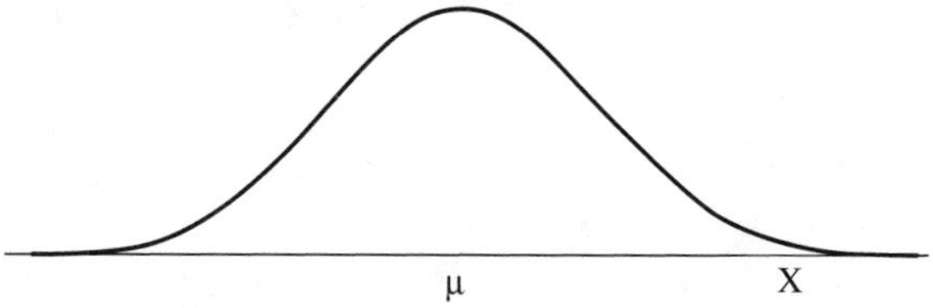

FIGURE 1.17 Normal Distribution with Large Standard Deviation is Shaped Like a Gentle Rolling Hill.

and the curve will appear like something of a spike. If σ is large, there will be less of a concentration of population values near μ, a larger proportion of measurements farther from μ, and the curve will resemble a gentle rolling hill (Figure 1.17).

A property of normal populations, the ***Empirical Rule,*** will prove helpful in understanding how control limits on many control charts are determined. Any population with a normal distribution satisfies the Empirical Rule. Large samples from a normal population also would be expected to satisfy the property. In fact, the Empirical Rule applies well to large samples from many reasonably symmetric mound-shaped distributions.

The Empirical Rule for populations with a normal distribution:

APPROXIMATELY

68% of the population measure within σ of μ

95% measure within 2σ of μ

99.7% measure within 3σ of μ

Figure 1.18 helps to illustrate the Empirical Rule. The large majority of the area under the curve, approximately 68% of it, lies within one standard

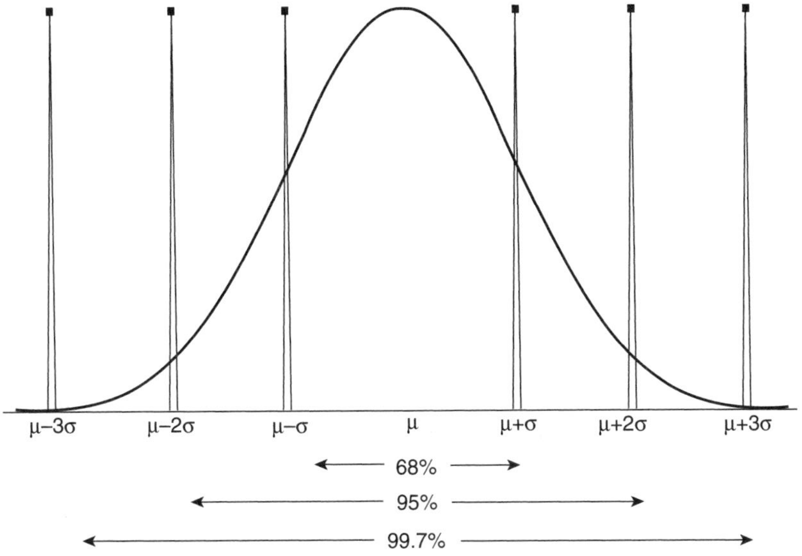

FIGURE 1.18 Graphic Illustration of the Empirical Rule.

deviation of μ; that is, between the vertical lines that intercept the x-axis at $\mu - \sigma$ and at $\mu + \sigma$. If we widen the interval to include all of the values within two standard deviations of μ, that is, between $\mu - 2\sigma$ and at $\mu + 2\sigma$, we pick up an additional 27% of the total area. That is, 95% of the total area under the curve lies between these lines. Finally, nearly every value in the population—99.7% of them, or 99.7% of the area under the curve—lies between $\mu - 3\sigma$ and at $\mu + 3\sigma$. In other words, almost every item in the population measures within three standard deviations of the mean. There would be an occasional population item that does not, but this would be extremely unusual.

Note in Figure 1.18 that all points on the curve lie above the x-axis, so that there is positive area under the curve to the right of $\mu + 3\sigma$ and to the left of $\mu - 3\sigma$. However, these areas combine to make up only .3% of the total area. Analogously, in a normal population, only 3 values in 1,000 fall more than 3σ from μ.

EXAMPLE

The resistance in ohms of the electrical components made by a company has an approximate normal distribution with mean 75.0 and standard deviation 3.0.

a. Roughly what percentage of components produced will have a resistance between 69 and 81 ohms?

b. Between 66 and 84?
c. Between 75 and 84?

Solution:

a. The values 69 and 81 are 6 ohms below and 6 ohms above the mean of 75 ohms, respectively. To address how far a value is from the mean statistically, we don't examine the distance from the mean in ohms. We consider how many standard deviations it is from the mean. Since $\sigma = 3$ ohms, being within 6 ohms from the mean is equivalent to being within two standard deviations from the mean. The Empirical Rule may be used to conclude that 95% of the components made will have a resistance between 69 and 81 ohms.
b. The values 66 and 84 are 9 ohms below and 9 ohms above the mean of 75 ohms, respectively. The question asked may be rephrased as "What percentage of components made should have a resistance within three standard deviations from the mean?" The Empirical Rule says that the answer is 99.7%.
c. From the symmetry of bell curves, the proportion of production having resistance between 75 and 84 ohms should be the same as the proportion of production between 66 and 75 ohms. Since both groups combined account for 99.7% of production, then each accounts for half of 99.7% of production, or 49.85%.

EXERCISE (FILL IN THE BLANK)

Unless the process mean and standard deviation change, only about _____ times in a 1,000 will a component's resistance differ from 75 ohms by more than 9 ohms. If 9 is replaced by 12 in the preceding sentence, how will the answer change?

Solution:

The question is "What percentage of production will not fall within three standard deviations from the mean?" Since 100% − 99.7% = .3%, we would expect only 3 times in 1,000 to have a component's resistance differ from 75 by more than 9.

If 9 is replaced by 12, we are asking what percentage of production falls more than 4σ from the mean. For most practical purposes, this should be regarded as 0% (the actual answer is .0064%).

That 99.7% of a population with a normal distribution will fall within 3σ from the mean is the basis for setting control limits on many control charts. Specifically, the upper and lower control limits will be located "3 sigma" on either side of the center line. The idea is that if a process is supposed to be centered at a value μ, and we either know σ or have a very good estimation of σ, and we get a data point lying more than 3σ from μ, this point has to be regarded as unusual for a process that is centered where it is supposed to be.

When a process is in control, only about 3 points in a 1,000 will lie outside the control limits, based on normal distributions.

Actually, "3 in 1,000" should be close to correct for many mound-shaped distributions, and "in the right ballpark" even for a wide variety of distributions. That this probability is so small is why such a point usually spurs an investigation for assignable cause—a reason the process may have changed.

There is a two-step arithmetic process, called *standardizing*, that can prove helpful when confronted with problems like those in the preceding example and exercise. The purpose of standardizing is to help answer the question "Is the particular value x just obtained from this process unusual?" The two arithmetic steps input the value x, and output the number of standard deviations x lies from the mean. The output value is called a *standardized score*, or sometimes a *z-score*. This value is computed using the following equation:

$$z = \frac{x - \mu}{\sigma}$$

In the previous example in which electrical devices had an average resistance of 75 ohms with a standard deviation of 3 ohms, we needed to compute how many standard deviations from the mean 81 ohms was. First we found the actual distance by subtracting the mean from 81, and then we divided the distance by the value of the standard deviation, 3. This is precisely what standardizing accomplishes. In this case, the z-score for a resistance of 81 is

$$z = \frac{x - \mu}{\sigma} = \frac{81 - 75}{3} = 2$$

The z-score tells us that 81 lies two standard deviations above the mean. On the other hand, 64 ohms has a z-score of

$$z = \frac{x - \mu}{\sigma} = \frac{64 - 75}{3} = -3$$

In other words, 64 is three standard deviations *less* than the mean.

Note that a z-score is unitless. The units in the numerator are ohms, the units in the denominator are ohms, and the units cancel.

RULE OF THUMB: A number is considered to be *unusual* when it is two or more standard deviations from the mean. A number is considered to be *extremely unusual* when it is three or more standard deviations from the mean.

This rule of thumb may be applied in many statistical arenas, and not just to normal distributions. In discrimination suits in a court of law, for example, it is very common for layoff, termination, or hiring outcomes that would be two or more standard deviations from the mean, were a company not treating a protected group differently than other individuals, to be considered as *significant*, and as circumstantial evidence against the defendant company. In an entirely different venue, whether water-heater lifespans are normally distributed or not, if the average lifespan of a hot water heater is 13 years with a standard deviation of 1.8 years, and a particular water heater developed a leak after only 9 years and 10 months, that water heater would not be viewed as having an unusually short life because the z-score is

$$z = \frac{x - \mu}{\sigma} = \frac{9\frac{10}{12} - 13}{1.8} = -1.76$$

While the lifespan is shorter than average, it is not more than two standard deviations short of the mean.

Control charts in business and industry typically use 3-sigma limits instead of 2-sigma limits because management doesn't want to disturb a functioning process or someone's routine unless an outcome is extremely unusual. Moreover, the laws of probability (see Chapter 5), can be used to show that if production is sampled, say, once daily, each sample results in a point on a control chart, 2-sigma limits are used for the control limits, and every out-of-control signal is investigated, then in roughly 75% of all 30-day work-months, staff will be futilely seeking an assignable cause for at least one out-of-control signal when there is nothing wrong with the process. When 3-sigma limits are used, the figure drops from 75% to 8%. No one wants to chase noise.

PROBLEMS

1. True or false: An out-of-control signal indicates that either the process mean has shifted or that the standard deviation has changed.
2. Historical data indicate that spot welds on a particular auto part have a mean shear strength of 492 lb and a standard deviation of 37.4 lb. Would a spot weld with a shear strength of 412 lb be unusual for this process? Would it fall outside 3-sigma limits?

3. A company's garage door opener has an average life span of 21 years with a standard deviation of 2.5 years. Should a garage door opener that lasts 24 years be viewed as having an unusually long life?
4. Assuming in Problem 3 that garage door life span has a normal distribution, what percentage of garage doors made would have life spans between 16 and 26 years?
5. In Problem 3, use the Empirical Rule to find two practical limits, a and b, where $a < b$, with the property that almost all garage door life spans fall between a and b. (Hint: Interpret "almost all" as meaning there might be an occasional door with a life span not between a and b, but such an event would be extremely unusual.)
6. If the minimum tension required to pull thread off a spool has a mean of 8.1 centinewtons (cn) and a standard deviation of .4 cn, find two practical limits, a and b, where $a < b$, with the property that almost all spool tensions fall between a and b. Assume that minimum tension has a normal distribution. (Hint: Interpret "almost all" as meaning there might be an occasional spool for which the required minimum tension is not between a and b, but such an event would be extremely unusual.)
7. In Problem 6, roughly what percentage of spools should have a minimum tension between 6.6 cn and 9.6 cn?
8. In Problem 6, roughly what percentage of spools should have a minimum tension between 7.7 cn and 8.5 cn?
9. Suppose the variable of interest, X, is the number occurrences of a given kind during a time period. Some examples would be the number of murders in Montana in a year, the number of births at Roper Hospital in a day, or the number of accidents at an industrial plant in a month. Let μ be the population mean for the number of occurrences during a time period. In other words, using the setting of accidents at a plant site to illustrate, μ represents the average number of accidents per month, computed across all conceivable months during which the plant would operate under the same safety procedures. It is common in such settings for σ, the standard deviation of the number of occurrences in a time period, to satisfy $\sigma \approx \sqrt{\mu}$. We will discuss the distribution of X in a later chapter. However, it is not a normal distribution nor even continuous. Nevertheless, it is common to apply the usual 2-sigma and 3-sigma rule of thumb for what constitutes an *unusual* or *extremely unusual* value, regardless of the distribution.

   If an industrial plant averages 2.3 accidents per month, would 4 accidents in a month be a) unusual? b) extremely unusual? c) Would 6 accidents in a month be unusual? d) extremely unusual?

10. See Problem 9. A newspaper article says that there were 17 murders in the county last year, 25 the year before that, and attributes the downward swing to the new sheriff. No other data are given. For a process having 25 as the population mean number of murders per year, should 17 murders in a year be considered unusually low?

1.4 RANDOM SAMPLING AND SAMPLE MEANS

As observed, we never really get to see the actual distribution of the variable of interest. However, we do get to see product samples taken periodically from production. To help ensure that a ***sample*** is as representative of the population as possible, it must be what statisticians call *random*, not just easy to obtain.

A ***random sample*** is a sample drawn in such a way that every member of the population has the same chance of being in the sample.

Consider vegetable soup that was heated in a pan on a stove and then allowed to sit for a minute or two. One could taste a soup-spoonful taken from the top of the soup in the pan. This would be an easy-to-get-at sample, but it wouldn't be representative of the soup in the pan because the vegetables would have settled to the bottom. But if the soup is stirred well and then just one spoonful is taken from the top, the spoonful should be representative of what is in the pan. That is, a small sample would give a reasonably faithful representation of the whole population. One doesn't have to eat all the soup in the pan to know what the whole population is like. By giving the soup a good stir, every member of the population has the same chance of being in the sample. If one is taking a sample of a liquid that just arrived in a tanker truck, knowledge of the product should resolve whether settling is an issue. The importance of knowledge of the product cannot be overstated. With knowledge of the product, one can answer the question "Can the way we sample be expected to give us a fair representation of the population we intend to be represented?"

A general rule of thumb for a process that produces distinct objects (e.g., ball bearings), is that the *objects in any sample should all be produced at as close to the same time as possible.* This is because one of the things done with sample data is to obtain estimates of process location and variation. Since the population can change over time (remember the term *out-of-control*), what usually is done is to regularly estimate what the population characteristics are at particular times. At 3 P.M., when five ball bearings that were all just made are taken off the production line, we *assume* that these five ball bearings constitute a random sample from the population of all ball bearings that could be made at 3 P.M. by the process before it degrades or fails. Out-of-control signals from charts on which these means and standard deviations are plotted over time can then be used to detect shifts and other changes, including process improvement.

A sample of objects produced at different times may not be representative of the process at any time.

Another rule of thumb is to *avoid mixing output from different workers, different machines, or different subprocesses into a single sample.* In the same sense that a "sample" of six copper wire spools made at six different times may not be representative of what the population is like, *ever*, six spools produced by six different machines, even if all are produced at the same time, may not be representative of any meaningful population. There may be circumstances that justify pooling together the output of different machines, different workers, and so on—for example, if the output from different machines necessarily would be, at most, marginally different. However, the decision to do such pooling should be made after much thought, in light of the fact that one purpose of maintaining control charts is to create signals when a process might be changing. It is easy to imagine that the output of one machine can be changing substantially, but because output from it is pooled with output from other machines, no unusual value results from the sample mean.

Sample size is the number of objects in one sample.

A sample size should be just large enough to aid in making good decisions about whether a population is changing, but not so large as to be "overkill," particularly when the cost for larger sample sizes is a factor. Actually, sample sizes of three, four, five, six, and even two, are very common in industry. In chemical batch processes, it is not uncommon to have a sample size of one—it is easy to imagine the situation in which one number, the result of a lab analysis of a single small portion taken from a well-mixed solution, would say everything that needs to be said about that solution. In fact, sometimes the analysis of multiple portions taken from that solution would produce essentially the same numbers over and over. In this case, the only purpose for using a sample size of more than one would be to measure the variation in the lab analysis procedure. Deciding whether the sample size should be 5, 6, or 20, for instance, should be made with the help of a statistician. It is not a topic addressed in this book.

In order to estimate the amount of common cause variation in some processes, it may be appropriate to have the values in a sample be obtained at different times. A possible example would be a process that produces a liquid on a continuous basis. Obtaining five concentrations of a chemical at nearly the same point in the flow of the liquid may be akin to measuring the diameter of the same ball bearing five times. This is fine if current interest is on variability of the measuring instrument.

The first chart we will examine is the $\overline{X}$-chart, which is updated by plotting the sample mean of the most recent sample. But before we can do this, we must address some facts about the statistical behavior of sample means.

When the sample size is greater than 1, the sample mean tends to be closer to μ than an individual sample value.

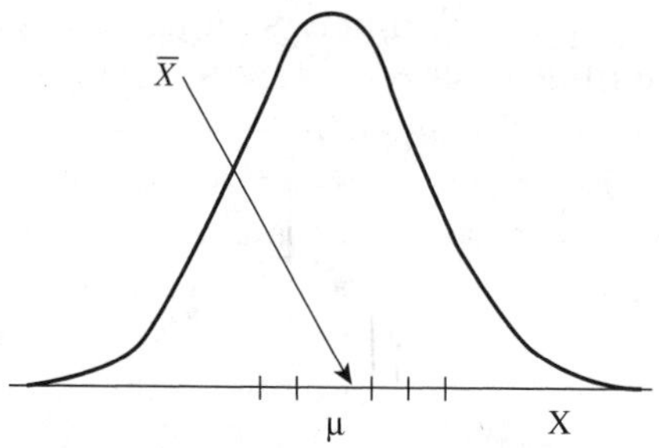

FIGURE 1.19 The Sample Mean Tends to be Closer to μ than Sample Values Are.

Figure 1.19 represents the individual values in a sample of size 5 from some population. Notice that the sample mean is closer to μ than are each of the five sample values. Of course, two different samples would be expected to have a different fallout and different means. But the key words in the previous statement are *tends to be closer to* μ. Sample means tend to be closer to the population mean than individual sample values are.

Figure 1.20 illustrates the same point. The top figure summarizes a random sample of 500 values taken from a population with mean $\mu = 40$ and standard deviation $\sigma = 10$. The bottom figure summarizes the means of 500 samples, each sample of size 5, and all samples taken from that very same population with $\mu = 40$ and $\sigma = 10$. Clearly the sample means are more concentrated near the population mean than individual values from the population are.

In order to set up $\overline{X}$-charts, we need to know more than just the fact that sample means tend to be less variable than individual sample values. To set up control limits, we need to quantify this lesser amount of variation that sample means possess.

Imagine two populations. The first is the one we are sampling from, taking samples of size n. This is the population that has mean μ and standard deviation σ. For each possible sample of n items from this population, imagine writing the value of the sample mean on a slip of paper. The second population consists of all of these slips of paper. Drawing a sample of n items at random from the first population and evaluating the mean is equivalent to drawing one slip of paper from the second population and seeing the number on it. To illustrate, consider the population that consists of the numbers 12, 4, 6, and 6. For this population

$$\mu = \frac{12 + 4 + 6 + 6}{4} = 7$$

and

$$\sigma = \sqrt{\frac{(12 - 7)^2 + (4 - 7)^2 + (6 - 7)^2 + (6 - 7)^2}{4}} = 3.$$

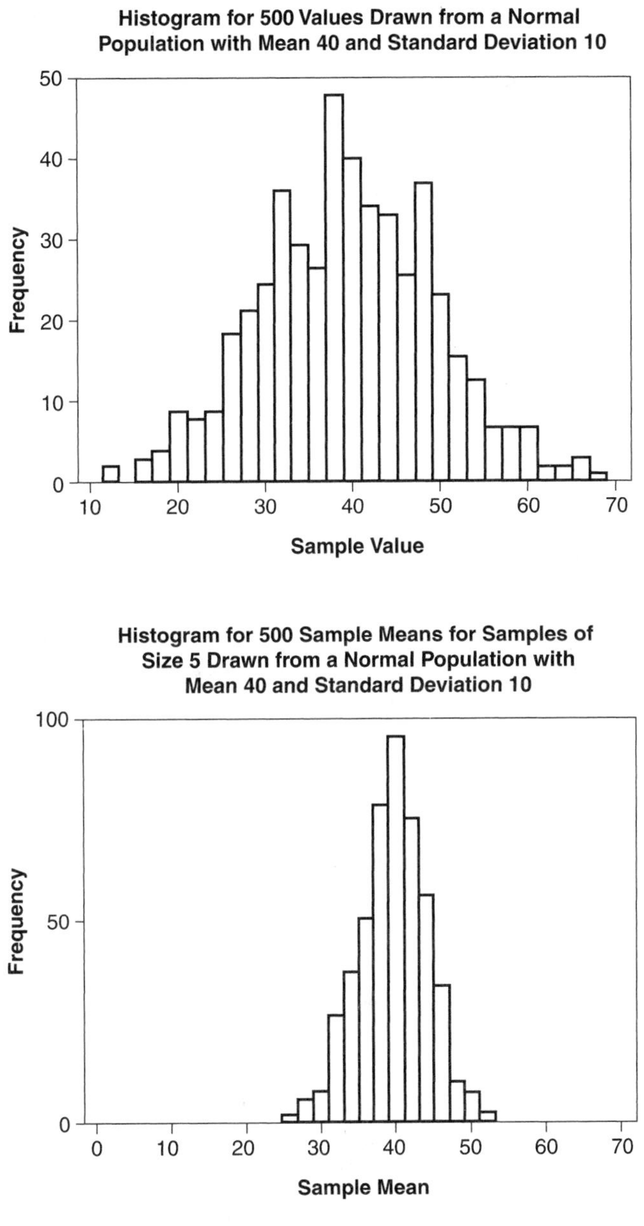

FIGURE 1.20 Sample Histograms.

Table 1.4 lists every possible sample of size 2. The samples are taken with replacement. This guarantees that each sample value comes from exactly the same population. To the right of each sample value is the mean for that

TABLE 1.4 Population of Means for Samples of Size 2 from the Population 12, 4, 6, 6.

X_1	X_2	$\overline{X}$	X_1	X_2	$\overline{X}$
12	12	12	6	12	9
12	4	8	6	4	5
12	6	9	6	6	6
12	6	9	6	6	6
4	12	8	6	12	9
4	4	4	6	4	5
4	6	5	6	6	6
4	6	5	6	6	6

sample. As discussed previously, there would be 16 slips of paper, one piece for each sample. We refer to the 16 numbers in the column as the *population of means.* The population mean and standard deviation for the population of means are:

$$\mu_{\bar{X}} = \frac{12 + 8 + 9 + 9 + \cdots + 9 + 5 + 6 + 6}{16} = \frac{112}{16} = 7 = \mu$$

and

$$\sigma_{\bar{X}} = \sqrt{\frac{(12 - 7)^2 + (8 - 7)^2 + (9 - 7)^2 + \cdots (6 - 7)^2}{16}} =$$

$$\sqrt{\frac{72}{16}} = \sqrt{\frac{9}{2}} = \frac{3}{\sqrt{2}} = \frac{\sigma}{\sqrt{n}}.$$

Notice that the population of means has a mean that is identical to μ, the mean of the population that is being sampled. Also notice that the population standard deviation of the population of means is σ divided by the square root of the sample size. One can show that this is true in general.

When drawing a random sample of size n from a population with mean μ and standard deviation σ, the population of means has mean $\mu_{\bar{X}} = \mu$ and standard deviation $\sigma_{\bar{X}} = \sigma/\sqrt{n}$.

These two facts give us information that justifies why $\overline{X}$-charts work. The population of means is centered at precisely the same place as the population being sampled. Additionally, we know not only that sample means are less variable than individual sample values (when $n > 1$), but also what the variation is—assuming that we know or have a good estimate of σ first.

Obtaining good estimates of μ and σ for a process that is in control will be important in setting up control charts. It will be helpful to know that when the sample size is large, the sample mean and the sample standard deviation provide good estimates of μ and σ, respectively. This fact may be shown by mathematical theory that we do not explore in this text. The matter is somewhat complicated because how "large" the sample size needs to be is tied to how close we want our estimates to be to the unknown μ and σ, to the value of σ, and to the kind of distribution the population has. Nevertheless, we state the following general truth:

When the sample size n is large, the sample mean $\overline{X}$ is a good approximation for μ and the sample standard deviation S for σ.

We illustrate by considering the 500 sample values and 500 sample means which gave rise to Figure 1.20. Recall that the population sampled had a mean $\mu = 40$ and standard deviation $\sigma = 10$. We now summarize some statistical facts about that data.

	sample mean	**sample st. dev.**
Sample of 500 values	39.436	9.980
500 sample means, $n = 5$	39.654	4.400
	$\frac{\sigma}{\sqrt{n}} = \frac{10}{\sqrt{5}} = 4.472$	

Notice that the mean of the random sample of 500 values is 39.436, very close to the known population mean of 40. The sample standard deviation of the 500 values is 9.980, very close to 10, the known population standard deviation. Had we not known the values of μ and σ, this is a case in which the sample size is large by practical standards, $\overline{X}$ would provide a good approximation of μ, and S would provide a good approximation of σ. We now consider the 500 sample means. Averaging the 500 means yields 39.654, which is closer to 40 than 39.436 is. Averaging the 500 means should give a good estimate of the mean of the population of means, which is also 40 (why?). The mean of sample means doesn't have to be closer to 40 than the mean of the 500 values, but there is a tendency for this to happen because the sample means come from a population with less variation: $\sigma/\sqrt{n}$ as opposed to σ. Finally, viewing the 500 sample means as a random sample from the population of means and using those values to compute a sample standard deviation should produce a good estimate of $\sigma/\sqrt{n}$, the standard deviation of the population of means. To illustrate the theory, we observe from our data that 4.400 is close to 4.472.

A final theoretical result is needed to understand why an $\overline{X}$-chart works. Recall from the Central Limit Theorem that as long as no inputs dominate, the more inputs there are that add together, the more similar to a normal distribution the output distribution will be. So when the sample size is large, we expect the population of means to be approximately normally distributed. The following result fine-tunes this point.

When a random sample is drawn from a population with a normal distribution, the population of means has a normal distribution.

If the population sampled has a normal distribution, the preceding theoretical result says that the population of means has a normal distribution, even when the sample size is small. Also, when the population sampled is simply mound shaped, it is common with small sample sizes for the population of means to have an approximate normal distribution. It is for this reason that $\overline{X}$-charts are based on normal distributions, even for small sample sizes. Additionally, it follows that the Empirical Rule has an application to sample means.

Empirical Rule for a process that is in control when the population sampled has a normal distribution:

After many samples have been drawn, approximately

68% of the sample means will be within $\frac{\sigma}{\sqrt{n}}$ of μ

95% will be within $2\frac{\sigma}{\sqrt{n}}$ of μ

99.7% will be within $3\frac{\sigma}{\sqrt{n}}$ of μ

Note: $\frac{\sigma}{\sqrt{n}}$ *is the appropriate standard deviation to use when answering probability questions about* $\overline{X}$.

EXAMPLE

A company manufactures an electrical component. The resistances in ohms of a daily sample of 16 components are measured and the mean computed and plotted on a chart. The process has been in control for a period long enough to estimate that the "process standard deviation" is roughly $\sigma = 3.0$ ohm (we will soon see how to do this). If the process mean is $\mu = 75.0$ ohm and the process stays in control, what is the approximate probability that the next sample mean is between 72.75 and 77.25?

Solution:

Let us assume that resistance has a normal distribution so that the population of means has a normal distribution. As when we were first introduced to normal distributions, we must determine how many standard deviations from μ each of 72.75 and 77.25 is. However, we must take care to use the proper standard deviation, for the question is not about the resistance of an individual component falling in a particular interval, but rather about the next $\overline{X}$ falling in the interval. The standard error of the mean (i.e., standard deviation of the population of means) is $\sigma_{\bar{X}} = \frac{\sigma}{\sqrt{n}} = \frac{3}{\sqrt{16}} = .75.$ The z-scores are $z = \frac{72.75 - 75}{.75} = -3$ and $z = \frac{77.25 - 75}{.75} = 3$. So the probability that the next sample mean will fall between 72.75 and 77.25 is .997.

PROBLEMS

1. The sample mean tends to be closer to the population mean than does an individual sample value when $n > 1$. What happens when $n = 1$?
2. Consider the small population consisting of the values 3, 4, and 8.
   a. Find μ and σ for this population.
   b. List all possible samples of size 3 with replacement and find the population of means, $\mu_{\bar{X}}$ and $\sigma_{\bar{X}}$.
   c. How do these values compare to μ and σ?
3. Figure 1.21a summarizes the number of hours that were needed to complete each of 500 jobs at an auto repair shop. As a matter of record keeping, it is decided in the future to group five consecutive jobs, find and record the average time spent per job for the group, group the next five consecutive jobs, find and record the average time spent per job for that group, and so on. Figure 1.21b summarizes the next 300 jobs, giving 60 averages. If the distribution for the amount of time that is needed to complete a job has not changed, explain why the shape of the distribution at the bottom of Figure 1.21 is not inconsistent with the shape of the distribution at the top of Figure 1.21.
4. A company spins yarn onto hundreds of spools simultaneously. It has been established that the weight of the yarn on a spool should have a

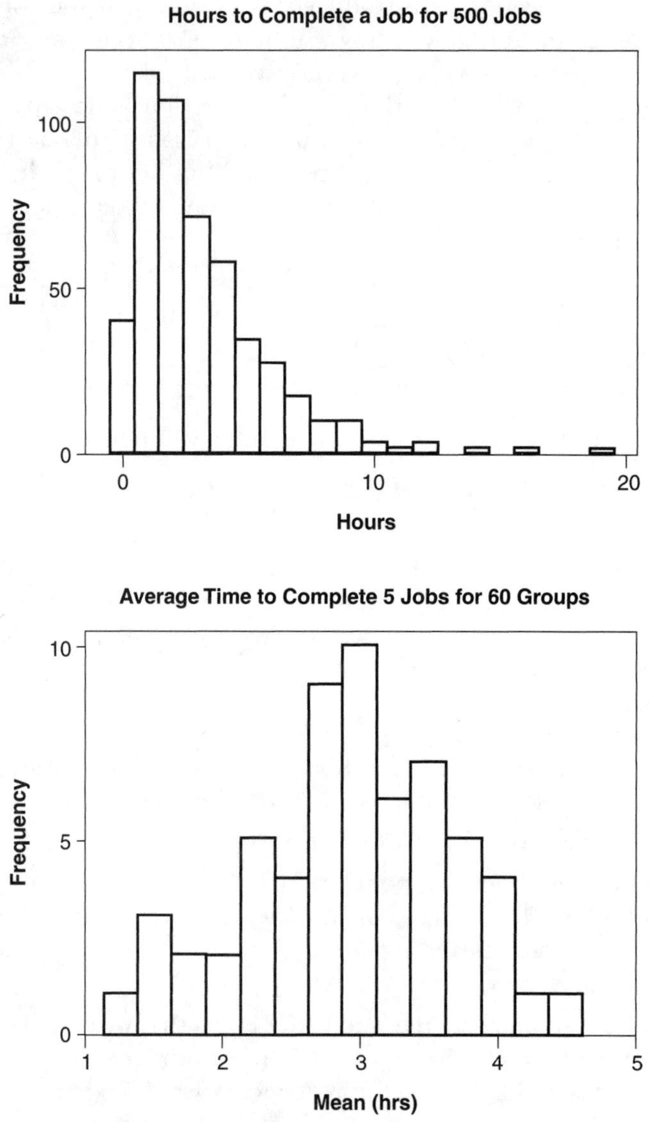

FIGURE 1.21 Histograms for Problem 3.

mound-shaped distribution when the process is in control, with mean $\mu = 1234$ g and standard deviation $\sigma = 6.4$ g. Periodically, four spools are sampled from production and weighed.

a. Determine practical limits a and b, where $a < b$, with the property that almost all sample means should fall between a and b when the process is in control.
b. What theoretical assumptions form the basis for your choices of a and b?
c. Suppose the theoretical assumptions are not quite met—the spool weight population sampled is simply "mound-shaped." Does this invalidate your bounds?
d. Out of 1,000 samples, roughly how many times should we expect the sample mean to fall outside the limits? To exceed b?

5. In Problem 3, suppose that the number of hours needed to complete a job at the auto repair shop has a mean of 3 hours and a standard deviation of 2.4 hours.

a. If for every fifth job, the mean completion time is computed for the last five jobs, determine practical limits a and b, where $a < b$, with the property that almost all sample means should fall between a and b when the repair shop is operating up to its capability.
b. Are these limits consistent with the histogram depicted at the bottom of Figure 1.21?
c. Out of 2,000 samples, roughly how many times should we expect the sample mean to exceed b when the repair shop is operating up to its capability?

2

Charting Sample Means and Variation

OBJECTIVES

- Explain how to recognize when a process exhibits apparent control
- Explain how to estimate the mean and standard deviation of a process that is in control
- Explain how to determine the center line and control limits for $\overline{X}$-charts, R-charts, and S-charts

2.1 ESTABLISHING THE CENTER LINE ON AN $\overline{X}$-CHART

To obtain tentative estimates for the *process mean* (μ) and *process standard deviation* (σ) on a new or existing process, a rule of thumb is that the process should appear to have been in control for 25 to 30 samples at a minimum. The reasoning is that this usually will provide enough data to generate reasonably good estimates of the process mean and standard deviation.

Consider a company that manufactures spools of copper wire. Machine settings were adjusted so as to produce spools of wire for a new product line having a target weight of 562.5 grams. Seven spools were sampled daily and the mean weight determined. Sample means and standard deviations for the first 35 days are plotted in Figure 2.1. The raw data are provided in Table 2.1. Based upon knowledge of customer needs and experience with its equipment, the company defined its own tentative ***upper spec limits (USL)*** and ***lower spec limits (LSL)*** for the weight of a spool:

USL=568.0 grams

LSL=557.0 grams

It is improper to mark USL and LSL values on the vertical axis when plotting sample means. Their values do not appear in Figure 2.1. This is because spec limits are designed to convey information about an individual copper wire spool—either the weight meets specs or it doesn't. If one were plotting the weights of individual spools, horizontal lines that identify upper and lower specs could prove helpful in identifying at a glance whether a large or small percentage of spools are outside specs. However, upper and lower spec limits are inappropriate landmarks for sample averages.

The chart on which mean values are plotted is not a control chart. There are no control limits. No center line is shown. However, the chart will be helpful in determining if the process appears to have been in control throughout the 35 days. Once we have data that we have decided reflects a process that is in control, we can use the data to estimate the level of common cause variation in the system as it currently functions. In other words, we will use the data to estimate σ.

In order to determine what data might reflect process control, we use the chart to help identify samples that should be culled out because there is evidence that the data in those samples are not representative of a process free of assignable cause problems. A glance at Figure 2.1 shows that soon after start-up—days 2 and 3—the sample means were large. A check of the event log for those days showed a pertinent entry by an operator who had reacted to a request by the quality manager. The latter had noted that on days 2 and 3, four of seven and two of seven spools exceeded USL. The quality manager had asked the operator to check certain machinery. An investigation revealed that a key machine part which affects the amount of wire

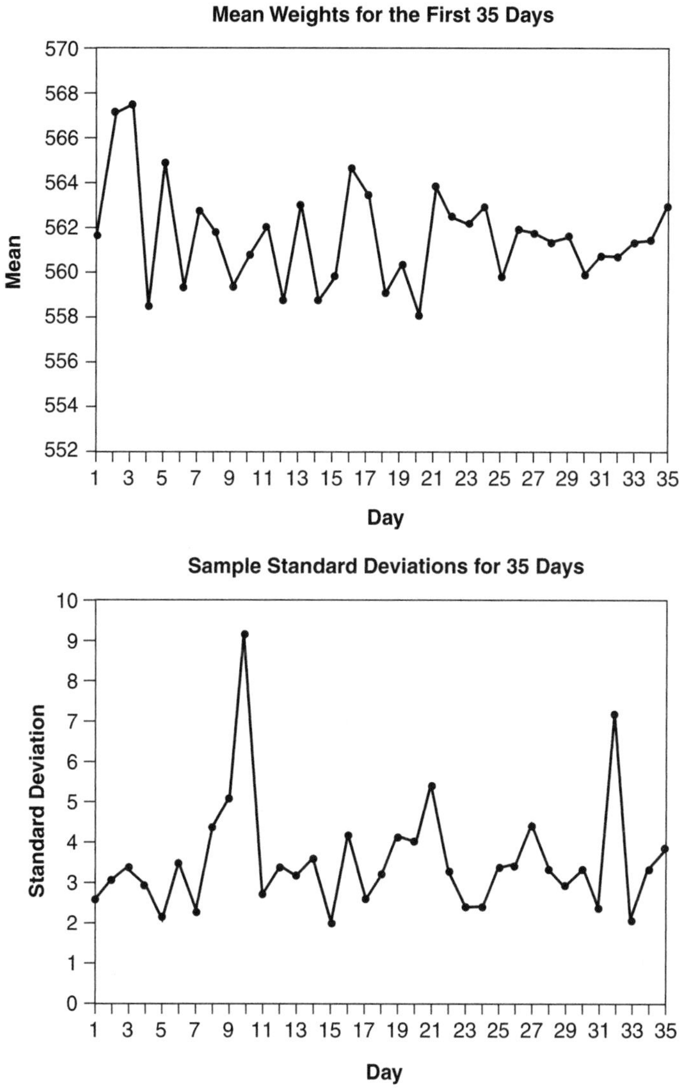

FIGURE 2.1 Sample Means and Standard Deviations for Spool Weights.

that goes on a spool was very worn. The part was replaced at the end of day 3 and the cause remedied.

Also notice that the standard deviation on day 10 is noticeably large. A check of the event log showed that this fact had not gone unnoticed by the operator on duty. Here, the machine part that would be the logical culprit when individual weights are erratic was found to be out of adjustment. It was reset immediately.

Note that the combination of charts in Figure 2.1 says something about day 10. The mean that day was a value near the center of the fallout of the

TABLE 2.1 Copper-Wire Spool Weights.

day	date	obs 1	obs 2	obs 3	obs 4	obs 5	obs 6	obs 7	mean	st dev	range
1	01-Apr	563	565.5	561.7	557.1	562.8	560.2	561.7	561.71	2.606	8.4
2	02-Apr	569.2	569.4	563.5	562.6	566	570.2	568.7	567.09	3.063	7.6
3	03-Apr	566.8	572.7	567.6	564.7	570.5	562.6	567.1	567.43	3.379	10.1
4	04-Apr	559.5	559.8	555.4	558.4	559.2	554.4	563	558.53	2.883	8.6
5	05-Apr	565.9	566.2	566.7	561.2	566.4	562.6	565	564.86	2.127	5.5
6	06-Apr	556.4	565.6	558.5	556.1	562.1	557	559.1	559.26	3.467	9.5
7	07-Apr	563.7	565.1	559.1	559.9	563.5	563.8	564	562.73	2.276	6
8	08-Apr	559.3	560.3	567.7	564.1	566.3	558.2	556.2	561.73	4.342	11.5
9	09-Apr	565.8	561	557.2	559.3	563	559.1	549.7	559.3	5.09	16.1
10	10-Apr	564.3	570.1	564.3	564.5	566.2	545.3	550.5	560.74	9.13	24.8
11	11-Apr	559.6	564	565.9	561.9	563.1	557.9	561.4	561.97	2.688	8
12	12-Apr	560.6	561.5	555.1	555.2	562.2	555.2	561.1	558.7	3.34	7.1
13	13-Apr	560.6	559.5	565.3	565.4	560.5	561.8	567.7	562.97	3.131	8.2
14	14-Apr	563.9	561.2	553.5	554.8	559.3	559.3	558.8	558.69	3.565	10.4
15	15-Apr	562.2	557.7	559	562.2	560.7	557.5	559.2	559.79	1.959	4.7
16	16-Apr	561.1	568.6	563	560.4	562.9	564.3	571.7	564.57	4.12	11.3
17	17-Apr	562.4	561.5	561.1	567.7	563.9	561.1	565.5	563.31	2.526	6.6
18	18-Apr	557.5	557	558.5	554.8	564.6	560.8	560	559.03	3.155	9.8
19	19-Apr	555.9	565.5	562.4	553.7	560.6	561.8	561.7	560.23	4.066	11.8
20	20-Apr	555.7	555.8	560.3	564.3	560.1	552.1	557.7	557.99	3.955	12.2
21	21-Apr	560.6	568.9	566.8	556.2	569.8	558.4	565.4	563.72	5.352	13.7
22	22-Apr	565.9	557	559.7	562.5	565.3	564.1	561.7	562.32	3.177	8.89
23	23-Apr	563.6	565.8	558.6	561.6	562.9	560.9	560.9	562.04	2.306	7.21
24	24-Apr	563	562	562.4	560.6	567.7	561.6	562	562.76	2.314	7.16
25	25-Apr	556.5	559.7	561.5	560.7	558.1	555.6	565.2	559.62	3.274	9.6
26	26-Apr	559.7	564.7	560.3	563.2	561.2	556.3	566.3	561.67	3.354	10
27	27-Apr	561.8	563.1	558.9	556.7	557	565	568.4	561.55	4.308	11.6
28	28-Apr	563.6	559	558.7	565.4	556.5	563.4	561.2	561.13	3.199	8.9
29	29-Apr	563.6	564.1	560.2	556.6	562.5	563.4	559	561.33	2.812	7.47
30	30-Apr	554.2	561	557.8	561.1	563.8	557.9	561.9	559.65	3.222	9.6
31	01-May	556.7	560.5	561.1	561.4	564.3	560.2	559.6	560.55	2.278	7.63
32	02-May	557	553.6	554.1	560.5	574.1	560	563.9	560.46	7.037	20.5
33	03-May	561.3	561.7	558.6	562.6	560.4	558.8	564	561.06	1.948	5.33
34	04-May	564.4	566.4	557.3	559.6	559.1	560.2	560.9	561.13	3.164	9.07
35	05-May	556.5	567.4	563.6	561.1	566.7	560.9	562.8	562.73	3.716	10.9

35 means—not an alarming value at all. If the average value for the sample is close to what the company would like it to be and the standard deviation is very high, then the sample must have at least one very heavy spool and at least one very light spool. An examination of Table 2.1 bears this out.

On day 32, variation seemed large, but no cause was found.

Data from days 2, 3, and 10 will be deleted from both charts in computing our estimates of μ and σ, as those data are not representative of how the process performs when working properly. We do not omit the data from day 32. Since no assignable cause could be found to explain the large standard deviation, we accept those data as representing something that occasionally would be expected to occur when this process is in control. This does not mean that there is not an assignable cause—only that none could be found. It is possible that someone very familiar with the equipment would know from experience what likely would have produced such a standard deviation and that someone probably fixed the problem but didn't log in the action taken, or that the value of S likely was just computed incorrectly. In a case like this, the decision to omit data from day 32 could be defended. Not all matters of judgment can be eliminated.

We now focus on the plot of successive standard deviations in Figure 2.1 for the points not omitted. For these points, variation appears to have been relatively stable. That is, there were no periods during which the points were obviously rising or obviously falling—the point cloud is approximately horizontal. Also, vertical swings do not become more and more (or less and less) severe. Even though we do not have a control chart for S yet, variation appears to have been in control.

We now similarly examine the plot of successive sample means. Swings from one point to the next are noticeably smaller for the last 10 days. As we shall see later, this usually is an indication that the process standard deviation, σ, has been reduced. However, the plot of standard deviations does not indicate that σ has gotten smaller. A potential explanation is that for the first three weeks, a particular operator was known to make occasional machine adjustments when he felt that the sample mean from the day before was *too far* from the target of 562.5 grams. The operator was applying overcontrol, which is known to inflate variation beyond natural levels.

The control limits and center line based on start-up data are viewed as tentative. Additionally, this is not as clean of a start-up situation as we would like to see. But these are all the data we have. Still, because variation has been stable and the average seems to have been centered at roughly the same place for 32 days, the process is taken to be in control, at least enough to obtain tentative estimates of μ and σ. With these estimates, we will be in a position to set up control charts.

At this company, the intention is to obtain monthly estimates of μ and σ. If future data make possible a clear judgment that σ has changed, control limits will be changed accordingly. Similarly, future data can indicate that a change in the center line is called for.

In order to determine a center line, one first must know if the intention is to keep the process centered as much as possible on some target value, or simply to monitor the process.

If interest is in controlling the process to keep measurements as close as possible to a target value T, then the y-intercept of the center line should be T.

If interest is just on monitoring, the y-intercept of the center line should estimate μ, the value at which the process has been centered lately.

Let k denote the number of samples upon which estimates of μ and σ will be based. In our example, $k = 32$. To estimate μ, we will find a mean of means. Let $\overline{X}_1$ be the mean of the first sample, $\overline{X}_2$ be the mean of the second sample, and so on, with $\overline{X}_k$ denoting the mean of the kth sample. The *mean of means* (or *grand mean*) is:

$$\overline{\overline{X}} = \frac{\Sigma \overline{X}_i}{k}.$$

For the copper wire data, this is

$$\overline{\overline{X}} = \frac{561.71 + 558.53 + 564.86 + \cdots + 562.73}{32} = 561.16.$$

Both location and variation need to be stable before attempting to establish a control chart for a new process.

Even if they are, the control chart should be viewed as tentative.

Figure 2.2 shows the average resistance in ohms of five electrical devices sampled from production four times a day for a new process. The 80 averages are plotted in sequential order. So are the sample standard deviations. The standard deviation plot shows that variation appears to be stable throughout the period—that is, "swings" from one point to the next are of the same order of magnitude before and after the drop in resistance. However, the process reflects lack of control because the plot of means shows an apparent shift downward in location sometime near sample 61. It would be inappropriate to establish a tentative control chart for this process because it has not met the rule-of-thumb guideline of having been in control for at least 25 to 30 samples. The process does appear to have been in control for the most recent samples, but only for about 20 samples, which is less than what is recommended by the rule of thumb.

Someone who wants to inappropriately *force* a control chart to be created out of these data might reason as follows. The process mean μ and

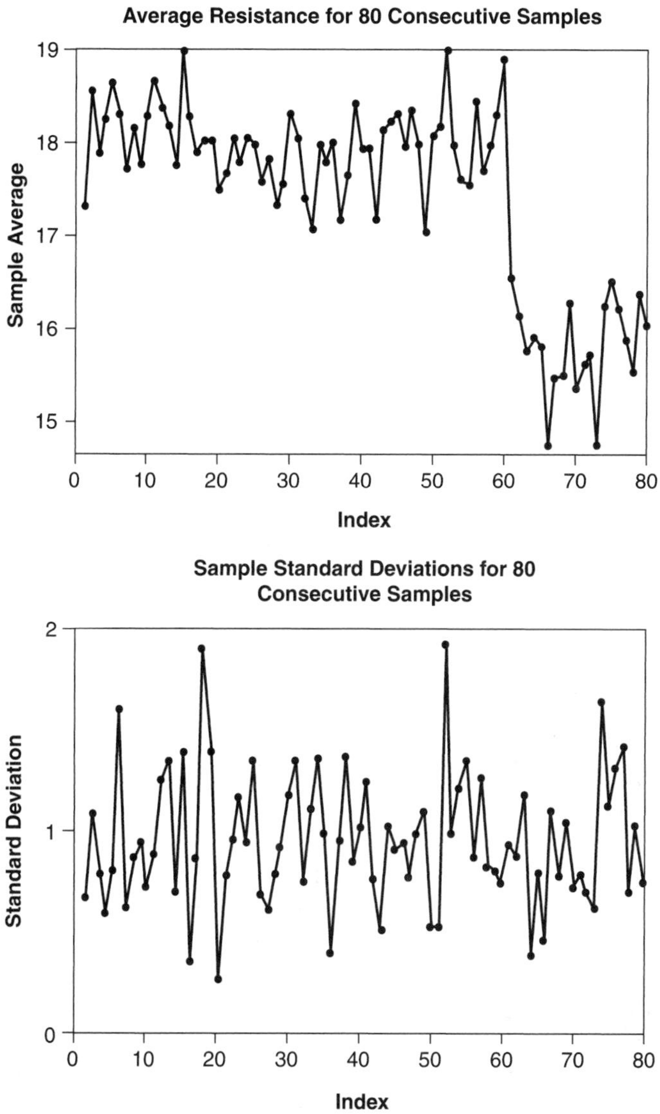

FIGURE 2.2 Means and Standard Deviations for 80 Samples of Electrical Devices.

process standard deviation σ obviously have been stable for about 20 samples, and due to our familiarity with the product and its manufacture, the process should stay stable, at least for a while. So let's make a tentative control chart. However, recall that the mean-of-means estimate of μ from the period of apparent control can be used to identify the location of the center line. Also, as we shall see in the next section, an estimate of σ from the period of apparent control will be needed to help determine the distance from either control limit to the center line. There must be enough data to

→	C1	C2	C3	C4	C5	C6	C7	C8	C9
	x1	x2	x3	x4	x5				
1	17.00	18.42	16.80	17.48	16.91				
2	19.06	18.36	17.31	17.87	20.11				
3	18.25	18.64	17.18	16.93	18.44				
4	18.53	17.70	18.30	17.65	19.03				
5	18.43	19.80	18.97	18.37	17.64				
6	20.04	18.71	16.47	19.52	16.76				
7	18.18	18.06	17.02	18.26	17.08				
8	17.44	17.31	19.30	18.77	17.95				
9	16.27	17.56	18.75	18.14	18.06				
10	18.48	18.75	17.75	19.06	17.35				

FIGURE 2.3 MINITAB Spreadsheet.

provide reasonably precise estimates of μ and σ. Otherwise, one cannot have faith in the control limits, center line, out-of-control signals based on the control chart, or on process and business decisions based on such a chart. The rule of thumb—a minimum of 25 to 30 samples—is based on experience and statistical theory. It would be an unwise statistical decision to violate the rule.

To obtain a plot of means such as is shown in Figure 2.2 using MINITAB, we assume that the sample data have been entered into MINITAB's spreadsheet (see Figure 2.3, which shows the data from the first 10 samples related to Figure 2.2). Here, the five sample values from the earliest sample occupy the first row of columns 1 through 5, sample values from the second sample occupy row 2 of columns 1 through 5, and so on. Neither graph in Figure 2.2 is a control chart, for neither has a center line or control limits. Each is a *time series plot,* meaning simply that values are plotted in the order in which the samples were obtained.

To obtain the first graph in Figure 2.2, we first have MINITAB compute the sample means. From the **Calc** menu, select **Row Statistics. . . .** A box will appear on the screen. In the box, from the list under **Statistic,** choose **Mean.** Click on the **Input variables** window in the box. Either type c1–c5 or, from a different window in the box that now displays which columns of your spreadsheet contain data, drag the mouse to highlight all of c1, c2, c3, c4, and c5, and hit **Select.** In either case, MINITAB now knows which columns contain the sample data. In the box, click on the window labeled **Store Results in:,** and type c6. Click **OK.** Note in the spreadsheet that in each row containing sample data, the sample mean has been entered in the cell where column 6 intersects that row. To name the column (*sample average* was used in Figure 2.2), click on the empty spreadsheet box just above the first sample mean of column 6, and type an appropriate name. If the column is not named, the y-axis of the plot we are about to produce will be labeled C6, indicating where the plotted values are from. From the **Graph** menu, choose **Time Series Plot . . . .** In the box that appears choose **simple** and click **OK.** Type c6 in the **Series** window (or in the window showing which columns contain data, either double-click on C6, or click once on C6 and hit **Select**). Hit the **Labels . . .** button and enter a title if you want to give the graph a title. Finally, click **OK** to produce the plot.

To obtain the second graph in Figure 2.2, first obtain the sample standard deviations in a manner similar to what was done for means. Namely, from the menus, choose **Calc,** then **Row Statistics . . . ,** and from the list under **Statistic,** choose **Standard Deviation.** In the **Input variables** box, you do not need to make changes from what was entered to produce means. Under **Store Results in:,** enter c7. Hit **OK.** To plot the standard deviations that are now listed in column 7, from the menus choose **Graph,** then **Time Series Plot .** Choose **simple,** and type c7 in the **series** window. Use the **Labels . . .** button to give the graph the proper title, and click **OK.**

PROBLEMS

1. If the 30 samples of ball-bearing diameters listed in Table 2.2 reflect a period of apparent control, estimate the current process mean μ. If the mean of sample 29 is determined to be an unusual value due to an assignable cause, what is the estimate of μ? The unit of measure is millimeters.
2. Enter the values of X_1, X_2, and X_3 from Problem 1 into columns 1, 2, and 3 of the MINITAB spreadsheet, and obtain a plot of sample means similar to that shown in Figure 2.2. Give the plot the title: Ball Bearing Sample Means from 0800 07-03-00 to 1400 07-04-00 in mm. Have the y-axis labeled: Meandiam. Similarly obtain a plot of successive sample standard deviations.
3. Should tentative control charts for the mean and standard deviation be established based on data that generated each of the pairs of plots shown in Figure 2.4? Give reasoning.
4. Refer to Problem 2. Some companies measure sample variation using the range rather than the standard deviation. From the **Calc** and **Row**

TABLE 2.2 Sample Ball-Bearing Diameters.

Sample	X_1	X_2	X_3	$\overline{X}$	S
1	8.01	7.96	7.91	7.960	0.050
2	7.91	8.10	8.05	8.020	0.098
3	8.30	7.88	8.01	8.063	0.215
4	8.08	7.94	8.17	8.063	0.116
5	7.94	7.92	7.82	7.893	0.064
6	8.02	7.93	7.81	7.920	0.105
7	8.02	8.10	8.19	8.103	0.085
8	8.05	8.14	8.01	8.067	0.067
9	7.95	8.08	7.89	7.973	0.097
10	8.11	7.84	8.03	7.993	0.139
11	8.02	8.23	8.12	8.123	0.105
12	8.13	7.91	8.04	8.027	0.111
13	8.09	7.99	7.89	7.990	0.100
14	7.80	8.12	8.10	8.007	0.179
15	8.06	8.13	8.06	8.083	0.040
16	8.07	8.04	8.03	8.047	0.021
17	8.06	8.20	8.10	8.120	0.072
18	8.07	8.14	7.94	8.050	0.101
19	8.17	8.23	7.91	8.103	0.170
20	8.11	8.00	8.01	8.040	0.061
21	8.06	7.96	7.94	7.987	0.064
22	7.99	8.18	7.72	7.963	0.231
23	7.92	8.16	8.01	8.030	0.121
24	7.98	7.94	7.82	7.913	0.083
25	8.16	7.95	8.14	8.083	0.116
26	8.12	7.94	8.06	8.040	0.092
27	7.96	8.01	8.07	8.013	0.055
28	8.01	7.97	8.05	8.010	0.040
29	8.17	8.23	8.10	8.167	0.065
30	8.00	7.72	8.04	7.920	0.174

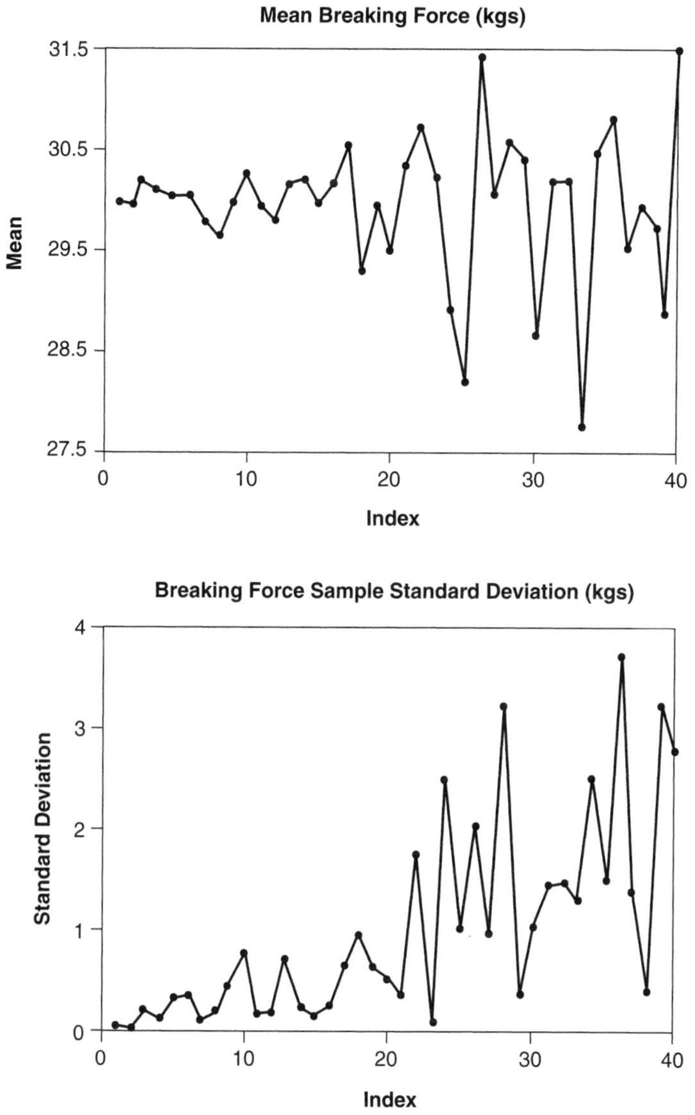

FIGURE 2.4 Sample Means and Standard Deviations.

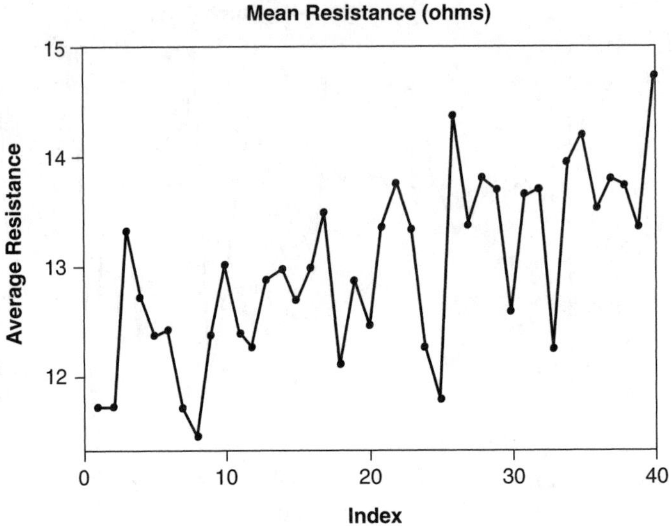

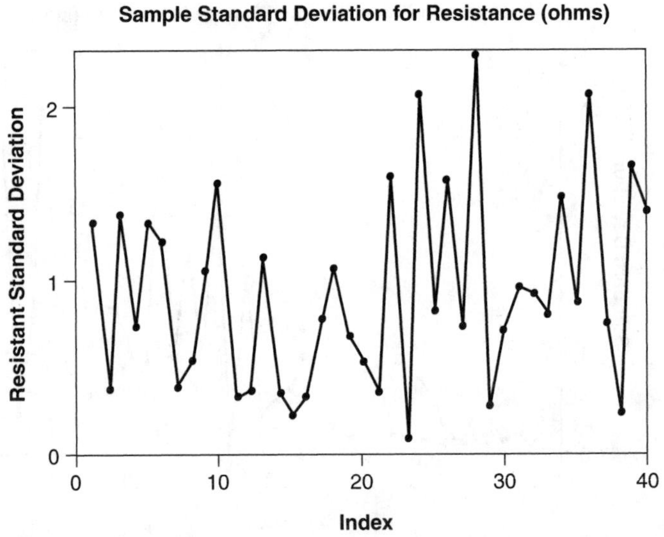

FIGURE 2.4 Continued.

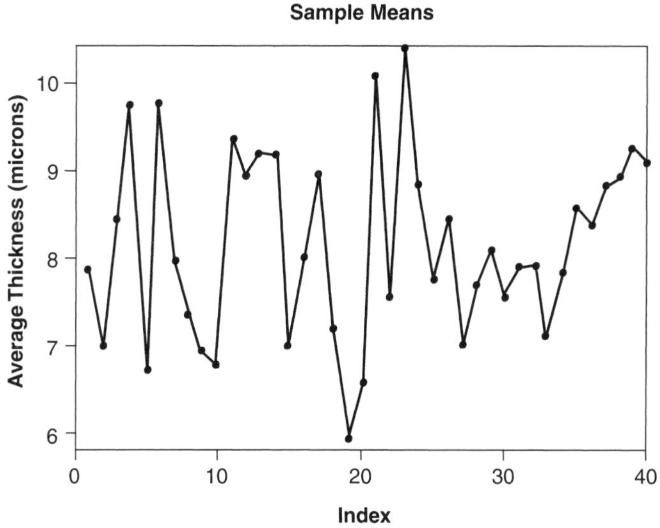

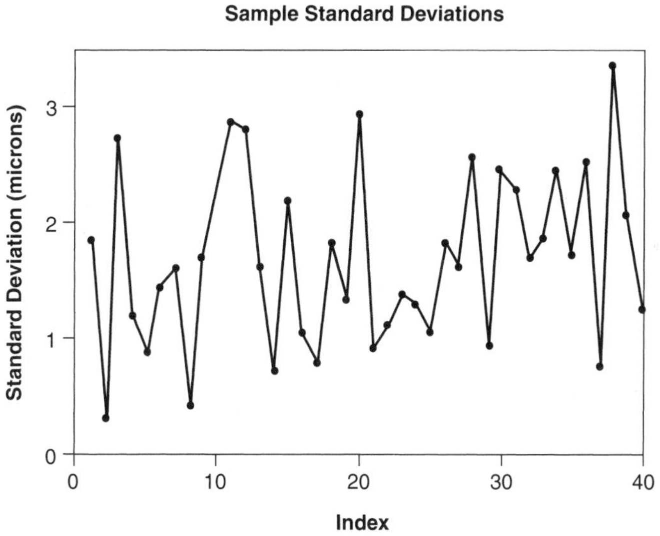

FIGURE 2.4 Continued.

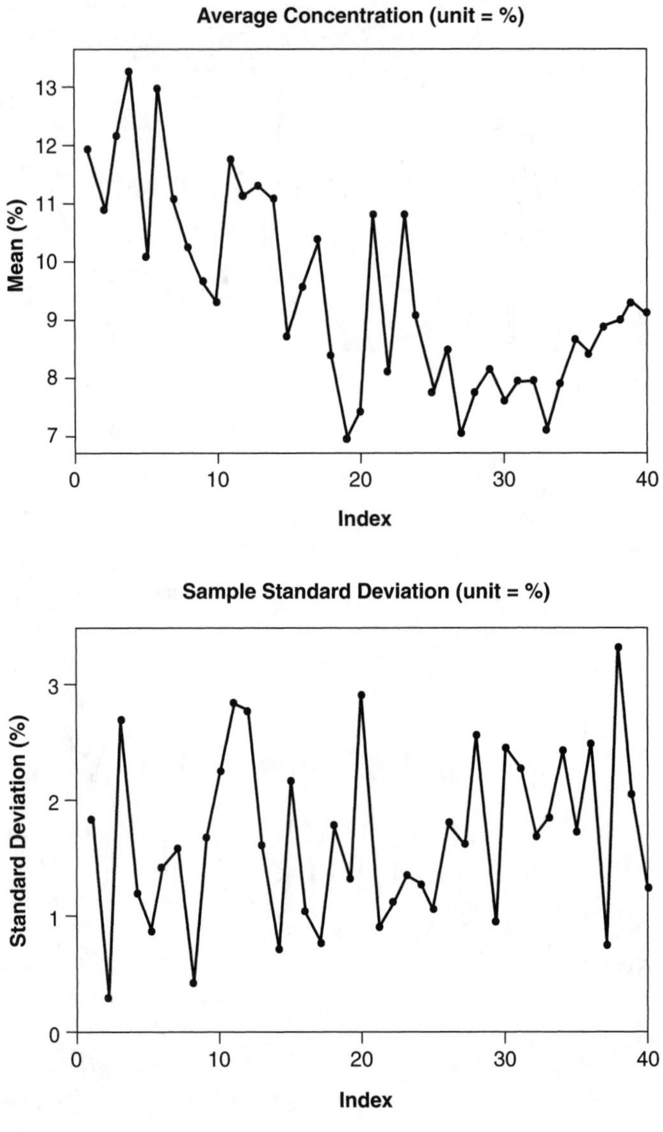

FIGURE 2.4 Continued.

Statistics ... features of MINITAB, or from a calculator or some other statistic package, obtain and graph the successive sample ranges. Does the graph indicate that variation is stable?

2.2 ESTABLISHING CONTROL LIMITS ON AN $\overline{X}$-CHART

Once variation and location have been stable and it has been decided whether a mean-of-means or target value will be used for the center line, it is time to determine the control limits for a tentative $\overline{X}$-chart. An underlying assumption for how we do this is that the values in the samples are *independent*, whether within or between samples. It is not the purpose of this book to introduce background material needed to be able to define independence in the precise language of statistical theory. Rather, we introduce the concept of independence from a practical, nontechnical point of view.

Namely, we take the view that the random sampling concept discussed in Section 1.4 applies. That is, we are getting all of our sample values *with replacement* from a conceptual population, a population with mean μ and standard deviation σ. Accordingly, two sample values in a row could be, say, substantially larger than either a target value or the historical mean μ, but this would be purely by chance, not because large values tend to be followed by more large values. An equivalent nontechnical meaning for independence within and between sample values is that any arbitrarily chosen collection of sample values from one or more samples does not influence nor is it influenced by other values in the same or other samples.

If sample values are independent within and between samples, it follows that sample means (or variances, standard deviations, or ranges) are independent. For example, two samples in a row could have sample means that are substantially larger than either the target or historical mean, but this would be purely by chance, not because large sample means tend to be followed by more large sample means. A more mathematically precise definition of independence may be found in Chapter 5 (Probability).

Independence is a very common assumption when charts indicate that fluctuation in sample values is due to common cause variation only.

Of course, an unusual, say large, sample mean would tend to be followed by more large sample means if the process mean has shifted upward. In such a case, sample means would not have the kind of fallout we would expect to see in relation to the center line and control limits on the $\overline{X}$-chart of a process that is in control. Out-of-control signals would be generated, and that is the whole idea.

An exception to independent samples is common in chemical industries when multiple batches of a product are made from chemicals stored in tanks. While the contents of a tanker truck filling a storage tank must meet specifications for that chemical, normally shipments of a given chemical will vary slightly from one tanker truck to the next. This variation is accepted as inherent in manufacturing with chemicals and has to be viewed as common cause variation. Alluding to an example used earlier, suppose the purity of Chemical A, an ingredient in the batch recipe of a chemical process, affects the hardness coefficient of the completed batch. Then hardness coefficients of consecutive chemical batches could tend to run somewhat low (or high) until the most recent shipment of Chemical A is used up, and batches are made using a new shipment of Chemical A. In this case, batch hardness is not independent from batch to batch. Rather, we say that batch hardness is *correlated* from batch to batch. We will see such an example in Chapter 4.

Meanwhile, back to out copper wire example, where we have seen that variation has been stable and we assume that samples are independent. We now consider several methods of estimating σ.

Pooling the Standard Deviations

Let $S_1, S_2, \ldots, S_k$ denote the standard deviations of the succession of samples obtained during the period when the process is in apparent control. Then

$$\sigma \approx S_p = \sqrt{\frac{\sum_{i=1}^{k} S_i^2}{k}}$$

The first estimator of σ we consider is S_p, commonly called the ***pooled standard deviation.*** A note on terminology is in order. S^2, the square of a sample standard deviation S, is called the sample ***variance.*** Based on statistical theory, the average of the k sample variances is what statisticians most often use to estimate σ^2. In fact, for a number of widely used theoretical statistical models, it can be shown that using $\sum_{i=1}^{k} S_i^2/k$ is the best method there is for estimating σ^2. It is natural then to take the square root of this average and use the result, S_p, to estimate σ.

Like all the methods used in this book to estimate σ, this method can be used even if μ is changing from sample to sample, so long as σ is not changing. Hence, if a plot of successive sample standard deviations reflects stability for 25 to 30 samples, then S_p may be used to estimate σ, regardless of whether the plot of sample means reflects control.

Dividing the Average Standard Deviation by a Constant

This method of estimating σ involves two steps. First we find $\overline{S}$, the average of the k sample standard deviations obtained during the period of apparent control:

$$\overline{S} = \frac{\sum_{i=1}^{k} S_i}{k}$$

Interestingly, using $\overline{S}$ itself to estimate σ is not a good idea. How well $\overline{S}$ estimates σ depends on the population being sampled and on the sample size. When the data come from a normal population, which is the usual idealized assumption and basis for the probabilities that get cited in connection with control charts, statistical theory shows that $\overline{S}$ tends to underestimate σ. The standard method of adjusting $\overline{S}$ in this case is to divide by a constant that is appropriate for the sample size n. The usual SPC notation for this constant is c_4. That is,

$$\sigma \approx \frac{\overline{S}}{c_4}$$

Values of c_4 may be found in Table I in the Appendix. For example, if during the period of apparent control $\overline{S} = 12.57$ mm and every sample consists of five widgets, then the estimate of σ is $\sigma \approx \frac{\overline{S}}{c_4} = \frac{12.57}{.9400} = 13.37$ mm. We soon will consider an example in which different methods of estimating are compared.

Dividing the Average Sample Range by a Constant

Let $R_1, R_2, \ldots, R_k$ denote the ranges of the succession of samples obtained during the period when the process is in apparent control. First we find $\overline{R}$, the average of these k sample ranges:

$$\overline{R} = \frac{\sum_{i=1}^{k} R_i}{k}$$

Then

$$\sigma \approx \frac{\overline{R}}{d_2}$$

where the value of d_2 is found in Table I in the Appendix. Again we assume that samples are drawn from a normal population. As was the case for c_4, d_2 depends on the sample size. In fact, the larger the sample size, the larger we expect the sample ranges, and hence $\overline{R}$, to be. It is not surprising, then, that as n increases, so does divisor d_2 (see Table in the Appendix I).

Consider once more the company that makes widgets and uses samples of size $n = 5$. Suppose $\overline{R} = 30.82$ mm. Recall that σ was estimated to be 13.37 mm based on $\overline{S} = 12.57$ mm. For an in-control process using samples of size 5 and having 12.57 mm as a typical value of $\overline{S}$, it happens that having an average range of 30.82 mm for those same samples would not be surprising. First, we expect ranges to be large in comparison to standard deviation values because the range is the distance between the two sample values that are farthest apart. Second, statisticians understand precisely how far apart these two values are expected to be, given the sample size. That these measures of variation are well understood usually is reflected in similar estimates of σ when different methods of estimation are used. The estimate of σ based on $\overline{R}$ here would be $\sigma \approx \dfrac{\overline{R}}{d_2} = \dfrac{30.82 \text{ mm}}{2.326} = 13.25$ mm similar to 13.37 mm, the estimate of σ based on $\overline{S}$.

On theoretical grounds, the sample range is the least preferred statistic because, by ignoring all but the two most separated sample values, it provides the most variable estimate of σ, except when $n = 2$. When $n = 2$, there is no statistical preference between $\overline{R}$ and $\overline{S}$. In fact when $n = 2$, both the sample range and sample standard deviation use all (both) sample values, and each statistic can be obtained from the other—algebraic manipulation shows that $R_i = \sqrt{2}S_i$ and $\overline{R} = \sqrt{2}\overline{S}$, always. In an era when manufacturing processes are automated and software packages commonly do the statistical calculations, S_p should be preferred when sampling from normal or near-normal populations, despite its more complicated formula. Software packages can do the heavy lifting. On the other hand, $\overline{S}$ and $\overline{R}$, particularly the latter, are easier for workers to understand. In particular, the statistical edge S_p has over $\overline{S}$ in estimating σ is slight. We cover all three methods; whoever sets up the control charts needs to select just one method.

We now see that the methods of estimating σ produce similar results for copper-wire spool weights (see Table 2.1). Table 2.3 shows the mean, standard deviation, variance, and range for each of the spool-weight samples. Samples 2, 3, and 10 are deleted for reasons discussed earlier. Recall that the sample size is 7. Later in this section, we discuss how to use MINITAB to obtain the mean of a column of numbers, including how to intentionally omit selected values from the computation. The bottom row of the table shows the values of $\overline{\overline{X}}$, $\overline{S}$, the mean of the sample variances, and $\overline{R}$. Estimates of σ using the three methods discussed with values from the bottom row of the table are fairly close:

TABLE 2.3 Copper-Wire Spool Weights: Mean, Standard Deviation, Variance, and Range for Period of Apparent Control.

Sample	Mean	S	S^2	Range
1	561.71	2.606	6.791	8.4
2	*	*	*	*
3	*	*	*	*
4	558.53	2.883	8.309	8.6
5	564.86	2.127	4.526	5.5
6	559.26	3.467	12.02	9.5
7	562.73	2.276	5.182	6
8	561.73	4.342	18.86	11.5
9	559.3	5.09	25.91	16.1
10	*	*	*	*
11	561.97	2.688	7.226	8
12	558.7	3.34	11.15	7.1
13	562.97	3.131	9.806	8.2
14	558.69	3.565	12.71	10.4
15	559.79	1.959	3.838	4.7
16	564.57	4.12	16.97	11.3
17	563.31	2.526	6.381	6.6
18	559.03	3.155	9.956	9.8
19	560.23	4.066	16.53	11.8
20	557.99	3.955	15.64	12.2
21	563.72	5.352	28.65	13.7
22	562.32	3.177	10.09	8.89
23	562.04	2.306	5.317	7.21
24	562.76	2.314	5.356	7.16
25	559.62	3.274	10.72	9.6
26	561.67	3.354	11.25	10
27	561.55	4.308	18.56	11.6
28	561.13	3.199	10.23	8.91
29	561.33	2.812	7.908	7.47
30	559.65	3.222	10.38	9.6
31	560.55	2.278	5.19	7.63
32	560.46	7.037	49.53	20.5
33	561.06	1.948	3.795	5.33
34	561.13	3.164	10.01	9.07
35	562.73	3.716	13.81	10.9
Average	561.16	3.336	12.27	9.48

$$\sigma \approx S_p = \sqrt{12.27 \text{ grams}^2} = 3.503 \text{ grams}$$

$$\sigma \approx \frac{\overline{S}}{c_4} = \frac{3.336 \text{ grams}}{.9594} = 3.477 \text{ grams}$$

$$\sigma \approx \frac{\overline{R}}{d_2} = \frac{9.48 \text{ grams}}{2.704} = 3.506 \text{ grams}$$

If the decision is made to use where the process has been centered lately (i.e., during the period of apparent control) to define the center line, then the location of the three defining lines for the $\overline{X}$-chart are:

$$\text{center line: } \overline{\overline{X}}$$

and

$$\text{(3 sigma) control limits: } \overline{\overline{X}} \pm 3\frac{\sigma}{\sqrt{n}}$$

If the company producing copper wire uses S_p to estimate σ, this places the center line at $\overline{\overline{X}} = 561.16$ grams and the upper and lower control limits at

$$\text{UCL} = \overline{\overline{X}} + 3\frac{\sigma}{\sqrt{n}} \approx \overline{\overline{X}} + 3\frac{S_p}{\sqrt{n}} =$$

$$561.16 + 3\frac{3.503}{\sqrt{7}} = 561.16 + 3.97 = 565.13 \text{ grams}$$

and

$$\text{LCL} = \overline{\overline{X}} - 3\frac{\sigma}{\sqrt{n}} \approx \overline{\overline{X}} - 3\frac{S_p}{\sqrt{n}} =$$

$$561.16 - 3\frac{3.503}{\sqrt{7}} = 561.16 - 3.97 = 557.19 \text{ grams.}$$

Figure 2.5 shows the control chart.

Note that while three samples were ignored in estimating μ and σ, in order to see the total production picture during this period, all 35 sample means have been plotted on the $\overline{X}$-chart. Due to the reasons samples 2 and 3 were ignored, it is not surprising that these samples each produce an out-of-control signal. It is also not surprising that the chart shows nothing unusual about the mean of sample 10. As we shall see, however, sample 10 will stand out on an *S*-chart or *R*-chart. *S*-charts and *R*-charts track process variation over time. In order to observe both process location and variation to-

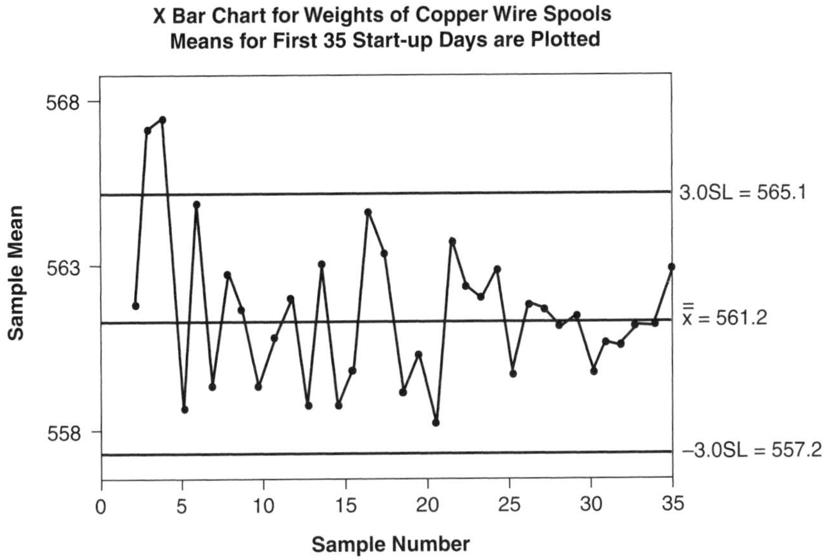

FIGURE 2.5 $\overline{X}$ Bar Chart for Weights of Copper Wire Spools (Means for First 35 Start-Up Days Are Plotted).

TABLE 2.4 Spool Weight Samples.

→	C1	C2	C3	C4	C5	C6	C7	C8
	x1	x2	x3	x4	x5	x6	x7	
1	563	565.5	561.7	557.1	562.8	560.2	561.7	
2	569.2	569.4	563.5	562.6	566	570.2	568.7	
3	566.8	572.7	567.6	564.7	570.5	562.6	567.1	
4	559.5	559.8	555.4	558.4	559.2	554.4	563	
5	565.9	566.2	566.7	561.2	566.4	562.6	565	
6	556.4	565.6	558.5	556.1	562.1	557	559.1	

gether, one of these charts usually accompanies an $\overline{X}$-chart. We examine these charts shortly.

Generating an $\overline{X}$-Chart with MINITAB

Suppose the copper-wire spool weight data from the 35 samples occupy the first 35 rows of columns 1 through 7. As a reference, the first six samples given in Table 2.1 are shown in Table 2.4.

To generate an $\overline{X}$-chart such as shown in Figure 2.5, from the menu choose **Stat,** then **Control Charts,** then **variables charts for subgroups,** and then **Xbar. . . .** A box containing buttons and windows will appear. In

TABLE 2.5 Copper-Wire Spool Weights—All Data Arranged in One Column.

→	C1	C2	C3	C4-D
			Sample value	Date
1			563.0	4/1/00
2			565.5	4/1/00
3			561.7	4/1/00
4			557.1	4/1/00
5			562.8	4/1/00
6			560.2	4/1/00
7			561.7	4/1/00
8			569.2	4/2/00
9			569.4	4/2/00
10			563.5	4/2/00
11			562.6	4/2/00
12			566	4/2/00
13			570.2	4/2/00
14			568.7	4/2/00
15			566.8	4/3/00

the box, we first describe how the data are arranged in the spreadsheet and specify which data will be used for the $\overline{X}$-chart. Data used for the chart can be arranged in several ways. If sample values occur in a one-sample-per-row format as they are in Table 2.3, then in the top-right window, choose **observations for a subgroup are in one row of columns.** Click on the window immediately below. Then either type c1–c7 in the window, which indicates that sample values occupy columns 1 through 7, or, in the window that identifies which columns contain data, drag the mouse to highlight the first seven columns and click on **Select.**

Alternatively, Table 2.5 shows another way in which the copper-wire spool data can be arranged in the spreadsheet. All sample values are in one column, C3 in this example. The top seven entries comprise sample 1, the next seven entries comprise sample 2, and so on. If data are arranged in this fashion, then in the top-right window, choose **All observations for a chart are in one column:** Click on the window immediately below and type c3. Then click on the window labeled **Subgroup sizes:** and type in a 7. If in another column there is an identifier that can be used to inform MINITAB which sample values belong together, such as the date on which the sample value was obtained (see Table 2.5), that information, rather than the sample size, can be entered in the **Subgroup size:** window. One would type c4 in the window (the **–D** suffix in the column name **C4-D** in the spreadsheet means that the column entries are dates). See a MINITAB manual for more details.

The next thing we do in the box is click the button labeled **Xbar Options. . . .** Under the **Parameters** tab, type 561.16 for the (historical) mean (recall that we chose to put the center line at 561.16, the mean-of-means with samples 2, 3, and 10 excluded). If nothing is typed in this window, MINITAB will use $\overline{\overline{X}}$, the mean-of-means for all the data just specified in the box, as the default center line location. In our case this would not be 561.16 because samples 2, 3, and 10 are included. If a target value is used for the center line, that is the value that should be typed in the window.

We elected to use 3.503 as the value for the process standard deviation. Click on the window labeled **Standard Deviation:** and type 3.503. If nothing is typed in this window, MINITAB will use S_p, the pooled standard deviation based on all the data just specified, as the default value for σ. Click **OK.** To give the control chart a title, click on the **Labels . . .** button and enter the desired title. Finally, click **OK,** which produces the $\overline{X}$-chart.

Computations and Omitting Samples

Based on time series plots of sample means and sample standard deviations, we decided to omit data from samples 2, 3, and 10 when computing $\overline{\overline{X}}$ and any of S_p, $\overline{S}$, and $\overline{R}$. In omitting data and doing the computation in MINITAB, it is a bad idea to delete data stored in their original columns. So we first copy the original data from those columns into an equal number of new columns. It is in the new columns where the data will be altered. Suppose the data occupy columns 1 through 7, as in Table 2.3. We will copy these data into columns 11 through 17 and omit rows 2, 3, and 10 in so doing. From the menu, choose **Data,** then **Copy Columns to Columns. . . .** The leftmost window inside the box that is produced identifies the columns that contain data. Drag the mouse to highlight columns 1 through 7, then click on **Select.** Alternatively, in the **Copy from Columns:** window, one can type: c1-c7. Next click on the **Store Copied Data in Columns** window and choose the option **In current worksheet, in columns.** Click on the window immediately below and type c11–c17. Since we want to omit some rows in the copy process, we now click on the **Subset the Data . . .** button, which produces a new box. Click on the circle labeled **Specify which Rows to Exclude,** then click on the circle labeled **Row numbers:** Type 2 3 10 in the window and click **OK.** MINITAB allows options such as omitting rows by sample date. The reader is referred to a MINITAB manual. In the first box, also click **OK.** Columns 11 through 17 will then appear as in Table 2.6.

Note that there are 32 rows of data instead of 35. Sample order has been preserved, but the data from rows 2, 3, and 10 in columns 1 through 7 have been removed. To obtain $\overline{\overline{X}}$, we first put sample means in column 18 using menus as was done in the preceding section: from the **Calc** menu, select **Row Statistics. . . ,** from the list under **Statistic,** choose **Mean,** for **Input variables** choose c11 through c17, under **Store Results in:** type c18, and click **OK.** Column 18 now contains the sample means. Next we find the mean of

TABLE 2.6 Spool Weights, Copied with Three Rows Omitted.

→	C10	C11	C12	C13	C14	C15	C16	C17
1		563.0	565.5	561.7	557.1	562.8	560.2	561.7
2		559.5	559.8	555.4	558.4	559.2	554.4	563.0
3		565.9	566.2	566.7	561.2	566.4	562.6	565.0
4		556.4	565.6	558.5	556.1	562.1	557.0	559.1
5		563.7	565.1	559.1	559.9	563.5	563.8	564.0
6		559.3	560.3	567.7	564.1	566.3	558.2	556.2
7		565.8	561.0	557.2	559.3	563.0	559.1	549.7
8		559.6	564.0	565.9	561.9	563.1	557.9	561.4
9		560.6	561.5	555.1	555.2	562.2	555.2	561.1
10		560.6	559.5	565.3	565.4	560.5	561.8	567.7
11		563.9	561.2	553.5	554.8	559.3	559.3	558.8
12		562.2	557.7	559.0	562.2	560.7	557.5	559.2
13		561.1	568.6	563.0	560.4	562.9	564.3	571.7
14		562.4	561.5	561.1	567.7	563.9	561.1	565.5
15		557.5	557.0	558.5	554.8	564.6	560.8	560.0
16		555.9	565.5	562.4	553.7	560.6	561.8	561.7
17		555.7	555.8	560.3	564.3	560.1	552.1	557.7
18		560.6	568.9	566.8	556.2	569.8	558.4	565.4
19		565.9	557.0	559.7	562.5	565.3	564.1	561.7
20		563.6	565.8	558.6	561.6	562.9	560.9	560.9
21		563.0	562.0	562.4	560.6	567.7	561.6	562.0
22		556.5	559.7	561.5	560.7	558.1	555.6	565.2
23		559.7	564.7	560.3	563.2	561.2	556.3	566.3
24		561.8	563.1	558.9	556.7	557.0	565.0	568.4
25		563.6	559.0	558.7	565.4	556.5	563.4	561.2
26		563.6	564.1	560.2	556.6	562.5	563.4	559.0
27		554.2	561.0	557.8	561.1	563.8	557.9	561.9
28		556.7	560.5	561.1	561.4	564.3	560.2	559.6
29		557.0	553.6	554.1	560.5	574.1	560.0	563.9
30		561.3	561.7	558.6	562.6	560.4	558.8	564.0
31		564.4	566.4	557.3	559.6	559.1	560.2	560.9
32		556.5	567.4	563.6	561.1	566.7	560.9	562.8
33								

the numbers in column 18—that is, we find $\overline{\overline{X}}$. From the **Calc** menu, select **Column Statistics. . . ,** choose **Mean,** use C18 as the **Input variable:,** and click **OK.** The value of $\overline{\overline{X}}$, 561.16, will appear in the session window (the window above the spreadsheet). $\overline{S}$ or $\overline{R}$ would be computed in a similar manner.

Obtaining the value of S_p takes several steps. We first need to put sample standard deviations in some column, say column 19. From the **Calc** menu, select **Row Statistics. . . ,** from the list under **Statistic,** choose **Standard deviation,** for **Input variables** choose c11 through c17, under **Store Results in:** type c19, and click **OK.** Column 19 now contains the sample standard deviations. Next we use the calculator feature of MINITAB to find the mean of the squares of the standard deviations, take the square root of this mean, and put the value in the top cell in column 20. This value is S_p. From the **Calc** menu choose **Calculator. . . ,** and in the **Store result in variable** window, type c20. In the **Expression** window type sqrt(mean(c19**2)). Note that ** is a calculator button whose role is to raise an expression to some power. The expression c19**2 squares all the values in column 19. Note that 3.50346, the value of S_p, now occupies the top cell in column 20.

An $\overline{X}$-chart that depicts only the 32 samples could be produced without computing $\overline{\overline{X}}$ or S_p in advance by using columns 11 through 17 as the input data and not entering values for μ and σ. MINITAB will use default values of $\overline{\overline{X}}$ and S_p for these.

If a company wishes to center the chart on a target value T, the three defining lines for the $\overline{X}$-chart are:

$$\text{center line: } T$$

and

$$\text{control limits: } T \pm 3\frac{\sigma}{\sqrt{n}}$$

For the copper-wire spool weights, this places the center line at $T =$ 562.50 grams and the upper and lower control limits at

$$UCL = T + 3\frac{\sigma}{\sqrt{n}} \approx T + 3\frac{S_p}{\sqrt{n}} = 562.50 + 3\frac{3.503}{\sqrt{7}} =$$

$$562.50 + 3.97 = 566.47 \text{ grams}$$

and

$$LCL = T - 3\frac{\sigma}{\sqrt{n}} \approx T - 3\frac{S_p}{\sqrt{n}} =$$

$$562.50 - 3\frac{3.503}{\sqrt{7}} = 562.50 - 3.97 = 558.53 \text{ grams}$$

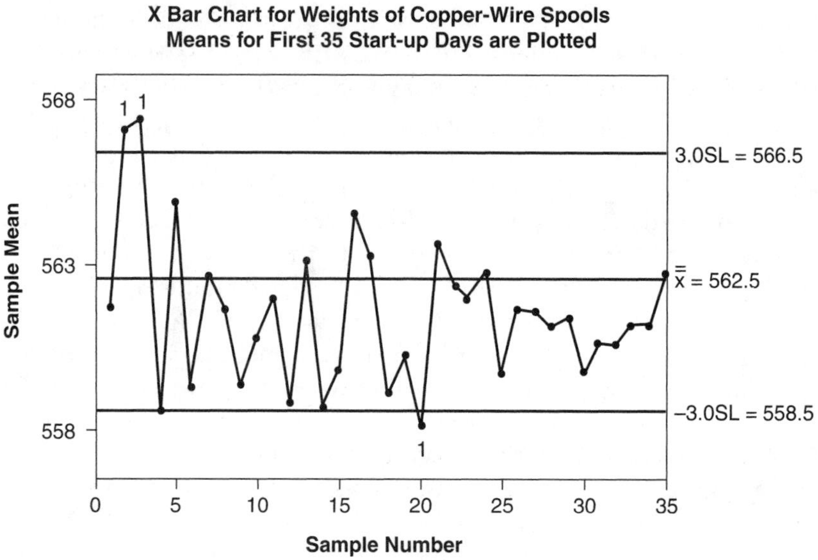

FIGURE 2.6 $\overline{X}$ Bar Chart for Weights of Copper-Wire Spools (Means for First 35 Start-Up Days Are Plotted).

Figure 2.6 shows the control chart. It is clear that the process is centered below target and that if it is important for the company to be centered on target, they must make adjustments to recenter the process. Sample 20 registers one out-of-control signal—a point below the lower control limit—and a number of other points are flirting with the lower control limit. We will soon be introduced to other kinds of out-of-control signals that would have told us before sample 20 that the process might be centered below target.

Control Limits for an $\overline{X}$-Chart Without Estimating σ First

Up to this point the procedure described for setting up an $\overline{X}$-chart has been to estimate the process standard deviation σ and then set the control limits at $\overline{\overline{X}} \pm 3\sigma/\sqrt{n}$ (or at target $\pm 3\sigma/\sqrt{n}$). Generally, those who want to *really know* their process would want to know or have an estimate for σ. As observed, the pooled standard deviation S_p estimates σ and may be substituted for σ in either formula to obtain the control limits. Alternatively, we have seen that we can substitute an estimate of σ based on either $\overline{R}$ or $\overline{S}$. However, for completeness, we point out that it is possible to compute the control limits directly from $\overline{R}$ or $\overline{S}$ using:

Control limits for an $\overline{X}$-chart: $\overline{\overline{X}} \pm A_2\overline{R}$ or $\overline{\overline{X}} \pm A_3\overline{S}$.

Values of A_2 and A_3 may be found in Table II in the Appendix. Because control limits may be obtained in different ways, one can conclude, appropriately, that there must be relationships between the constants that are involved with different methods. For example, $\overline{\overline{X}} \pm 3\sigma/\sqrt{n}$ with σ replaced by $\overline{R}/d_2$ should yield the same values as $\overline{\overline{X}} \pm A_2\overline{R}$. It follows that $A_2\overline{R}$ must equal

$\dfrac{3\overline{R}}{d_2\sqrt{n}}$, and then that $A_2 = \dfrac{3}{d_2\sqrt{n}}$.

For example, when $n = 4$, $\dfrac{3}{d_2\sqrt{n}} = \dfrac{3}{2.059(2)} = .7285 = A_2$.

Do the Sample Values Come from a Population That Is Approximately Normally Distributed?

Notice from Figure 2.5 that the points not related to special-cause problems all fall between the control limits (sometimes referred to as 3-sigma limits because they are set three standard deviations [of $\overline{X}$] from the mean). Recall that based on properties of normal distributions, 99.7% of the points should be expected to fall between the upper and lower 3-sigma limits and .3% to not do so. Since .3% of 32 = .003(32) = .096, we wouldn't expect any of the 32 points to fall outside the control limits.

However, the value .3% is based on the theoretical assumption that the sample means from this period of apparent control come from a population with a normal or close to normal distribution. Recall that if the sample values come from a normally distributed population, then the sample means also come from a normally distributed population. That the sample means come from a normal or near-normal distribution is the theoretical model behind $\overline{X}$-charts.

> Once it is decided from time series plots for either $\overline{X}$ and S or for $\overline{X}$ and R that there is apparent control, that is, that all sample values could reasonably have come from the same population, there needs to be some kind of check as to whether sample means could reasonably have come from a population having a normal or near-normal distribution.

If sample means don't appear to have come from a near-normal distribution, then the reader should consult a statistician to design an appropriate control chart, because the standard practices associated with $\overline{X}$-charts can lead to making inappropriate decisions about the process. At a minimum we suggest a simple visual check of two histograms. One is the histogram of all sample values during the period of apparent control. The other is a histogram of *z*-scores of sample means.

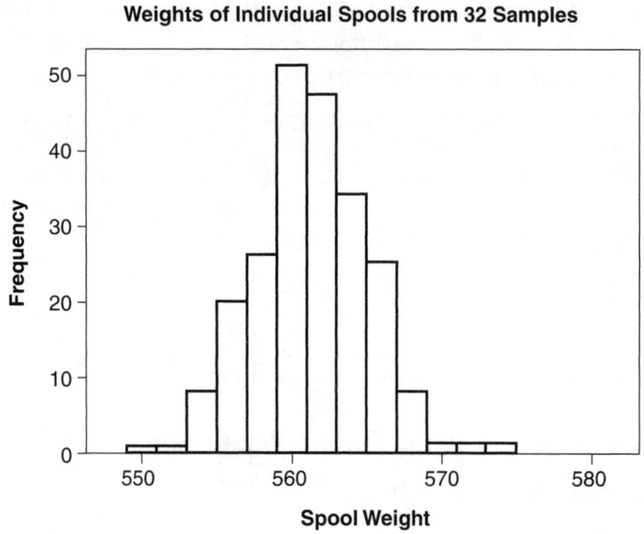

FIGURE 2.7 Weights of Individual Spools from 32 Samples.

Figure 2.7 is a histogram of the 224 spool weights from the 32 samples taken when the process was in apparent control. Note the mound shape, which is key. The nonmathematically stated version of the Central Limit Theorem given in Chapter 1 says the following about an output variable that is a sum or average of independent inputs: as long as no inputs dominate, the more independent inputs there are, the more similar to a normal distribution the output distribution will be. As a practical guide, if the sample size is 30 or more, one is justified in treating the sample means as if they come from a normal distribution, regardless of the shape of the distribution the sample values come from. That is, when $n \geq 30$, individual sample values can come from an odd-shaped distribution and, still, the sample means will come from a near-normal distribution, so that an $\overline{X}$-chart can be used.

In many manufacturing settings, it is common to have small sample sizes, as small as 4 or 5, or even 2. One of the beautiful and practical truths about the Central Limit Theorem is that if the sample values come from a distribution that is mound shaped to begin with, then a large sample size as large as 30 or more is not required in order that sample means come from a near normal distribution. The histogram in Figure 2.7 is mound shaped and even reasonably symmetric. In this case, we can anticipate that sample means should come from an approximate normal distribution, even for a sample size as small as 2 or 3.

Before examining a histogram of z-scores of sample means, we discuss how to produce a histogram of the sample values using MINITAB. First we need to arrange all data values in one column if they are not already arranged that way.

Assume that the data are arranged as portrayed in Table 2.6, where the sample values occupy columns 11 through 17. We will *stack* all 224 spool weights into column 21 with the numbers from column 11 occupying the first 32 rows of column 21, the numbers from column 12 occupying the next 32 rows of column 21, and so on. From the menus, choose **Data,** then **Stack/Unstack,** and then **Stack Columns. . . .** This produces a box. In the window labeled **Stack the following columns:,** type c11-c17. In the **Store the stacked data in:** window, **the one labeled column of current worksheet,** type c21. Then click **OK.** The data are now stacked in column 21 as described. To obtain a histogram, go to the menus, choose **Graph,** then **Histogram. . . .** A box will appear. Click on the option **simple** and click **OK.** In the top cell of the **Graph variables:** window, type **the one labeled column of current worksheet.** c21 and click **OK.** The histogram will appear. It is possible to control how many rectangles there are in the histogram and where they are centered. The reader is referred to a MINITAB manual for details.

We now examine a histogram of z-scores of the sample means. Recall that based on 224 sample values from 32 samples, we estimated the process mean to be $\mu = 561.16$ grams (the value of $\overline{\overline{X}}$) and $\sigma = 3.503$ grams (the value of S_p). So sample means come from a population with approximate mean 561.16 grams and approximate standard deviation $\sigma/\sqrt{n} \approx 3.503/\sqrt{7} = 1.324$. We now compute a z-score for each sample mean:

$$z = \frac{\text{mean} - \overline{\overline{x}}}{\sigma/\sqrt{n}} = \frac{\text{mean} - 561.16}{1.324}$$

To do this in MINITAB, first recall that the sample means occupy column 18. We will put their z-scores in column 21. From the **Calc** menu, choose **Calculator. . . .** In the box that appears, type c21 in the **Store result in variable:** window, and in the **Expression:** window type (c18-561.16)/1.324. Click **OK.** The desired z-scores are now in column 21. Figure 2.8 shows a histogram of these 32 z-scores.

How do we use Figure 2.8 to help us decide whether sample means could have come from a normal or near-normal distribution? Based on the Empirical Rule, if the population of sample means has a normal or near-normal distribution, we would expect most sample means to be near μ, and only .3% of the means to fall more than $3\sigma/\sqrt{n}$ from the mean. In other words, we would expect most z-scores for sample means to be near zero, and virtually all z-scores to fall between -3 and 3, as they do in Figure 2.8. So, using an $\overline{X}$-chart for this process is justified.

Figure 2.9 shows sample data from a process that has no assignable cause problems but which occasionally produces values far from the process mean. Figure 2.10 shows the distribution of z-scores for the sample means. It is not the case that virtually all z-scores of sample means of this process *live* between -3 and 3. This is a process that inherently will produce points outside 3-sigma control limits at a rate far in excess of the customary .3% rate

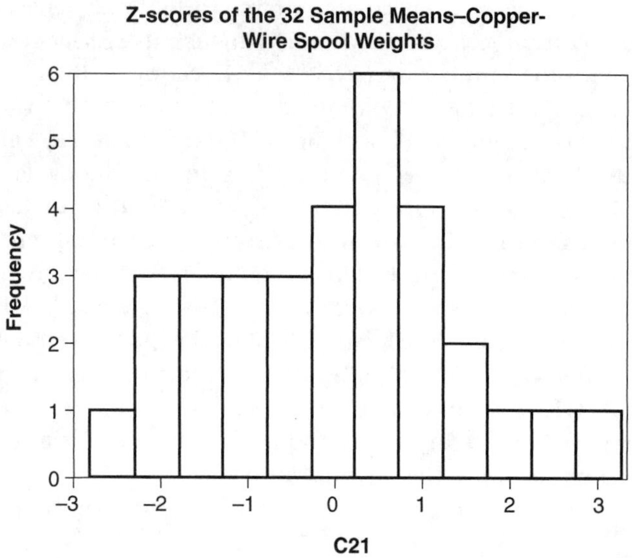

FIGURE 2.8 z-Scores of the 32 Sample Means—Copper-Wire Spool Weights.

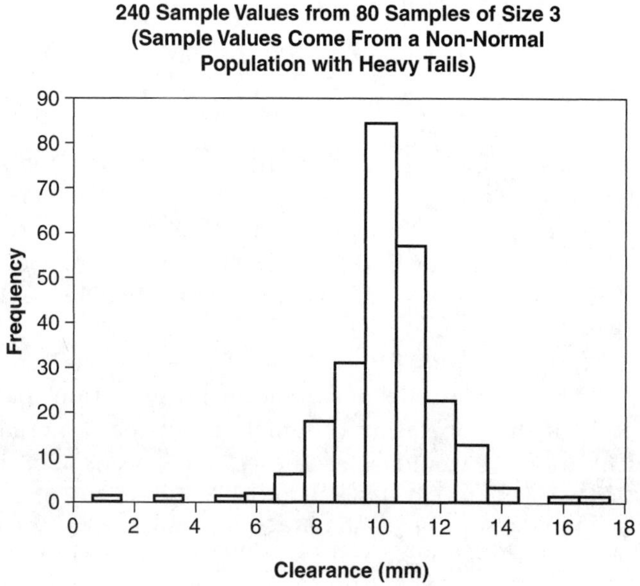

FIGURE 2.9 240 Sample Values from 80 Samples of Size 3 (Sample Values Come from a Non-Normal Population with Heavy Tails).

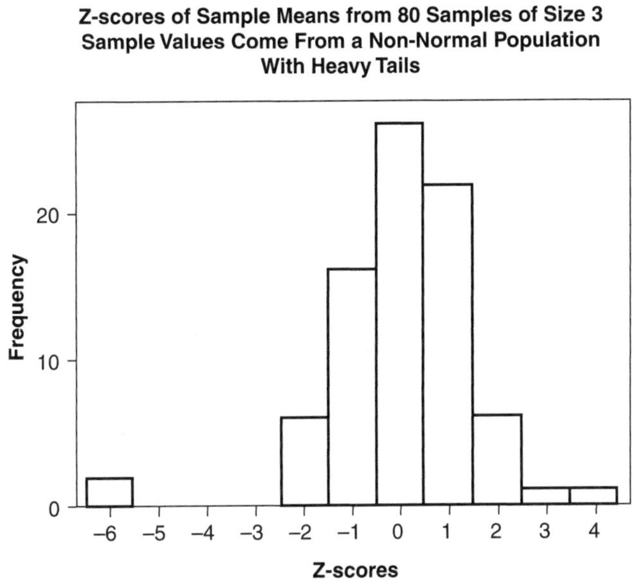

FIGURE 2.10 z-Scores of Sample Means from 80 Samples of Size 3 (Sample Values come from a Non-Normal Population with Heavy Tails).

that pertains when sampling is from a normal population. Management has two choices. The not-so-good choice is to force an $\overline{X}$-chart on the sample means anyway, and accept the fact that when no assignable causes are present, people will not infrequently be *chasing noise*—that is, out-of-control signals will cause them to look for assignable causes that aren't there. A better choice is to work with a statistician who can decide upon an appropriate data transformation. With an appropriate transformation, original sample values could be transformed into values that come from a near-normal distribution. An $\overline{X}$-chart could be kept on the transformed data. Such transformations are beyond the scope of this text.

Nevertheless, it is important that management continually check for normality of process data to ensure that the control chart being used is the proper tool. We would be in remiss not to mention that statistics as a discipline has developed some wonderful statistical tests for normality and that monitoring of process data can be done in a very sophisticated manner, should management wish to do so. The histograms we have discussed represent a practical minimum approach.

PROBLEMS

1. Enter into a spreadsheet the three columns of data as given in Problem 1 of Section 2.1. Use statistical software to:

   a. Obtain the column of sample means, the column of standard deviations, and the column of sample ranges.
   b. Find $\overline{\overline{X}}$, $\overline{S}$, $\overline{R}$, and S_p.
   c. Estimate μ.
   d. Compare the three estimates of σ that are based on $\overline{S}$, $\overline{R}$, and S_p.
   e. Find the center line and control limits for an $\overline{X}$-chart (based on $\overline{S}$).
   f. Find the center line and control limits for an $\overline{X}$-chart (based on $\overline{R}$).
   g. Find the center line and control limits for an $\overline{X}$-chart (based on S_p).

2. Control charts are kept on $\overline{X}$ and S for the diameters of holes (in centimeters) drilled in metal plates. The sample size is 4. Based on 34 samples during which the process was in apparent control, $\overline{\overline{X}} = 1.203$ cm and $\overline{S} = .019$ cm. Estimate σ and identify the center line and 3-sigma limits for an $\overline{X}$-chart.

3. Control charts are kept on $\overline{X}$ and S for the contents of a soft drink product sold in cans. The sample size is 6. Based on 30 samples during which the process was in apparent control, $\overline{\overline{X}} = 354.4$ mL and $S_p = 2.03$ mL. Estimate σ and identify the center line and 3-sigma limits for an $\overline{X}$-chart.

4. Control charts are kept on $\overline{X}$ and R for the resistance in resistors produced by a manufacturing process. The sample size is 5. Based on 44 samples during which the process was in apparent control, $\overline{\overline{X}} = 112.7\ \Omega$ and $\overline{R} = 2.74\ \Omega$. Estimate σ and identify the center line and 3-sigma limits for an $\overline{X}$-chart.

5. A company is starting a new product run of spools of thread ordered by a customer. Two spools are sampled every half hour and tested to find the minimum tension (in centinewtons) needed to make thread pull from the spool without snagging. Data for the first 26 samples, along with time series plots of means and standard deviations, are given in Table 2.7 and Figure 2.11.

   a. Does the process appear to be in control? (See the plots of $\overline{X}$ and S values.)
   b. If the answer to the previous question is yes, estimate μ and σ (use $\overline{S}$ to estimate σ).
   c. Find the center line and 3-sigma control limits for an $\overline{X}$-chart for this process, assuming that the company elects to use where the

TABLE 2.7 Tension Data: 26 Samples of Size Two.

Sample	1	2	3	4	5	6	7	8	9	10	11	12	13
Obs 1	7.42	6.94	6.82	7.15	7.16	7.22	7.05	7.06	6.65	7.43	7.33	7.54	6.75
Obs 2	7.47	7.18	7.24	7.38	7.13	7.41	7.51	7.85	7.35	7.02	7.45	7.15	7.43

Sample	14	15	16	17	18	19	20	21	22	23	24	25	26
Obs 1	6.86	7.10	6.66	6.99	7.08	6.79	7.38	7.55	7.84	7.31	7.52	6.48	7.60
Obs 2	7.16	7.36	7.52	7.51	7.69	7.54	7.19	7.42	7.48	7.13	7.08	7.23	6.71

process seems to be centered as the center line of the control chart on which future sample means will be plotted.

6. Use a statistics software package to enter and display the existing data of Problem 5 on an $\overline{X}$-chart. Does the chart indicate that the process was in control?
7. In starting up a new process, three product units are sampled every half hour and a key measurement made from each unit. Start-up data for the first 15 samples are shown in Table 2.8.
   a. Does the process appear to be in control?
   b. If the answer to the previous question is yes, estimate μ and σ (use S_p to estimate σ).
   c. Find the center line and 3-sigma control limits for an $\overline{X}$-chart for this process, assuming that the company elects to use where the process seems to be centered as the center line of the control chart on which future sample means will be plotted.
8. In starting up a new process, three product units are sampled every half hour and a key measurement made from each unit. Start-up data for the first 30 samples are shown in Table 2.9.
   a. Does the process appear to be in control? (See Figure 2.12)
   b. Estimate μ, if appropriate.
   c. Estimate σ, if appropriate, using S_p.
   d. Is it appropriate to set up a tentative control chart on which to plot means of future samples?
9. An $\overline{X}$-chart has 814.20 as the upper 3-sigma limit and 795.75 as the lower 3-sigma limit. The sample size is 4. Estimate the process mean and standard deviation.
10. Table 2.10 shows product unit measurements for the first 28 samples of size 2 from production of a new product. Enter these values into

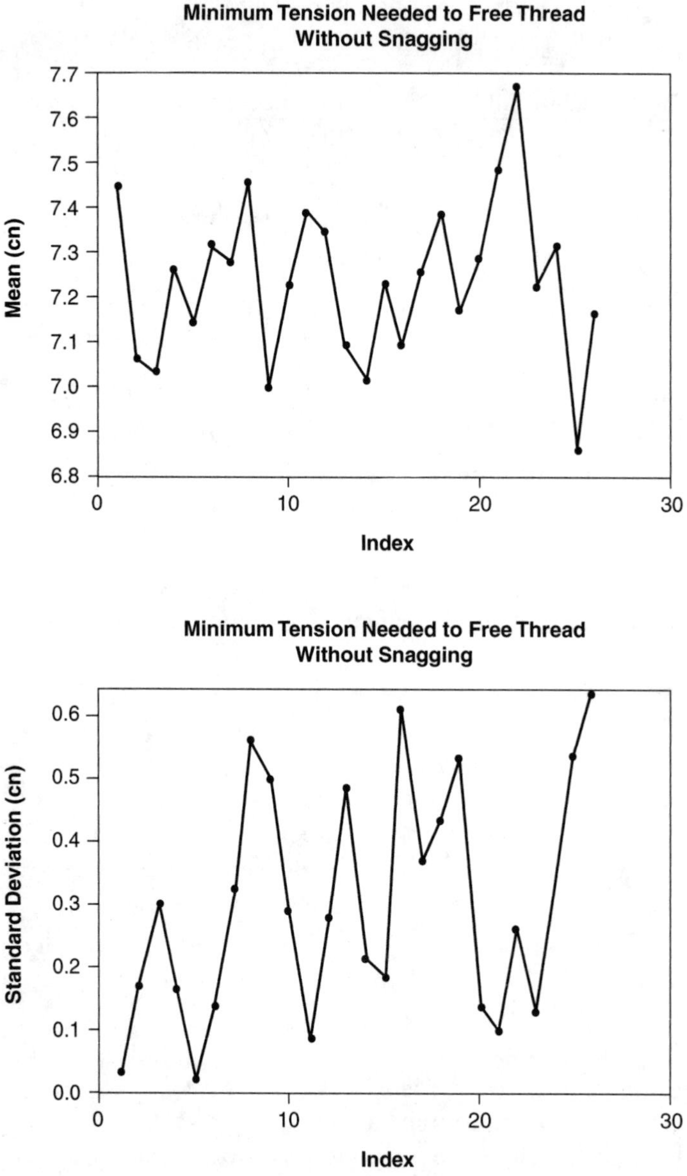

FIGURE 2.11 Minimum Tension Needed to Free Thread Without Snagging.

TABLE 2.8 Start-Up Data—Fifteen Samples.

Sample	1	2	3	4	5	6	7	8	9	10	11	12	13	14	15
Obs 1	9.40	9.73	9.96	9.40	9.73	9.96	9.40	9.73	9.96	9.40	9.73	9.96	9.40	9.73	9.96
Obs 2	9.95	9.95	9.38	9.95	9.95	9.38	9.95	9.95	9.38	9.95	9.95	9.38	9.95	9.95	9.38
Obs 3	9.82	9.29	9.60	9.82	9.29	9.60	9.82	9.29	9.60	9.82	9.29	9.60	9.82	9.29	9.60

TABLE 2.9 Start-Up Data—Thirty Samples.

Sample	1	2	3	4	5	6	7	8	9	10	11	12	13	14	15
Obs 1	9.35	9.12	9.84	9.35	9.12	9.84	9.35	9.12	9.84	9.35	9.12	9.84	9.35	9.12	9.84
Obs 2	9.13	9.83	9.95	9.13	9.83	9.95	9.13	9.83	9.95	9.13	9.83	9.95	9.13	9.83	9.95
Obs 3	9.92	9.37	9.15	9.92	9.37	9.15	9.92	9.37	9.15	9.92	9.37	9.15	9.92	9.37	9.15

Sample	16	17	18	19	20	21	22	23	24	25	26	27	28	29	30
Obs 1	8.98	8.72	8.52	8.98	8.72	8.52	8.98	8.72	8.52	8.98	8.72	8.52	8.98	8.72	8.52
Obs 2	8.98	8.77	8.18	8.98	8.77	8.18	8.98	8.77	8.18	8.98	8.77	8.18	8.98	8.77	8.18
Obs 3	8.53	9.14	8.78	8.53	9.14	8.78	8.53	9.14	8.78	8.53	9.14	8.78	8.53	9.14	8.78

two MINITAB columns or into the spreadsheet of a comparable software package. Obtain time series plots of $\overline{X}$ and of S.

a. Does the process appear to be in control?
b. Some investigation reveals that data from samples 16 and 18 are unreliable because an operator did not follow measurement protocol on those samples. Copy the data into two new columns either omitting samples 16 and 18, as discussed earlier, or copy without omitting, but when copying is completed, replacing the 4 sample values with asterisks (MINITAB recognizes asterisks as missing values). Estimate μ and σ (using S_p) for this process.
c. Find an appropriate center line and 3-sigma limits for an $\overline{X}$-chart upon which future means can be plotted.
d. Obtain an $\overline{X}$-chart for the start-up data with samples 16 and 18 removed. Does the chart seem to reflect control? Be sure to do Problem 11 next. Problem 11 indicates that there may be a problem in using an $\overline{X}$-chart for this process.

11. Using the data from Problem 10 with samples 16 and 18 deleted, obtain a histogram of sample values and a histogram of the z-scores of

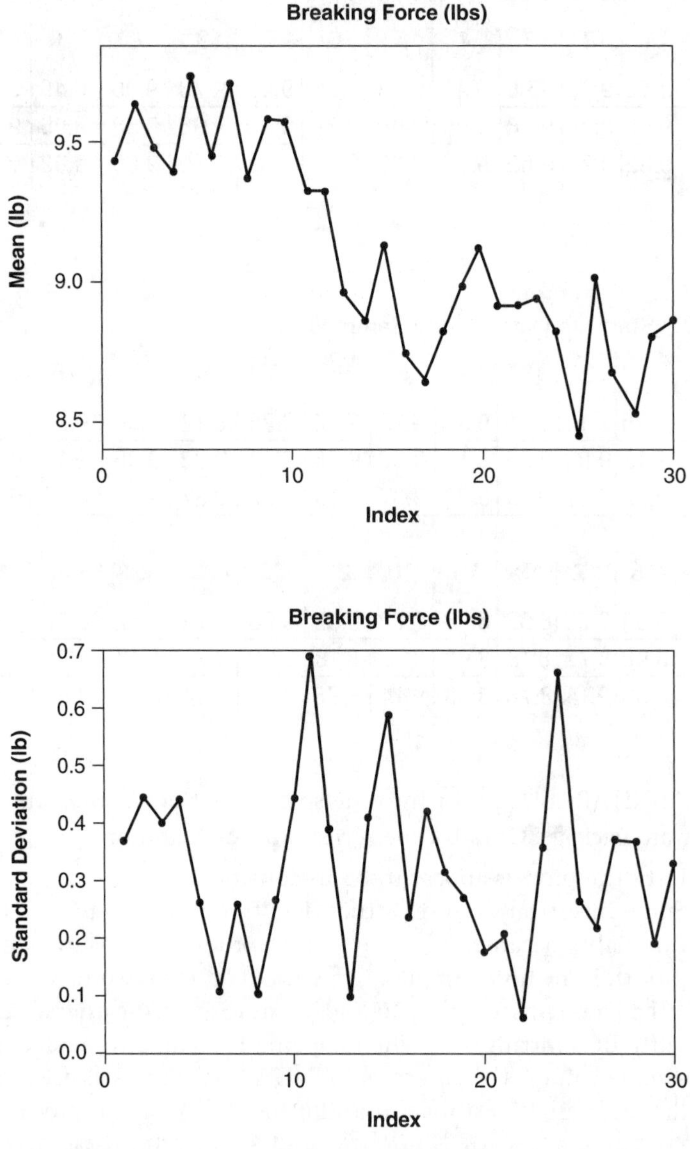

FIGURE 2.12 Sample Means and Standard Deviations for Breaking Force Start-up Data.

TABLE 2.10 Product Unit Measurements—28 Samples of Size Two.

X_1	203.66	204.83	203.66	204.83	203.66	204.83	203.66	204.83	203.66	204.83
X_2	207.94	200.93	207.94	200.93	207.94	200.93	207.94	200.93	207.94	200.93
X_1	203.52	209.80	203.52	209.80	203.52	209.80	203.52	209.80	203.52	209.80
X_2	203.10	205.04	203.10	205.04	203.10	205.04	203.10	205.04	203.10	205.04
X_1	202.16	202.48	202.16	202.48	202.16	202.48	202.16	202.48		
X_2	204.51	203.99	204.51	203.99	204.51	203.99	204.51	203.99		

sample means as discussed in this section. The fallout of *z*-scores is not what we would expect to see if sample means come from a near-normal distribution. Instead of *z*-scores having a mound shape and *living* on the interval from -3 to 3, the *z*-scores are concentrated essentially between -1 and 1. This is also indicated by the control chart obtained in part d. of Problem 10, in that all the points *hug* the center line. The sample data are skewed to the right, and a sample size of 2 may not be enough to ensure that sample means come from a near-normal distribution. More samples should be taken from this process before an $\overline{X}$-chart is implemented. Why?

12. A company wants to center its light bulb manufacturing process on the target value of 100 watts. Based on experience with the process, it is known that the process standard deviation is .25 watts. What should the center line and control limits be for an $\overline{X}$-chart if the sample size is 4?

2.3 CHARTING VARIATION ON *S*-CHARTS OR *R*-CHARTS

We will examine a number of control charts, all of which have a center line and control limits that are three appropriate standard deviations from the center line. The two we examine in this section are used for assessing process variation.

S-Charts

An *S*-chart is updated by adding to the plot the sample standard deviation of the most recent sample. This chart allows us to detect changes in variation over time, either sudden or gradual. An $\overline{X}$-chart is usually accompanied by an *S*-chart (or an *R*-chart), and, used in concert, can give information that

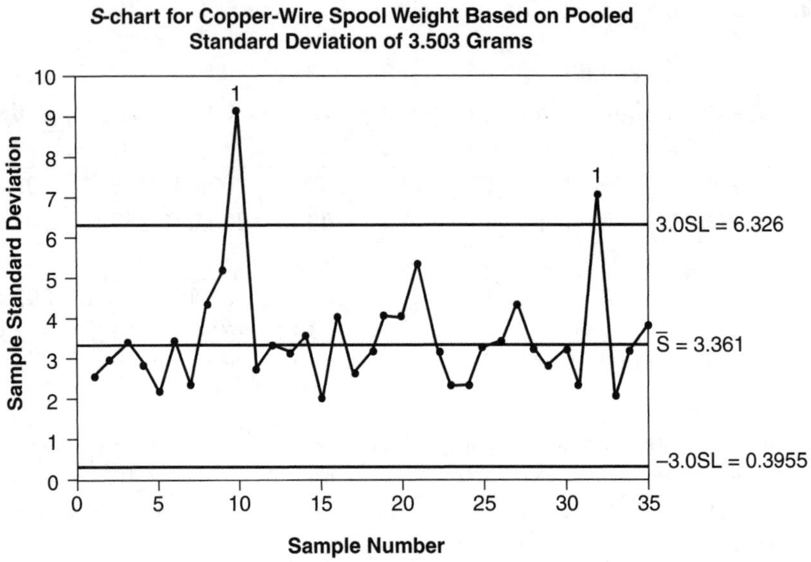

FIGURE 2.13 S-Chart for Copper-Wire Spool Weight (Based on Pooled Standard Deviation of 3.503 Grams).

neither chart would reveal by itself. If σ is known or there is a good estimate of σ, then the defining lines of an *S*-chart are:

center line: $c_4\sigma$

$LCL_S = B_5\sigma \qquad UCL_s = B_6\sigma$

Values for B_5 and B_6 may be found in Table III in the Appendix.

Recall from the 32 samples of copper-wire spool weights that $\overline{S} = 3.336$ and $S_p = 3.503$. If S_p is used to estimate σ, the center line and control limits of an *S*-chart for this process are

center line: $.9594(3.503) =$ 3.361 grams
$UCL_S = 1.81(3.503) =$ 6.340 grams
$LCL_S = .11(3.503) =$ .385 grams

The *S*-chart is shown in Figure 2.13. A comparison of the values computed by formulas such as those just given to those shown in software-produced control charts sometimes reflects disagreement after the first few significant digits. This disagreement is slight and is simply the result of roundoff error and the fact that software packages may know the value of constants like B_5 to a different number of decimal places than is given in a book.

If one wishes to use $\overline{S}$ rather than S_p as the basis of an *S*-chart, then the defining lines of an *S*-chart are:

center line: $\overline{S}$

$LCL_S = B_3\overline{S}, \quad UCL_s = B_4\overline{S}$

where values for B_3 and B_4 may be found in Table III in the Appendix. For the spool weights, the center line and control limits of an *S*-chart for this process would be:

center line:	3.336 grams
$UCL_S = 1.88(3.336) =$	6.272 grams
$LCL_S = .12(3.336) =$	.400 grams

values similar to the ones obtained for an *S*-chart based on S_p. It is common for statistical software packages to allow the user to choose either the pooled standard deviation or $\overline{S}$ as the basis for the defining lines on an *S*-chart.

All 35 sample standard deviations are shown in Figure 2.13. Recall that samples 2, 3, and 10 did not figure into estimates of μ and σ because of assignable causes, but sample 32 did because no assignable cause was found (which is different than saying that there was no assignable cause). In any case, we are reminded that points outside the control limits do occur occasionally when there is nothing wrong with the process.

Note that when the sample size is 5 or less, $B_3 = B_5 = 0$, which makes $LCL_S = 0$. Consider the case in which the control limits are a function of $\overline{S}$. The upper and lower 3-sigma limits on an *S*-chart are designed to be equally distant from the center line (located at $\overline{S}$). The smaller the sample size, the greater this distance is supposed to be. An examination of the values of B_3 and B_4 as n decreases confirms this. While it is possible for the values in a sample to be so extremely dispersed when the sample size is 5 or less that the standard deviation falls beyond a given distance in excess of $\overline{S}$, it is impossible for a standard deviation to fall beyond that distance from $\overline{S}$ in the other direction because a standard deviation cannot be negative. Therefore the only points outside the 3-sigma limits when $n \leq 5$ must be above the upper 3-sigma limit. Some people prefer to say that a lower 3-sigma limit does not exist rather than to say it equals zero.

This raises a natural question when $n \geq 6$. What could be wrong with having a sample standard deviation so small that it falls below the lower 3-sigma limit? After all, aren't small standard deviations supposed to be good? There are several reasons to look for an assignable cause when such an out-of-control signal occurs. Something accidental, beneficial to the process, and having an assignable cause could have occurred. For example, suppose a company's regular supplier of component parts could not meet current demand and so additional parts were ordered on a temporary basis from a different vendor, and it is discovered that those parts are what has reduced variation in the end product. The company could choose to make the new company its sole supplier. On the downside, the small variation in sample values could be the result of a measuring device being stuck or broken or even of someone "cooking" the data.

R-Charts

An *R*-chart is updated by adding to the plot the range of the most recent sample. If σ is known or there is a good estimate of σ, then the defining lines of an *R*-chart are:

center line: $d_2\sigma$

$LCL_R = D_1\sigma$

$UCL_R = D_2\sigma$

Values for D_1 and D_2 may be found in Table IV in the Appendix.

Using $S_p = 3.503$ for σ, the center line and control limits of an *R*-chart for this process would be

center line: $2.704(3.503) =$ 9.472 grams

$UCL_R = 5.20(3.503) =$ 18.22 grams

$LCL_R = .20(3.503) =$ .70 grams.

The *R*-chart is shown accompanying the $\overline{X}$-chart in Figure 2.14, where $\overline{\overline{X}} = 561.16$ grams was used for the historical mean and 3.503 grams for the historical standard deviation.

It is routine for an $\overline{X}$-chart to be accompanied by either an *S*-chart or *R*-chart.

An *S*-chart or *R*-chart used in concert with an $\overline{X}$-chart conveys information that might not be noticed otherwise. For example, Figure 2.14 shows that the means of samples 2 and 3 are exceptionally large because all the values in each sample are large, and not because of one isolated large value in the sample (since the variation within each sample is typical for this process). In sample 10 there is nothing unusual about the mean, but the standard deviation is large, which says there must be both exceptionally large and exceptionally small values in the sample. This also is true of sample 32 although variation is not so extreme.

One need not prescribe values for the historical standard deviation or mean to set up an *R*-chart. In MINITAB, the default for the historical mean is $\overline{\overline{X}}$ and the default method for estimating σ is to use $\overline{R}$. However, if default values are used, samples 2, 3, and 10 must be omitted because all selected data are used in computing these means.

If in setting up an *R*-chart the estimate of σ is to be based on $\overline{R}$, then the defining lines of an *R*-chart are:

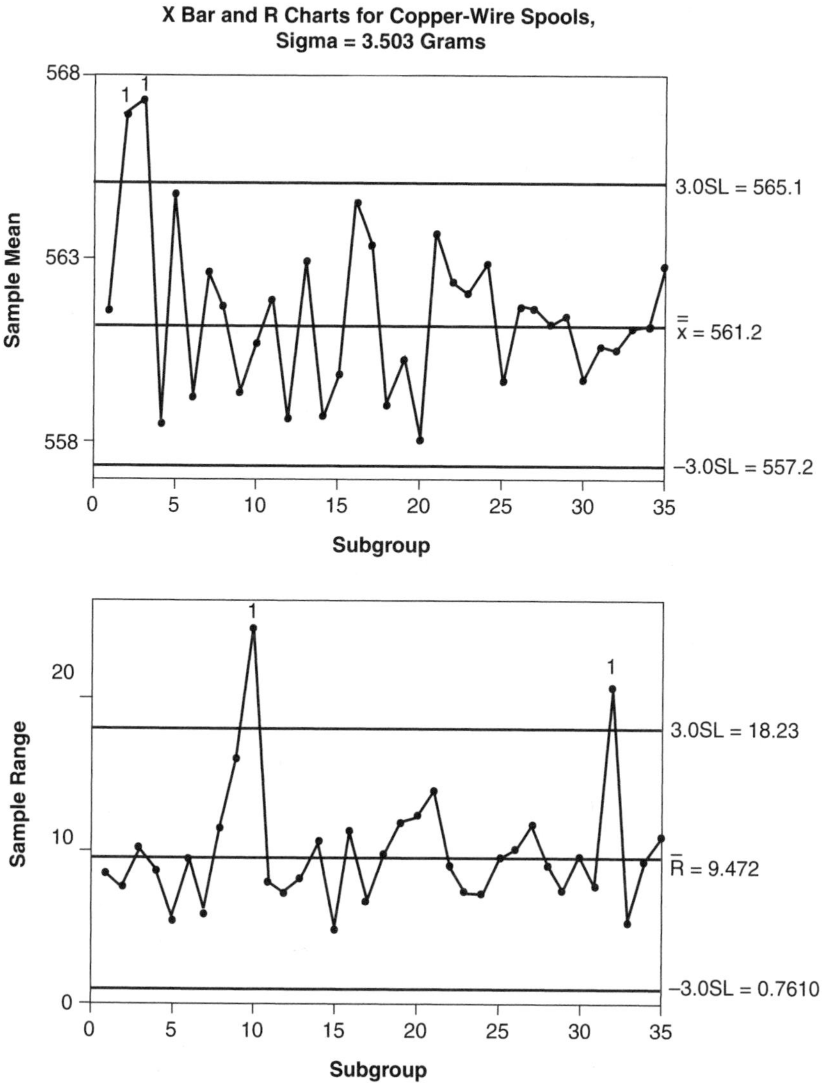

FIGURE 2.14 $\overline{\mathrm{X}}$ Bar and *R*-Charts for Copper-Wire Spools, Sigma = 3.503 Grams.

center line: $\overline{R}$

$LCL_R = D_3\overline{R}$

$UCL_R = D_4\overline{R}$

Values for D_3 and D_4 may be found in Table IV in the Appendix.

Minitab Procedures

The procedure for obtaining S-charts, R-charts, or either of these with an $\overline{X}$-chart is similar to what was done to obtain just an $\overline{X}$-chart. From the **Stat** menu, choose **Control Charts,** and then one of **R. . . , S. . . , Xbar-R. . . , Xbar-S.** A box will appear. In the same sequence of steps used with $\overline{X}$-charts, specify how the data are arranged and which column(s) contain the data. Click on the rightmost **options. . .** button and another box will appear. To estimate Sigma Select the **estimate** tab and select an option (e.g., **Sbar, Rbar, Pooled standard deviation**). Click **OK.** As before, click on the **Label** button if it desired to give the control chart(s) a specific title.

Control Charts Can Reveal Improper Subgrouping

Suppose control charts are kept on $\overline{X}$ and S for the thickness of metal motor mounting plates. Each subgroup (i.e., sample) consists of three thickness measurements, one at each of the three corners of a single plate. We shall see why this plan, not a good one, is an example of what is called *improper subgrouping.* We examine what can be expected to happen with such a sampling plan. Based on the 27 samples summarized in the Table 2.11, we have $\overline{\overline{X}} = .7689$ cm and $S_p = .01941$ cm. Hence for the $\overline{X}$-chart, the center line is given by .7689 cm and the 3-sigma limits by

$$\overline{X} \pm 3\frac{\sigma}{\sqrt{n}} \approx .7689 \pm 3\frac{.01941}{\sqrt{3}} = .7689 \pm .03362, \text{ or } .7353 \text{ cm and } .8025 \text{ cm.}$$

For the S-chart the center line is at $c_4\sigma \approx .8862(.01941) = .01720$ cm, the upper 3-sigma limit at $B_6\sigma \approx 2.28(.01941) = .04425$ cm, and the lower 3-sigma limit is viewed as either zero or not existing (why?). The $\overline{X}$- and S-charts appear in Figure 2.15. (Why would values just obtained for a

TABLE 2.11 Sample Means and Standard Deviations for Plate Thickness (cm).

Sample	1	2	3	4	5	6	7	8	9
$\overline{X}$	0.800	0.780	0.733	0.747	0.770	0.807	0.780	0.807	0.760
S	0.017	0.020	0.012	0.015	0.020	0.031	0.010	0.025	0.026
Sample	**10**	**11**	**12**	**13**	**14**	**15**	**16**	**17**	**18**
$\overline{X}$	0.727	0.783	0.740	0.760	0.747	0.760	0.753	0.793	0.787
S	0.006	0.025	0.000	0.020	0.012	0.035	0.025	0.015	0.038
Sample	**19**	**20**	**21**	**22**	**23**	**24**	**25**	**26**	**27**
$\overline{X}$	0.790	0.767	0.757	0.797	0.753	0.753	0.727	0.803	0.780
S	0.010	0.012	0.021	0.006	0.015	0.006	0.021	0.015	0.000

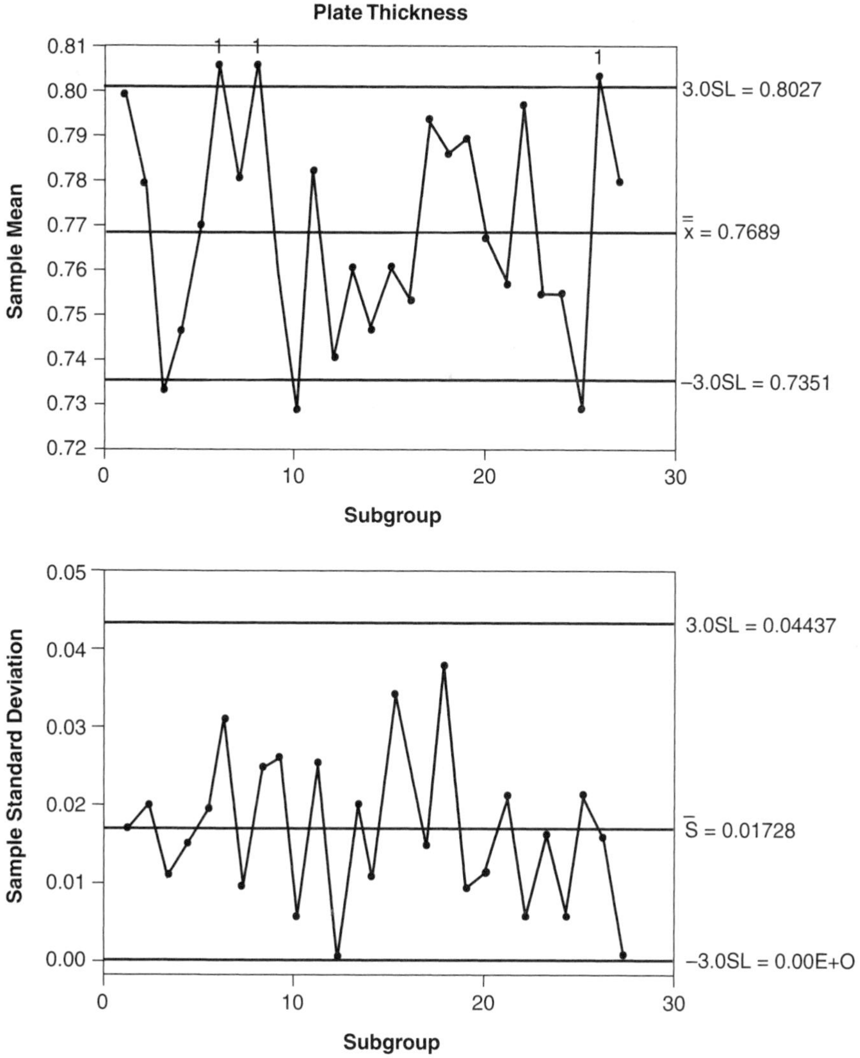

FIGURE 2.15 *X*-and *S*-Charts for Plate-Thickness Data.

center line or 3-sigma limits be expected to show slight roundoff error in comparison to values given in Figure 2.15, which is based on 81 sample values?) Variation appears stable, as does process location (imagine these charts without 3-sigma limits, as a time series plot would appear). Yet despite the apparent stability, the $\overline{X}$-chart reflects lack of control. This is because the proper σ for the $\overline{X}$-chart is being underestimated. A sample standard deviation based on the three measurements on a given plate assesses thickness consistency on that plate. So, because of the way the subgroups are organized, the pooled standard deviation S_p is an aggregate measure of within-plate variation. However, thickness could vary substantially from plate to plate, even if all plates are made by the same machine. Also, the plates could be made by different machines, different shifts, or with different workers at the controls. The S_p being measured here does not assess plate-to-plate variation. In fact, if thickness within plates is extremely uniform, the three measurements off a plate come close to being three attempts to measure the same value. One should not be surprised, then, when a plot of the average plate thicknesses of different plates doesn't fit nicely between the 3-sigma limits being used.

What is needed for an $\overline{X}$-chart on plate thickness is a subgrouping method according to which each mean plotted on the chart is the mean thickness of the several plates in each subgroup, with one thickness value being obtained from each plate. Every hour, for example, four plates could be sampled from production, the thickness measured at the center of each plate, and $\overline{X}$ would be the mean on those four values. There could be circumstances in which it makes sense for each of the four measurements to be the average of three measurements taken from the same plate. In either case, $n = 4$, and whether 4 is the best sample size to use for this particular process depends on various costs not presented and on how quickly management wants to be able to react to location and variation shifts of a size they consider to be serious. Management should consult with a statistician on selecting an optimal sample size. This example serves only to call attention to an obvious improper subgrouping situation.

If excessive within-plate variation can cause problems and the risk of it is real, then keeping an S-chart to monitor within-plate variation should be considered. In this case, the subgrouping method deemed improper for the $\overline{X}$-chart may be right for this particular S-chart.

PROBLEMS

1. Control charts are kept on $\overline{X}$ and S for the thickness of metal motor mounting plates. One measurement is made at the center of the plate and each sample consists of three plates. Based on 49 samples during which the process was in apparent control, $\overline{\overline{X}} = .451$ cm and $S_p = .0049$ cm.
   a. Estimate σ.
   b. Identify the center line and 3-sigma limits for an $\overline{X}$-chart.
   c. Identify the center line and 3-sigma limits for the accompanying S-chart.

TABLE 2.12 Sample Data for First 27 Days for Weights of Bags of Sugar.

Sample	1	2	3	4	5	6	7	8	9
$\overline{X}$	902.41	905.42	907.32	907.21	907.56	907.24	907.10	907.29	907.47
S	0.226	0.311	0.299	0.288	0.352	0.394	0.335	0.278	0.326

Sample	10	11	12	13	14	15	16	17	18
$\overline{X}$	907.46	907.27	907.36	907.36	907.31	907.25	907.28	907.24	907.40
S	0.196	0.230	0.371	0.327	0.345	0.190	0.296	0.148	0.328

Sample	19	20	21	22	23	24	25	26	27
$\overline{X}$	907.21	907.18	907.18	907.28	907.38	907.20	907.29	907.10	907.32
S	0.363	0.272	0.329	0.321	0.241	0.251	0.149	0.290	0.331

2. Work Problem 1, given that $\overline{\overline{X}} = .451$ cm and $\overline{S} = .0044$ cm.
3. Control charts are kept on $\overline{X}$ and R for the contents of a soft drink product sold in cans. The sample size is 4. Based on 36 samples during which the process was in apparent control, $\overline{\overline{X}} = 354.4$ mL and $\overline{R} = 2.28$ mL.
   a. Estimate σ.
   b. Identify the center line and 3-sigma limits for an $\overline{X}$-chart.
   c. Identify the center line and 3-sigma limits for the accompanying R-chart.
4. A packaging process is now using new equipment to put sugar in paper bags. Nine bags are weighed each day. The target weight for a bag is 907 grams. It is easy to adjust the process mean. The process standard deviation could be related to the process mean. The standard deviation in bag weight is unknown for the new process. Table 2.12 summarizes sample data for the first 27 days after start-up. After each of the first two samples, adjustments were made in an effort to bring bag weight closer to target. No adjustments were made subsequently.
   a. After the first two days, does the process appear to be stable?
   b. Estimate σ using the pooled standard deviation.
   c. Identify the center line and 3-sigma limits for an $\overline{X}$-chart.
   d. Identify the center line and 3-sigma limits for the accompanying S-chart to be used for the next month.
   e. What amount of adjustment should be made in an effort to recenter the process at 907.00 grams?

5. A company uses a process that makes a paper product out of wood chips. One critical quality measure is the thickness of the paper. At the end of the production process, a hot liquid is spewed onto rollers, dries while rolling forward, and becomes a 20-foot wide sheet of paper. At 5-foot intervals, four thickness measurements are taken across the 20-foot width at a point where the spew has dried into paper. These four measurements constitute a sample of size 4, and $\overline{X}$ and S charts are kept. The target thickness is .400 mm. Control charts for 25 samples appear in Figure 2.16.

   On the S-chart, variation appears stable, most points concentrate near the center line, very few points are near the 3-sigma limits, and a histogram of the standard deviation values would occupy most of the 6-sigma span from the lower 3-sigma limit to the upper 3-sigma limit. However, while points on the $\overline{X}$-chart reflect stability, they *hug* the center line, which is not what is expected from a stable process. The $\overline{X}$-chart is *too* good. The data that produced these charts appear in Table 2.13.

   a. Find the mean and confirm the standard deviation of the 25 values obtained from each measurement position. Are the four means similar? Are the four standard deviations similar?
   b. How do the sample standard deviations in the table (from which $\overline{S}$ or S_p is computed) compare to those four standard deviations?
   c. Why do the points in the $\overline{X}$-chart hug the center line?
   d. What might be a better subgrouping method?

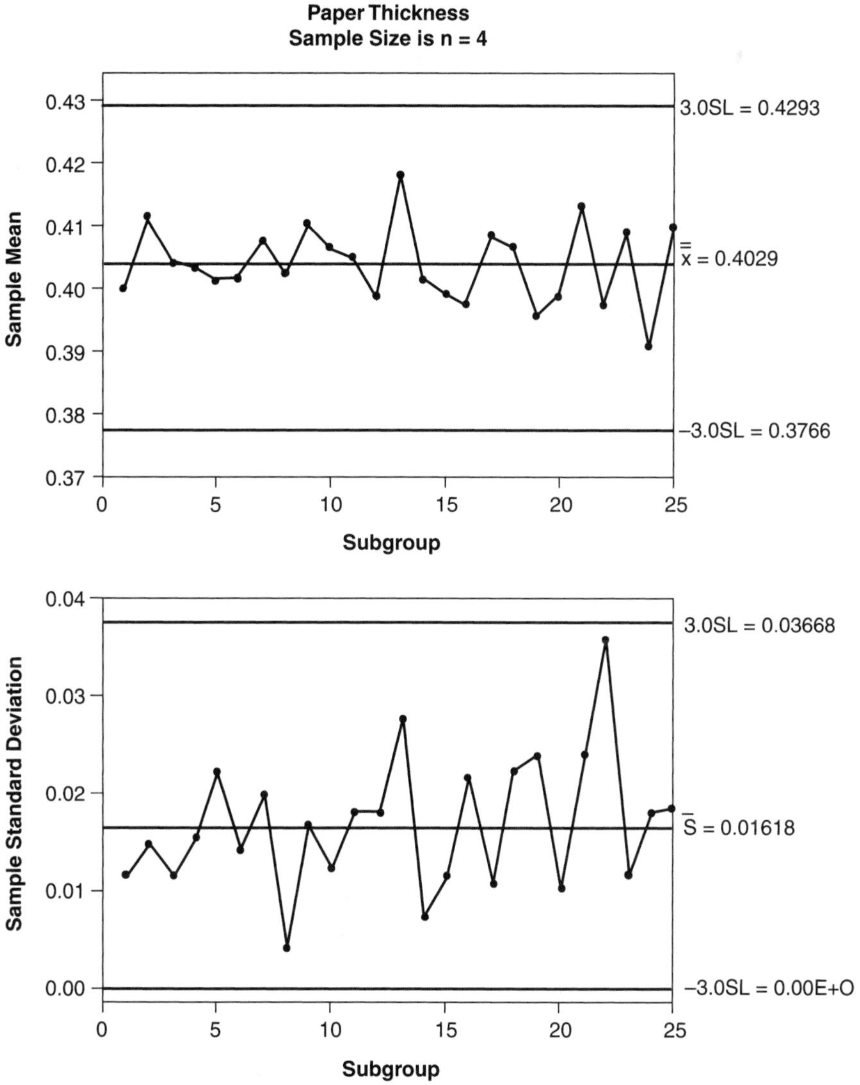

FIGURE 2.16 Paper Thickness (Sample Size is $n = 4$).

TABLE 2.13 Paper Thickness (mm).

Sample	Position 1	Position 2	Position 3	Position 4	Sample St. Dev.
1	0.405	0.386	0.412	0.394	0.0115
2	0.397	0.410	0.431	0.404	0.0147
3	0.416	0.389	0.406	0.406	0.0112
4	0.412	0.402	0.415	0.382	0.0149
5	0.416	0.381	0.422	0.383	0.0215
6	0.407	0.398	0.416	0.383	0.0141
7	0.394	0.419	0.427	0.387	0.0193
8	0.397	0.402	0.402	0.406	0.0037
9	0.403	0.417	0.427	0.391	0.0158
10	0.413	0.410	0.412	0.388	0.0119
11	0.409	0.409	0.419	0.379	0.0173
12	0.384	0.405	0.419	0.383	0.0174
13	0.440	0.422	0.427	0.378	0.0269
14	0.405	0.408	0.395	0.395	0.0068
15	0.397	0.390	0.414	0.392	0.0109
16	0.419	0.399	0.398	0.369	0.0206
17	0.419	0.413	0.399	0.399	0.0101
18	0.423	0.391	0.425	0.382	0.0220
19	0.388	0.383	0.429	0.379	0.0231
20	0.407	0.390	0.404	0.390	0.0090
21	0.424	0.405	0.433	0.384	0.0217
22	0.391	0.398	0.439	0.354	0.0348
23	0.410	0.402	0.421	0.396	0.0108
24	0.404	0.376	0.405	0.374	0.0171
25	0.399	0.399	0.434	0.399	0.0175
Column St. Dev.	.0127	.0122	.0124	.0120	

3

Signals and Measures Used in Assessing Control and Quality

OBJECTIVES

- Explain how to recognize basic out-of-control signals
- Explain how to diagnose likely causes of out-of-control signals
- Explain the basic methods used to measure process capability

3.1 TELL-TALE CONTROL CHART PATTERNS

When a process is in control, there are several things we expect to see in each of the control charts for the processes we have discussed up to now. One is an apparent random balance of points above and below the center line. The control chart of Figure 3.1 illustrates this balance. If 40.00 mm is the target for bolt diameter, it is clear that the process was centered close to target during the period when these samples were taken.

A second thing one expects to see is that the points concentrate near the center line and become less and less frequent as one moves away from the center line in either an upward or downward direction. This can be seen in the control chart of Figure 3.1. Basically, this means that we would ex-

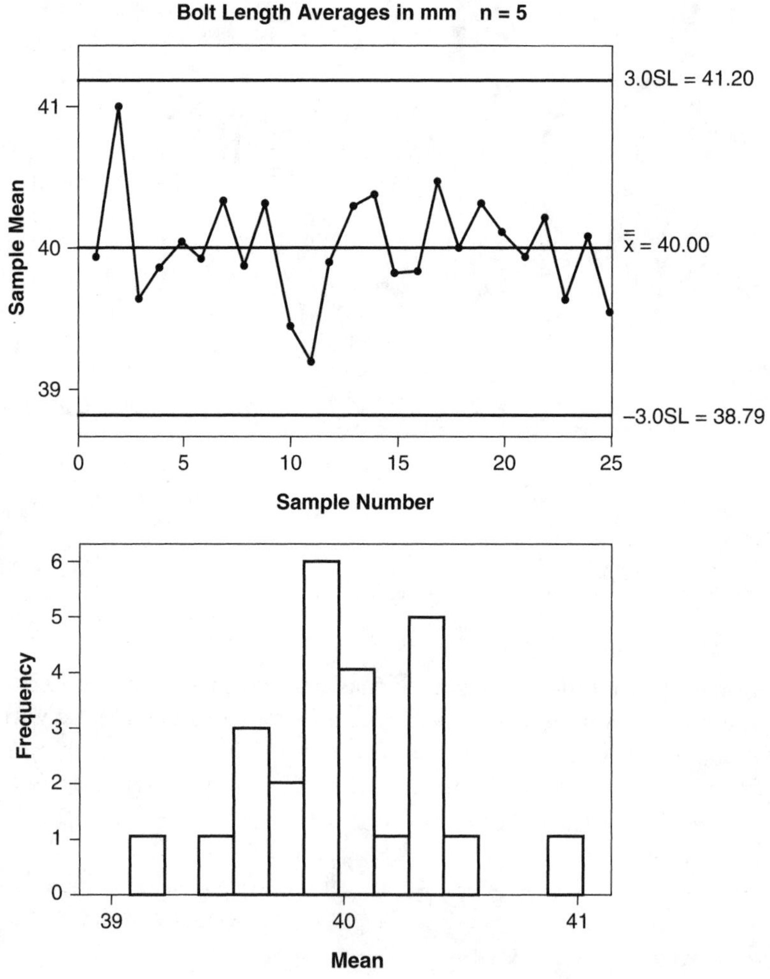

FIGURE 3.1 Ideal Control Chart Pattern and Histogram.

pect a histogram of the plotted values to approximate the shape of a normal distribution. The histogram of sample means shown in Figure 3.1 illustrates this concept. The histogram indicates that the process is generating sample means from a distribution that is approximately normal and that the process mean is near 40.00 mm. However, unlike the control chart, the histogram does not take into account the time sequence in which sample means were obtained. If the same sample means that generated the $\overline{X}$-chart are rearranged in a different time order so that the means with a value less than 40.00 mm are plotted first and greater means are plotted last, the histogram would look the same, but the control chart would not exhibit the random fallout above and below the center line as previously described. Such a nonrandom pattern could mean a shift in the process mean. We consider such patterns shortly. In any case, it is the control chart that should be examined for random fallout with respect to the center line, not the histogram.

A third thing one expects to see from a process that is in control is that the plotted values "live" in a region that uses up most of the interval between the lower and upper 3-sigma limits. One can see from either the control chart or the histogram that this process should be expected to generate an occasional sample mean near 38.79 mm or 41.20 mm. Remember that when a process is in control, the upper and lower 3-sigma limits give practical upper and lower bounds for the plotted values and a small fraction of the points would be expected to be near these limits. A control chart that exhibits this characteristic gives reassurance that σ is being estimated properly. If a noticeable percentage of the points fall outside the 3-sigma limits, something may be wrong. Also, if the points hug the center line, something may be wrong. Both are telltale patterns we discuss in this section, along with a number of other patterns one would not expect to see on a control chart from a process that is in control. For each telltale pattern, we also will discuss potential assignable causes to troubleshoot.

That third thing one expects to see from a process that is in control—that the plotted values "live" in a region that uses up most of the interval between the lower and upper 3-sigma limits—relates to the first of several indices we consider.

Consider the individual values that make up the samples. Recall that when a process is in control, practically all sample values will be within $\pm$ 3σ of μ, assuming that the population sampled is approximately normally distributed. For normal distributions, the expected percentage is 99.7%. For many other types of mound-shaped distributions, 99.7% will be in the ballpark. It follows that nearly all of the sample values generated by a process may be expected to occupy an interval of length 6σ, so long as the process remains in control. This interval length is a process characteristic important enough to have a name:

The **natural tolerance of the process** is 6σ.

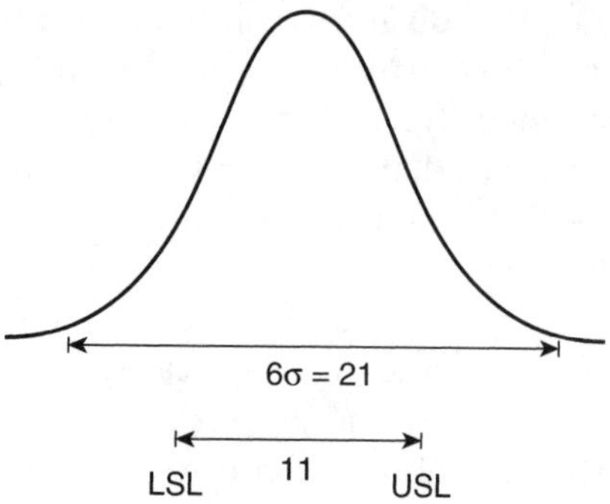

FIGURE 3.2 When the Spec Range is Less than 6, a Large Percentage of Defective Items Will Be Produced.

For the company that manufactured copper wire spools (see Section 2.1), the process standard deviation was estimated as 3.503 grams. This value was obtained during a period when the process was in apparent control using one of the standard methods of estimating σ. The natural tolerance of this process would be $6\sigma \cong 6(3.503) = 21.0$ grams, approximately. Unfortunately, suppose the spec range for this product is 562.5 ± 5.5 grams, a spread of only 11.0 grams. Figure 3.2 illustrates this state of affairs. Clearly, a sizeable percentage of spools produced will have weights that do not meet specifications.

The company in this predicament has two strategies to fix or lessen the problem. One is to check whether product specs are needlessly tight. Interestingly, it is not unusual in manufacturing to have product specs that can be changed. Sometimes it is the case that product specs have not been carefully thought out, and thus widening them would not be a problem. It may be the specs are internal and out of date, in light of changes made in the product to adapt to the market. Or, if the specs are the customer's and the customer won't widen the specs, the manufacturer still might be able to work a deal with a customer who can find a use for off-spec items.

A second strategy is to reduce σ. Reducing σ should be a never-ending struggle and philosophy. This is the statistical way of saying that increasing product consistency is a never-ending struggle and philosophy. At every turn, management and the employee-mind-set that management creates should be thinking about how to reduce variation in every part of the process. This will make the ultimate product be more consistent, and hence be of higher quality. The goal is to make σ so small (and hence 6σ so small) that the quality variable being measured has a distribution that fits comfortably between the spec limits, and to continue working toward making σ even smaller. Some of the business consequences of reducing σ are:

- a more competitive product,
- happier customers,

- the company stays in business,
- employees continue to have jobs, and
- society has a higher quality of life than it would without the product.

The goal of reducing σ has been around since Walter Shewhart introduced the statistical point of view through control charts in the 1920s. In the late 1990s, people in the quality community began to see articles about the *6-sigma* concept, a movement promoting the use of statistics. In particular, the term *6-sigma* used in this context means something different than the natural tolerance of a process. Namely, a process is *6-sigma* when it is centered 6σ from any product specs. A process could be also a 4-sigma process, a 5-sigma process, a 4.6-sigma process, and so on. For any of these processes, however, the natural tolerance of the process is still 6σ.

In addition to the two strategies for dealing with the state of affairs shown in Figure 3.2, a company also will have to deal with an unfortunate short-term consequence if production must be sold as top quality to a customer who can't budge on product specs. Namely, 100% inspection. Every item must be inspected prior to shipping in order to ferret out items that don't meet specs. This is necessary to avoid customer refunds, customer ill will, loss of repeat business, and weakening of product reputation.

Other Kinds of Out-of-Control Signals

Up to this point, the only kind of out-of-control signal considered has been for a single point—namely, the point that has fallen outside of the 3-sigma limits on a control chart. Because the sequence of points on a control chart has a time chronology attached, there are trend patterns consisting of two or more consecutive points that commonly are viewed as out-of-control signals, even when no single point in the pattern indicates lack of control. We now consider some of these.

Sudden Shifts in $\bar{X}$-, R-, or S-Charts

Figure 3.3 illustrates a process that was centered near the historical mean of 40.00 mm until some time near sample 15, 16, 17, or 18. After that, the process mean shifted to some smaller value. Activity logs should be examined to identify potential assignable causes that could produce a sudden shift. Some potential causes could be:

- new part supplier,
- new machine or operator,
- new measuring device,
- new inspector,
- new chemical shipment,

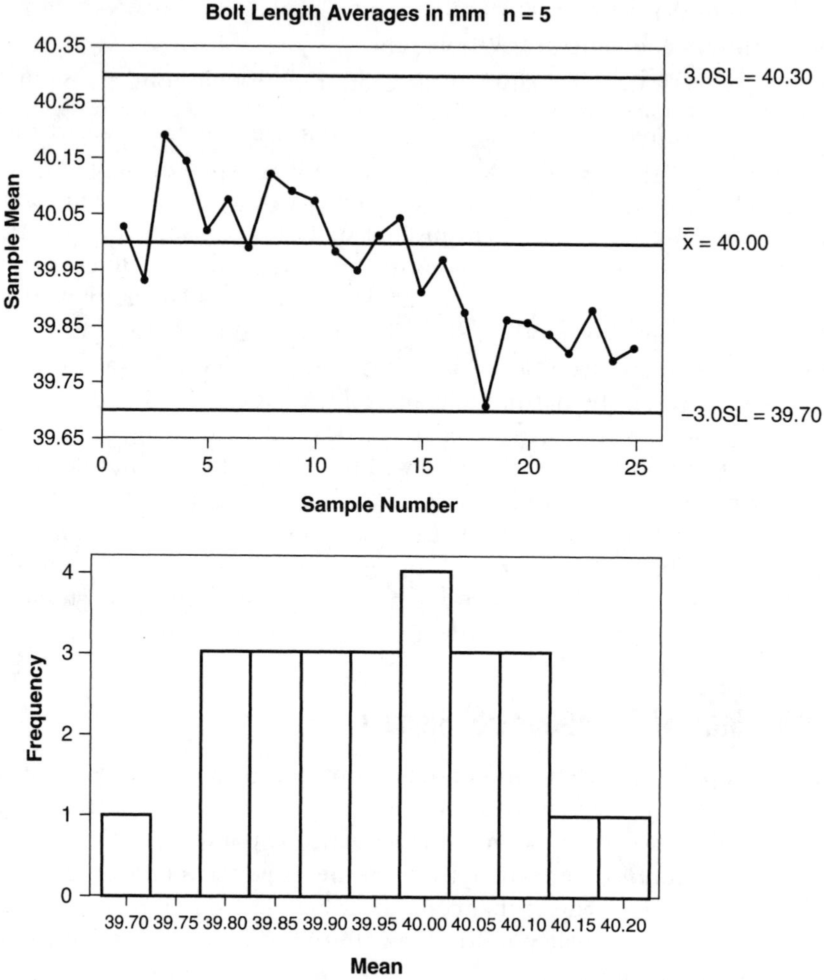

FIGURE 3.3 Process with a Mean that Shifts Downward at Some Time Near Sample 15, 16, 17, or 18.

- equipment adjustment,
- equipment component failure, or
- intentional change in the process.

Note that not all sudden shifts are necessarily unwanted. Suppose Figure 3.3 represents an *S*- or *R*-chart, in which case a drop in the process standard deviation would be desirable. If a search for an assignable cause shows that the sudden drop in variation is due to a temporary changeover to a new part or material supplier or to a temporary change in a process standard operating procedure (SOP), this is something management would want to know and consider continuing.

While it is likely that the control chart would be noticed first, the histogram of Figure 3.3 shows the effect of mixing the two kinds of data. Unlike a normal distribution, this particular histogram is flat in the middle because this middle represents a pooling of a small proportion of sample means from the lower tail of the earlier distribution with a small proportion of sample means from the upper tail of the more-recent distribution.

High Percentage of Points Near or Beyond Control Limits on $\bar{X}$-, *R*-, or *S*-Charts

Figure 3.4 shows a control chart with a high percentage of points near the control limits and, consequently, a smaller percentage of points near the center line. The accompanying histogram tells the same story in its own way, showing high frequencies near the 3-sigma limits. Some possible causes could be:

- overcontrol,
- John and Mary (or two processes) both being tracked on one chart,
- different measuring systems being used,
- swings in quality of incoming raw material, or
- error in computing control limits.

This $\bar{X}$-chart is a classic picture of the consequences of overcontrol. Notice how often a point far from the center line immediately follows a point also far from the center line, but in the opposite direction. It is easy to imagine when a compulsive tinkerer had hands on the controls, did not like the location of a particular sample mean (which could have been noise), and adjusted the process to "compensate."

Another possibility is that two workers, John and Mary, say, both have produced samples that are tracked on one chart, and if each were tracked alone, it would be clear that they have different personal means. Analogously, the explanation could be one chart tracking output from different machines or different shifts. In the same spirit, component parts from different suppliers could be the explanation if a lot from one supplier is installed in product units until the lot is depleted. It would be common, then, for all of the product units in one sample to contain component parts from the same supplier. If the plotted points summarize the results of chemistry lab analyses, the explanation could be that as John and Mary do the analyses, they do not follow the same lab procedures. The common thread in these potential explanations is that the system is allowing something to happen in two or more ways that could affect a quality measure. Management needs to be comfortable that different machines, people, component parts,

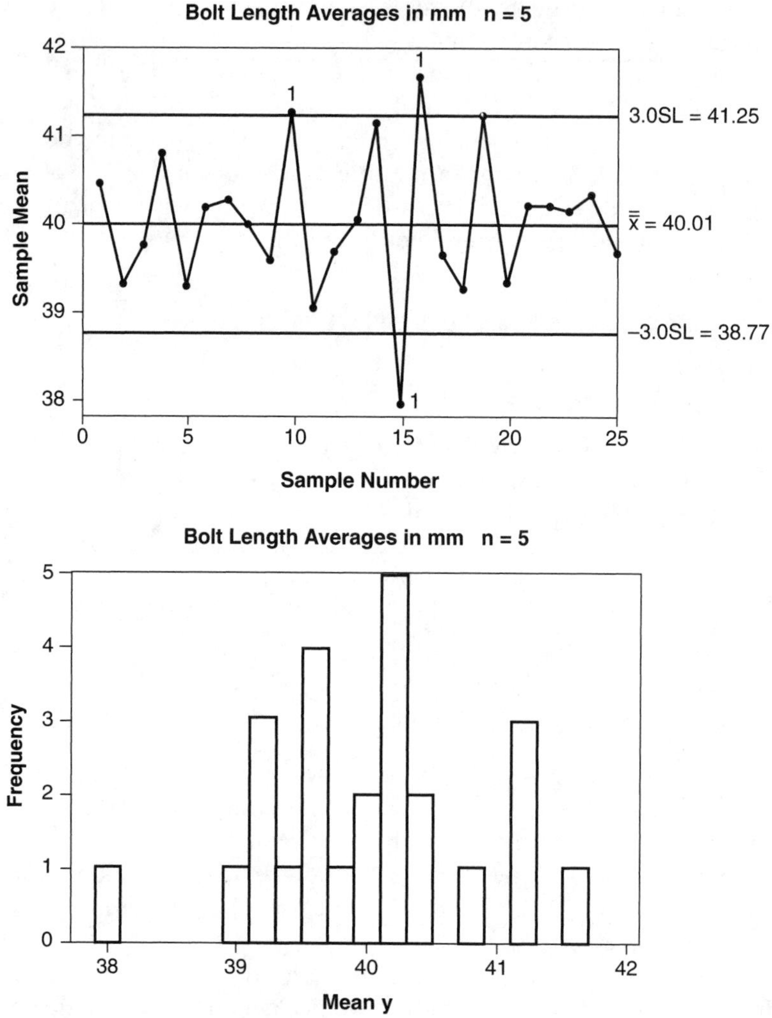

FIGURE 3.4 Low Percentage of Points Near the Center Line, and a High Percentage Near the Control Limits.

and so on, generate comparable output before pooling the data onto one control chart. The wisest course may be to keep separate control charts for each machine, shift, and so on.

The fact that the control limits could be incorrect merits a comment. Many companies know they should be practicing SPC. Many seek certification from bodies that require proof that the company uses statistics and control charts in its decision making. While anyone may make a mistake in setting up 3-sigma limits for a control chart, it would be truly unfortunate

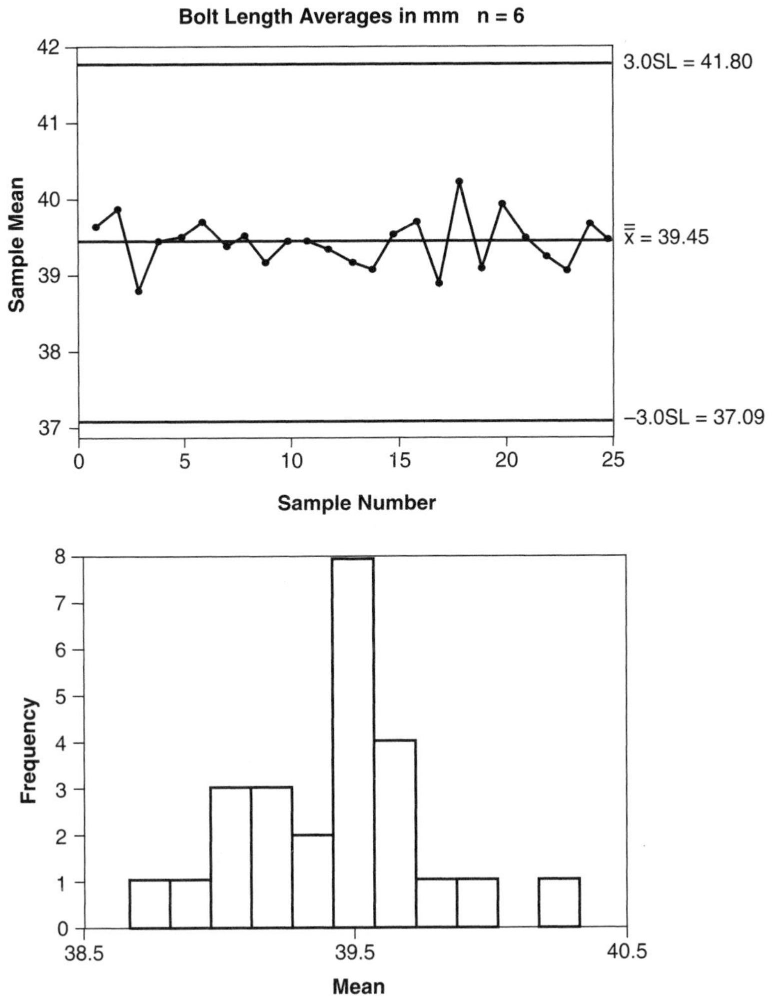

FIGURE 3.5 When Points Hug the Center Line, There is a Problem.

for a company to invest in expensive SPC software but not have access to someone who can set the charts up properly.

Points Hug the Center Line on $\bar{X}$-, R-, or S-charts

Figure 3.5 shows an $\bar{X}$-chart with a pattern that would not be expected from a process that is in control. The plotted points would be expected to use up most of the interval between the 3-sigma limits, whereas these points are exceptionally tight on the center line.

Some possible causes could be:

- output from different machines, workers, or merges systematically make up every sample;
- error in computing control limits; or
- process change has reduced σ, but the control chart was set up using the old value as the user-defined sigma.

The first of these possible causes warrants a comment. Just as tracking output from different machines on one control chart can produce *too many* points near the 3-sigma limit, another kind of mixing of output from different machines can produce a chart with a relative absence of points outside the immediate vicinity of the center line. Suppose every sample systematically combines items from both machines and the machines have different personal means.

To illustrate, suppose that every sample consists of three bolts produced by Machine 1 and three bolts produced by Machine 2, that Machine 1 always produces bolts of length near 39.8 mm, and that Machine 2 produces bolts of length near 39.1 mm. The bolt lengths from two typical samples might look like:

Sample 1	39.7, 39.9, 39.8, 39.0, 39.1, 39.2
Sample 2	39.7, 39.7, 39.7, 39.2, 39.3, 39.0

With samples structured as described, sample means should always be near 39.45 mm. However, with three sample values in each sample far from the other three values in the sample, the sample ranges or standard deviations used to estimate the process standard deviation σ will produce 3-sigma limits far from 39.45 mm. If separate control charts were kept on each machine, it would be obvious where the output distribution of each is centered and easier to know when and how much to adjust each machine.

$\overline{X}$ Chart Is Correlated with an *S*- or *R*-Chart

Figure 3.6 illustrates an $\overline{X}$-chart that is positively correlated with its *R*-chart. That is, when a sample has a large mean, it tends to also have a large standard deviation; when a sample has a small mean, it tends to also have a small standard deviation. The overall picture is that the graphs tend to rise and fall together. Figure 3.7 illustrates an $\overline{X}$-chart that is negatively correlated with its *R*-chart. In this case, when a sample has a large mean, it tends to have a small standard deviation, and when a sample has a small mean, it tends to have a large standard deviation. The graphs will be approximate mirror images.

Some possible causes are:

- the population being sampled is skewed to the right (positive correlation),

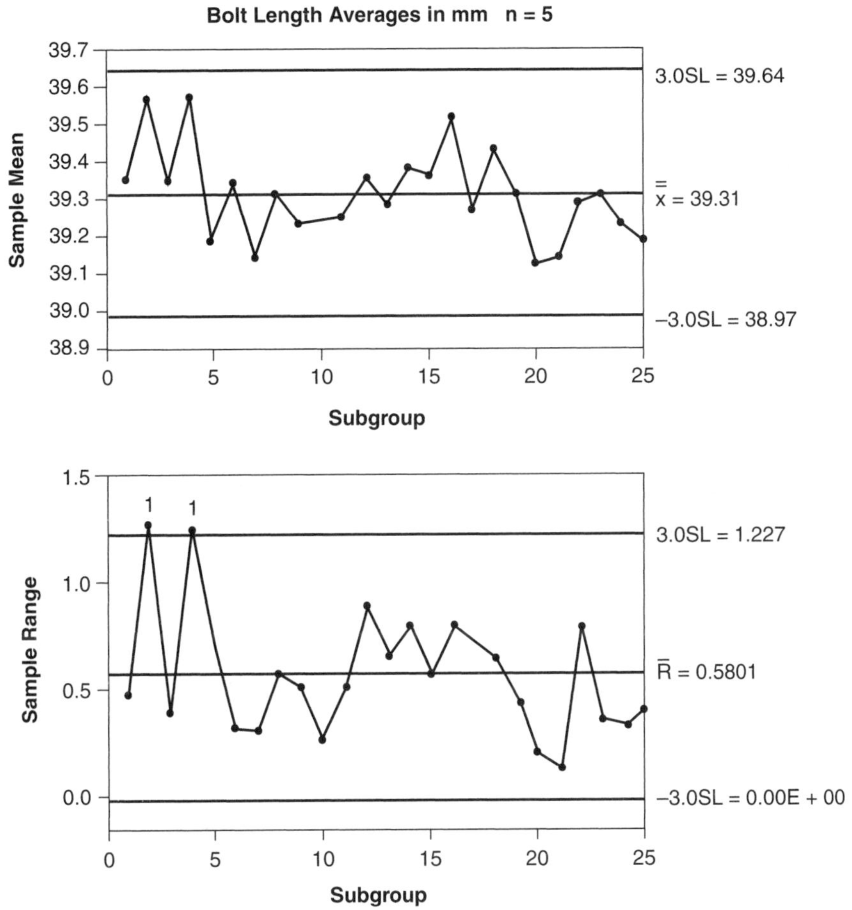

FIGURE 3.6 $\overline{X}$-Chart Positively Correlated with *R*-Chart.

- the population being sampled is skewed to the left (negative correlation), and
- causal relationships inherent in the product.

Figure 3.8 shows a histogram of the 125 sample values whose means are plotted on the $\overline{X}$-chart in Figure 3.6. The graph indicates that the population being sampled is skewed to the right. For most samples from such a population, all five sample values will fall near the mound. However, some samples will have one or two sample values from the right tail, in which case the mean, range, and standard deviation of the sample will be large. Only rarely would all sample values fall in the right tail, close enough together to produce both a large mean and a small standard deviation. The explanation for why one should expect negative correlation between sample location and variation when the population is skewed to the left is similar.

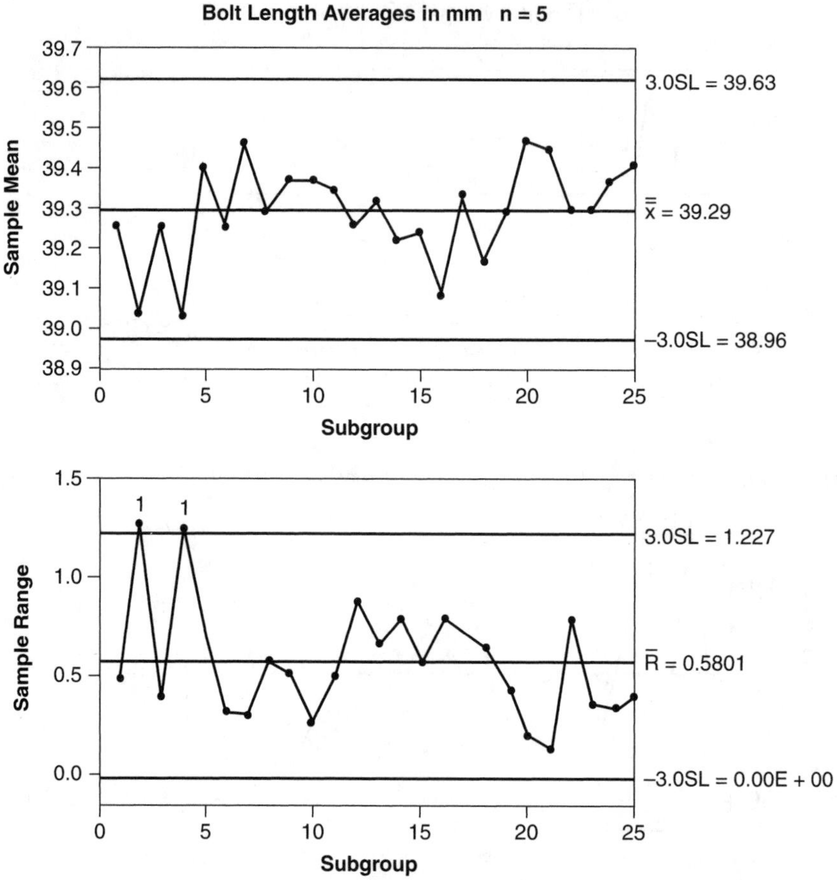

FIGURE 3.7 $\overline{X}$-Chart Negatively Correlated with R-Chart.

The potential explanation of causal relationships inherent in the product can be illustrated by an example. Suppose the company in question manufactures thread, which it winds onto spools that are then sold to customers who manufacture fabrics. Periodically, spools are sampled from production in order to measure various quality characteristics of the thread. One of the characteristics measured is the minimum tension needed on the free end to make the thread pull freely from the spool without snagging. To measure the minimum tension on a spool, the spool is immobilized on a frame in such a way that the cylindrical hole in the core is vertical and the free thread-end hangs down. The company wants to know if where the loose thread contacted the spool affected the minimum tension measurement. The company examined four ways the loose thread could hang down: from the top (where the loose end contacts the spool and the thread winds downward), from the middle (where the thread winds downward), from the middle (where the thread winds upward), and from the bottom (where the

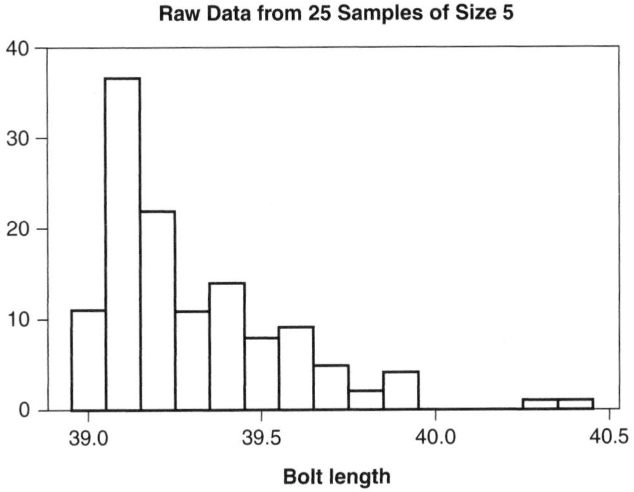

FIGURE 3.8 Histogram of Sample Values Used to Obtain Figure 3.6.

Four Ways Thread Can Hang from the End of a Spool

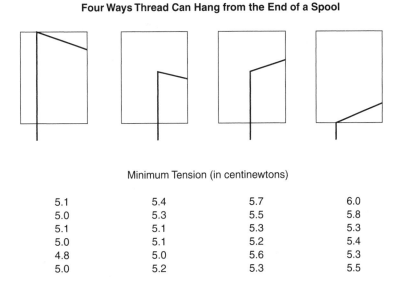

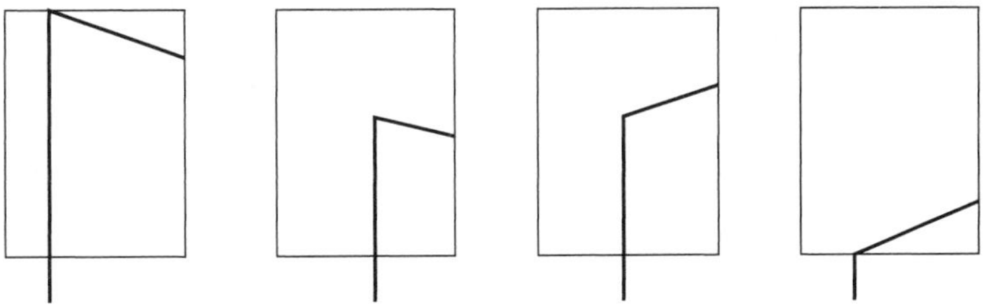

Minimum Tension (in centinewtons)

5.1	5.4	5.7	6.0
5.0	5.3	5.5	5.8
5.1	5.1	5.3	5.3
5.0	5.1	5.2	5.4
4.8	5.0	5.6	5.3
5.0	5.2	5.3	5.5

FIGURE 3.9 Minimum Tension Needed to Pull Thread from a Spool.

thread winds upward). Figure 3.9 illustrates this example. To measure the minimum tension, gradually incremented downward pull is applied to the loose end of the thread until the point is reached when the thread falls freely from the spool. The suspicion was that it would take less pull to bring the thread down if the thread wound downward from the contact point with the spool (in a sense, was ready to fall) than if the thread wound upward (there would be some drag to overcome). In particular, the suspicion was that minimum tension would increase in the left-to-right order shown in Figure 3.9.

So that differences between spools wouldn't mask differences between positions, all 24 measurements were obtained from one spool, 6 from each position. The loose thread would be pulled until the contact point matched one of the four positions and a minimum tension obtained. The thread pulled off in obtaining the measurement would be cut off and the process repeated to obtain the next measurement. To avoid the masking of differences between positions that might occur if all six measurements for one position were to come from one possibly quirky section of thread, the measurements were obtained by rows in the order shown in Figure 3.9. Descriptive statistics for these data were:

$\overline{X}$:	5.00	5.18	5.43	5.55
S:	.110	.147	.197	.288

A statistical test called *analysis of variance* was used to show that position did affect minimum tension, and did so in the suspected way. It is not within the scope of this book to examine statistical testing. However, this example does illustrate a causal relationship inherent in the product that is reflected in a positive correlation between $\overline{X}$ and S. Prior to this experiment, employees who did the minimum tension tests had paid no attention to how the loose thread contacted the spool. Still, management had been aware of the positive correlation between $\overline{X}$ and S. The experiment showed that thread position helped explain what was happening. Incidentally, as a result of the experiment and in the spirit of consistency, minimum tension measurements were obtained thereafter using the *hang from the top* position.

Alternating Patterns in $\overline{X}$-, *R*-, or *S*-Charts

The histogram shown in Figure 3.10 is bimodal, indicating that there is a mix of large and small values plotted. However, the $\overline{X}$-chart shows that the mix is not random over time. There is an alternating pattern.

Some possible causes could be:

- two samples from A.M. shift alternating with two samples from the P.M. shift,
- temperatures in A.M. and P.M. affect product,
- daytime inspector and nighttime inspectors are different,
- same machine with rotating operators, and
- worker fatigue (four samples from one shift).

Rising or Falling Pattern on $\overline{X}$-, *R*-, or *S*-Chart

The $\overline{X}$-chart in Figure 3.11 shows a rising pattern over time. This is a picture of common cause error superimposed on a steadily increasing process mean.

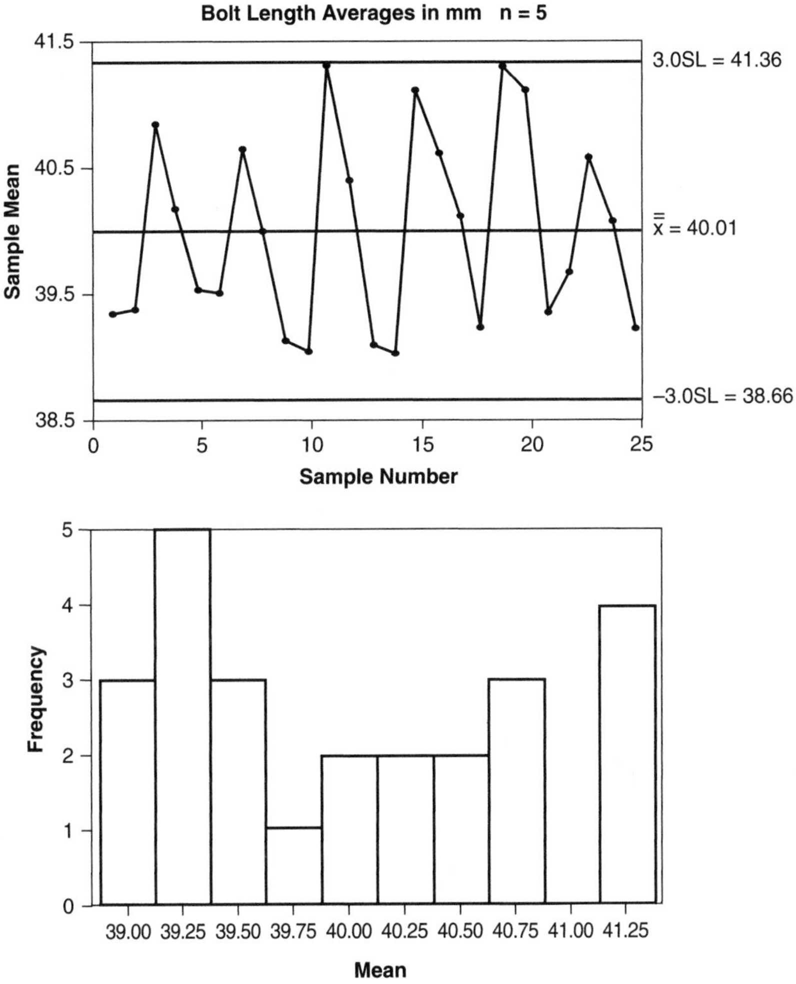

FIGURE 3.10 Control Chart with Cycles.

The accompanying histogram is somewhat flat as a result. Some possible causes could be:

- tiring worker,
- machine deterioration,
- increasing substitution of component parts from new vendor for parts purchased from former vendor, and
- gradual replacement of old chemical shipment with new.

The key word here is *gradual*. If the cause is a tiring worker or machine deterioration, one would expect increasing sample variation over time, hence a steadily rising *S*- or *R*-chart.

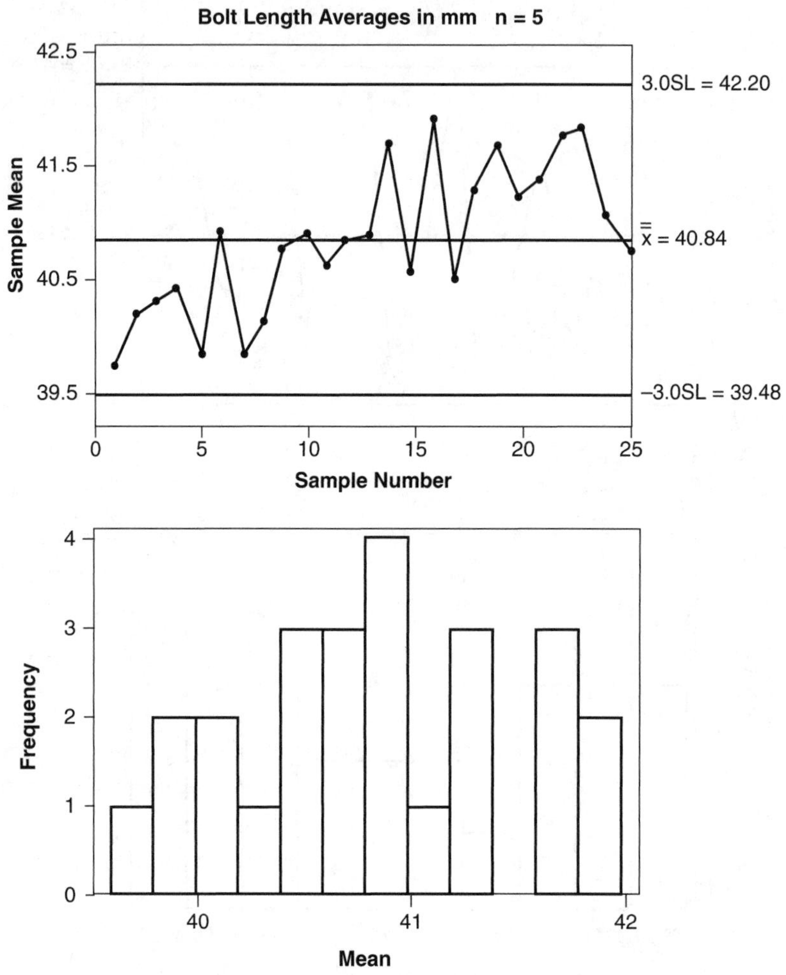

FIGURE 3.11 Steadily Increasing Process Mean.

PROBLEMS

In Problems 1 and 2, the estimate of the process mean μ and standard deviation σ is based on a period when the process was in control and a histogram of all individual sample measurements combined is mound shaped and roughly symmetric (i.e., the data are assumed to come from a normal population). For Problems 1 and 2: a) Find the natural tolerance of the process. b) If the process mean is approximately 3σ from both spec limits, the company will do frequent sampling in order to detect as quickly as possible whether the process mean has shifted, even slightly. If the process mean is closer than 3σ to either spec limit, the company will do 100% inspection (inspect all items produced) in order to keep nonconforming items from reaching the customer. Otherwise, the company will do routine sampling to monitor the process. Given the specification limit(s) and/or target value, tell whether the company should do routine sampling, frequent sampling, or 100% inspection.

1. Estimate of

	μ	σ	LSL	USL	Target
a.	80	.20	78.1	81.4	none
b.	75	.32	74.0	76.0	75
c.	27	.40	26.1	28.3	27.2

2. Based on 40 samples of size 3 during a period of control, the average downward pressure a company's plastic pipette can tolerate before cracking is $\overline{\overline{X}} = 184.6$ lb and $\overline{S} = 2.31$ lb. The lower spec limit is 170.0 lb.

3. In Problem 2, suppose a customer wants to purchase a large order of pipettes and the lower spec limit is 180.0 lb. The company can adjust the process mean upward in order to get the customer's business, although σ will not change. What is the smallest value, approximately, of μ for which there hardly ever will be a nonconforming pipette?

In Problems 4 through 8, identify the most likely potential explanation for the pattern shown in the control chart from the following list: a) tiring machine operator; b) subassembly from temporary supplier put to use and appears in units sampled beginning with sample 18; c) daytime and nighttime inspectors are different; d) every sample contains product units made by two different machines; e) no problem is indicated.

4. In Figure 3.12, the historical mean and standard deviation have been prescribed by the user.

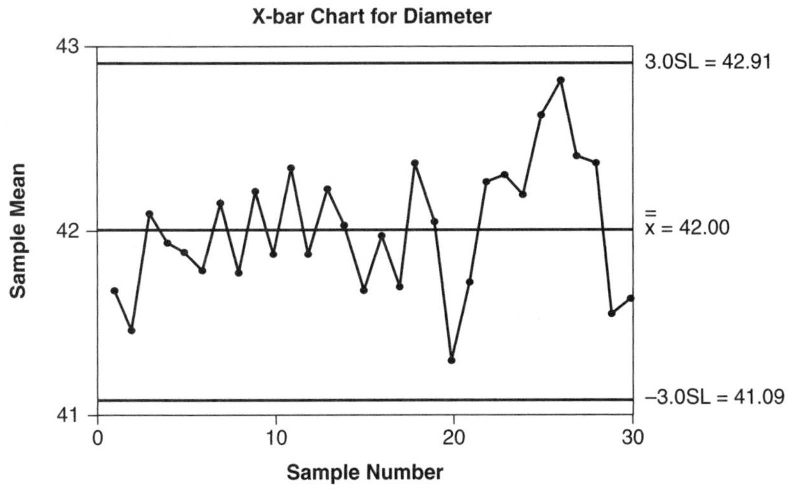

FIGURE 3.12 Control Chart for Problem 4.

5.

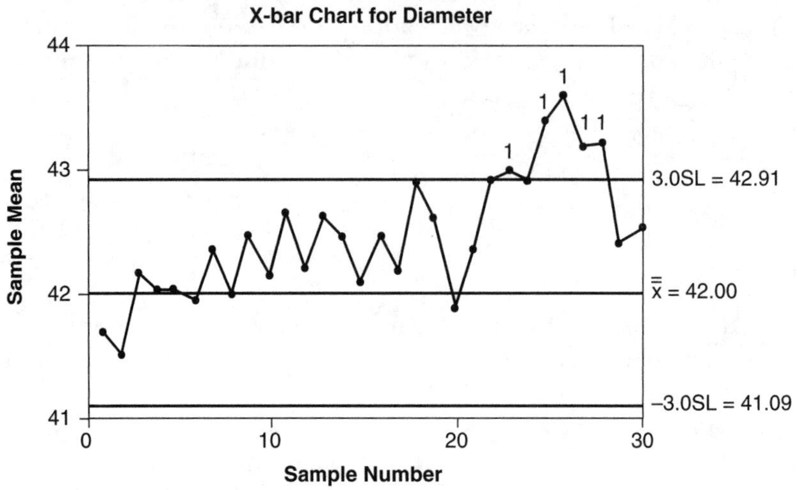

FIGURE 3.13 Control Chart for Problem 5.

6.

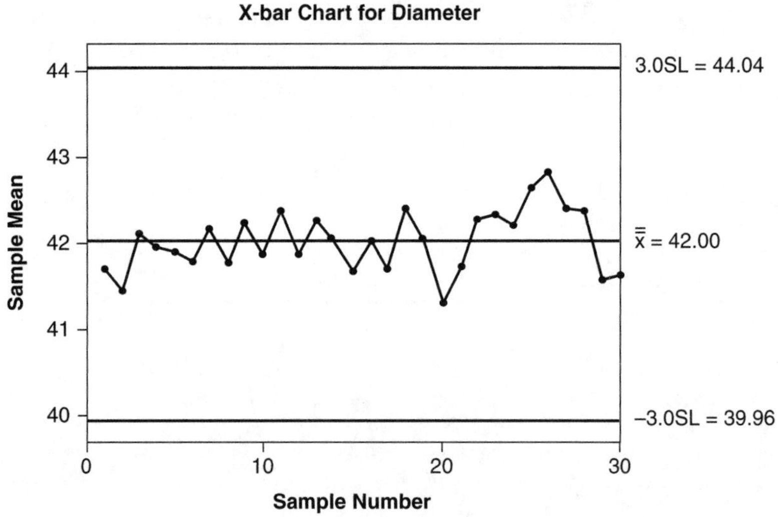

FIGURE 3.14 Control Chart for Problem 6.

7.

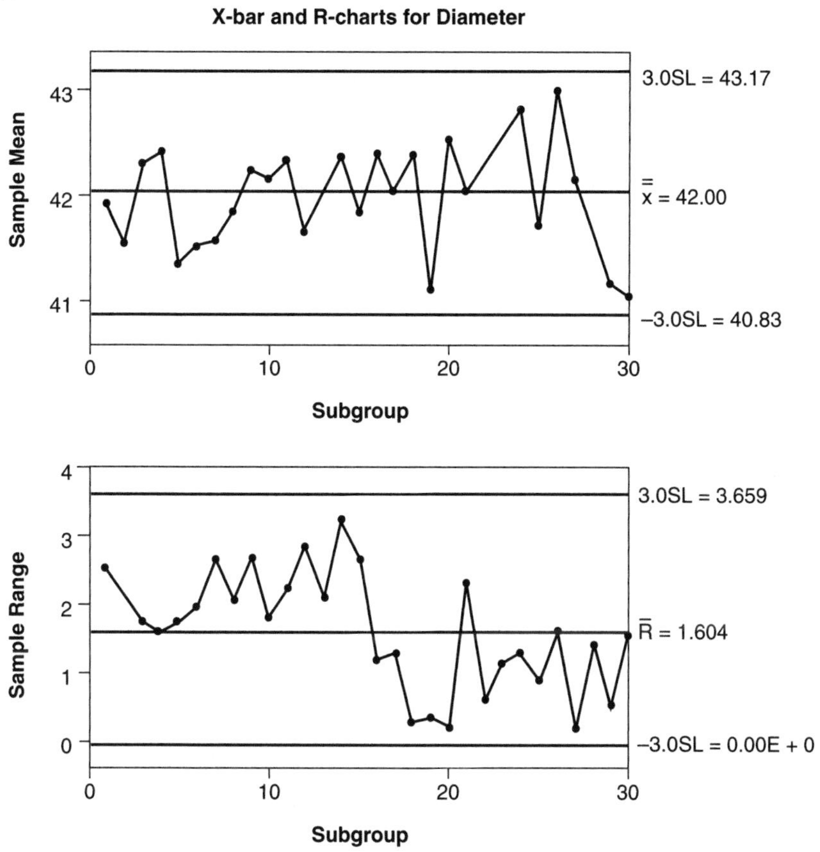

FIGURE 3.15 Control Chart for Problem 7.

8.

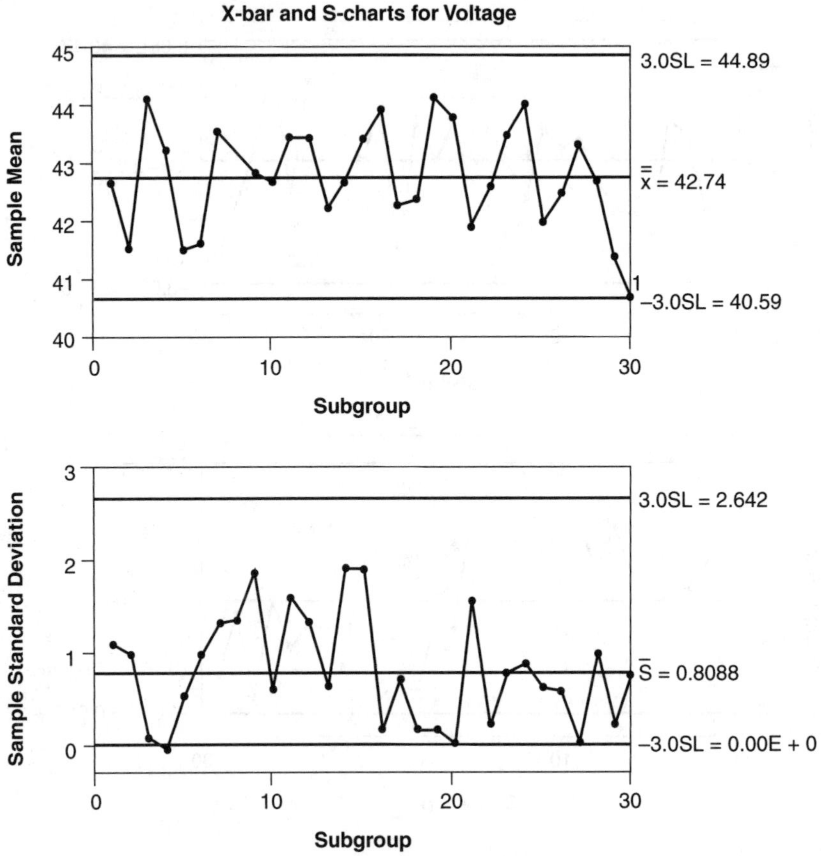

FIGURE 3.16 Control Chart for Problem 8.

3.2 OTHER OUT-OF-CONTROL SIGNALS

The preceding section examined several general patterns consistent with lack of control and some possible assignable causes that might explain the presence of those patterns. We did not examine how much of a pattern needs to be present in order to be able to say, "We have an out-of-control signal *now.*" That is the focus of this section. So far, we have met only one out-of-control signal that lets us know *when* to look for an assignable cause—the single point that falls beyond a 3-sigma limit. We now augment our arsenal with a number of other out-of-control signals that enable us to know when to look for assignable causes without having to wait for the single extreme point.

There is a statistics principle behind how the out-of-control patterns presented in this section were arrived at. Without going into the statistical de-

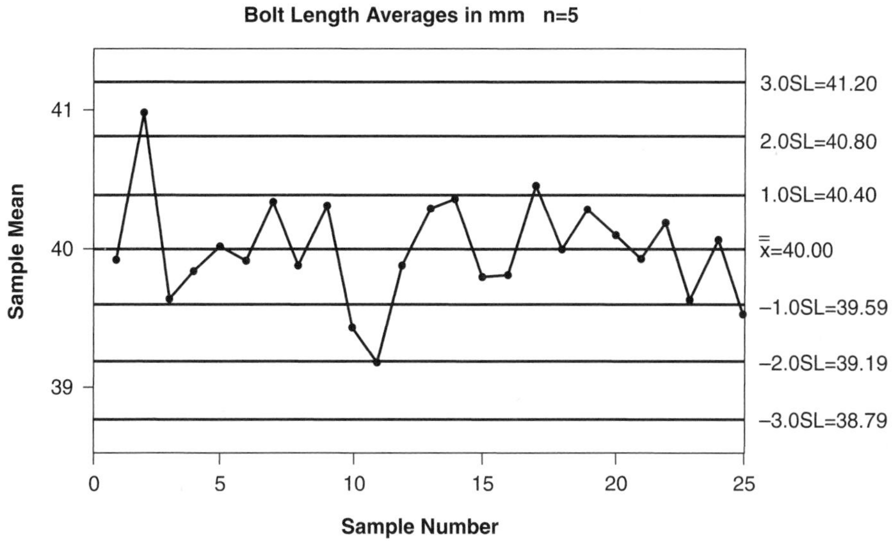

FIGURE 3.17 Zones Used to Define Out-of-Control Signals Based on a Succession of Points.

tails, this is that each pattern should be seen rarely by pure chance when a process is in control, but should be seen frequently after the process has changed in a particular way. In some sense, these signals are like Geiger counters. They are sensitive to trends.

To understand how these signals work, we first need to partition a control chart into finer gradations. The interval between the 3-sigma limits is divided into six zones of equal width. We illustrate on an $\overline{X}$-chart. Since the upper and lower and lower 3-sigma limits are each $3\frac{\sigma}{\sqrt{n}}$ from the center line, this means that each zone has width $\frac{\sigma}{\sqrt{n}}$. See Figure 3.17, which gives the center line and 3-sigma limits for a process centered at 40.00 mm and having a process standard deviation $\sigma = .8944$ mm. Each sample consists of 5 bolts. The zone width is $\frac{\sigma}{\sqrt{n}} = \frac{.8944}{\sqrt{5}} = 40\ (mm)$. The lines separating the zones are denoted ± 1.0SL, ± 2.0SL, and ± 3.0SL.

Some of these commonly used out-of-control signals are shown in the following list. Each signal pertains to a consecutive group of points, enough of which fall above the center line that the event is unusual. In order not to double the length of the list, it is understood that the symmetric pattern with the same number of points falling below the center line also would constitute an out-of-control signal. To avoid ambiguity, we treat any point that falls exactly on any of the zone boundaries ± 1.0SL, ± 2.0SL, and ± 3.0SL as falling in whichever adjacent zone is nearer to the center line.

Some Commonly Used Out-of-Control Signals for a Sequence of Points

- One point above the upper control limit
- Of three consecutive points, two are above 2.0 SL
- Of five consecutive points, four are above 1.0 SL
- Eight consecutive points are above the center line

We should point out that many SPC texts identify other signals that are in common use. For example, another is: of 11 consecutive points, 10 are above the center line. With this kind of signal, one could have the first four points fall just above the center line, the fifth point just below it, the next six points just above it, and an out-of-control signal would register on that 11th point. Such a pattern of points near the center line can occur without any of the four signals identified in the previous list occurring first. But in this case, such a signal on an $\overline{X}$-chart must be considered very weak, in the sense that the points are near the center line. It is common for those who implement control charts to get to choose from a list of signals precisely which signals to use. SPC software packages often are designed to let the user choose. Recall that the key to a good out-of-control signal is that the particular combination of points is rarely seen when a process is in control, but is seen frequently when the process has changed in a particular way. If the mean has shifted upward by more than a modest amount, any of the four signals mentioned in the previous list are more likely to register first. Further, if all of the 11 points are near the center line, the particular fallout described may well be within the routine fluctuation (common cause variation) inherent in a process that is in control. Nobody likes to chase noise.

A reasonably thorough list of standard out-of-control signals can be found in Hoyer and Ellis (1996). In the view of this author, a wise course for management would be to adopt just a few signals, no more than three or four, for the company to use. A compelling case for holding down the number can be made by considering how often one would be looking for assignable causes if many different out-of-control signals are adopted. Each individual signal is rare when the process is in control. But if many different signals are adopted, out-of-control signals are no longer rare when the process is in control. Again, nobody likes to chase noise.

Consider bolt lengths for 25 samples from a process that has the mean gradually shifting upward from the target of 40 mm (see Figure 3.18). Notice that the process standard deviation does not appear to have changed. In the $\overline{X}$-chart, the SPC package, MINITAB, announces when and what kind of signal has occurred. Can you guess what kind of signals correspond to 1, 2, and 6 in Figure 3.18?

In answer to the question posed: Regarding the first plotted point identified with a 6, notice that it is the last of 5 consecutive points, 4 of which are above 1.0 SL. The next point also is identified with a 6 because it is the

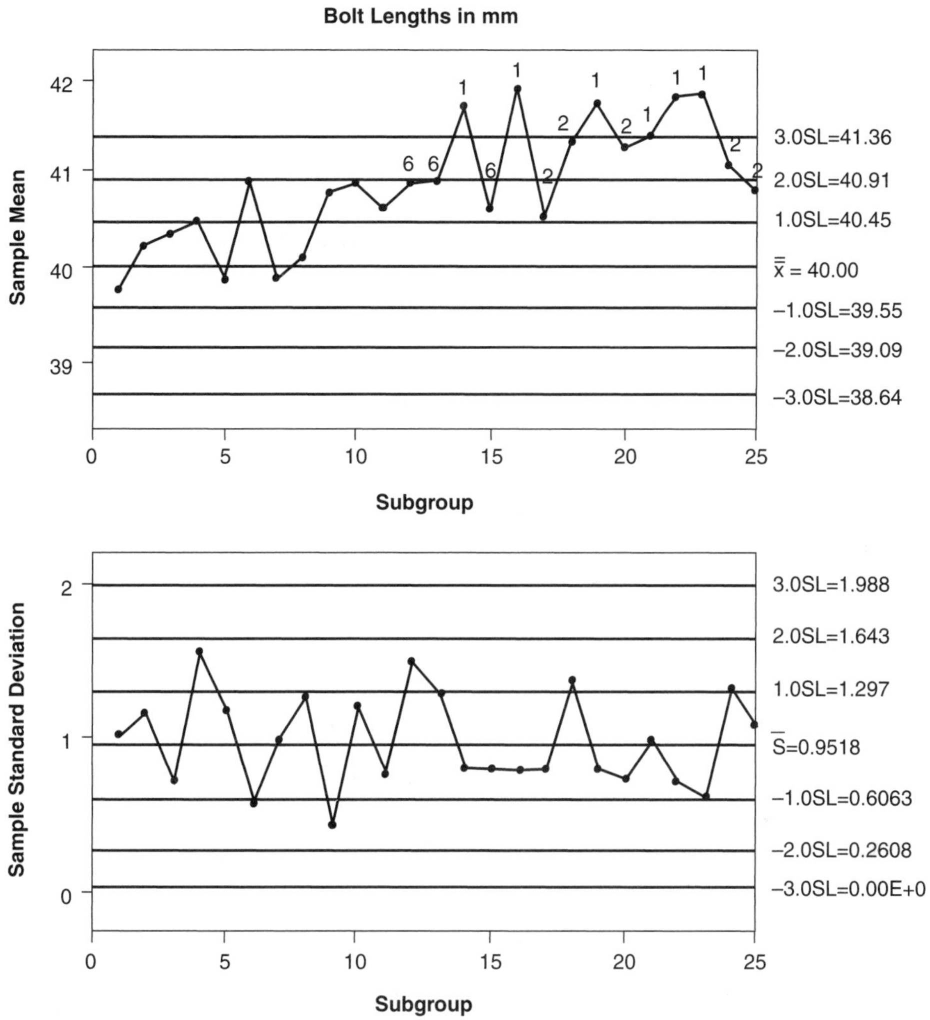

FIGURE 3.18 Various Types of Out-of-Control Signals Register for a Gradually Increasing Process Mean.

fifth of 5 points in a row, the last 4 of which are above 1.0 SL. The next point also has this property, but is identified with a 1, which designates that a single point beyond 3.0 SL has occurred. Regarding the first plotted point identified with a 2, MINITAB is telling us that the process has had 9 points in a row above the center line (MINITAB counts 9 points in a row on one side of the center line as an out-of-control signal, rather than 8).

MINITAB registers out-of-control signals in a way that allows them to overlap. Also, MINITAB is similar to a number of SPC software packages that allow the user to define which signals are to be noticed.

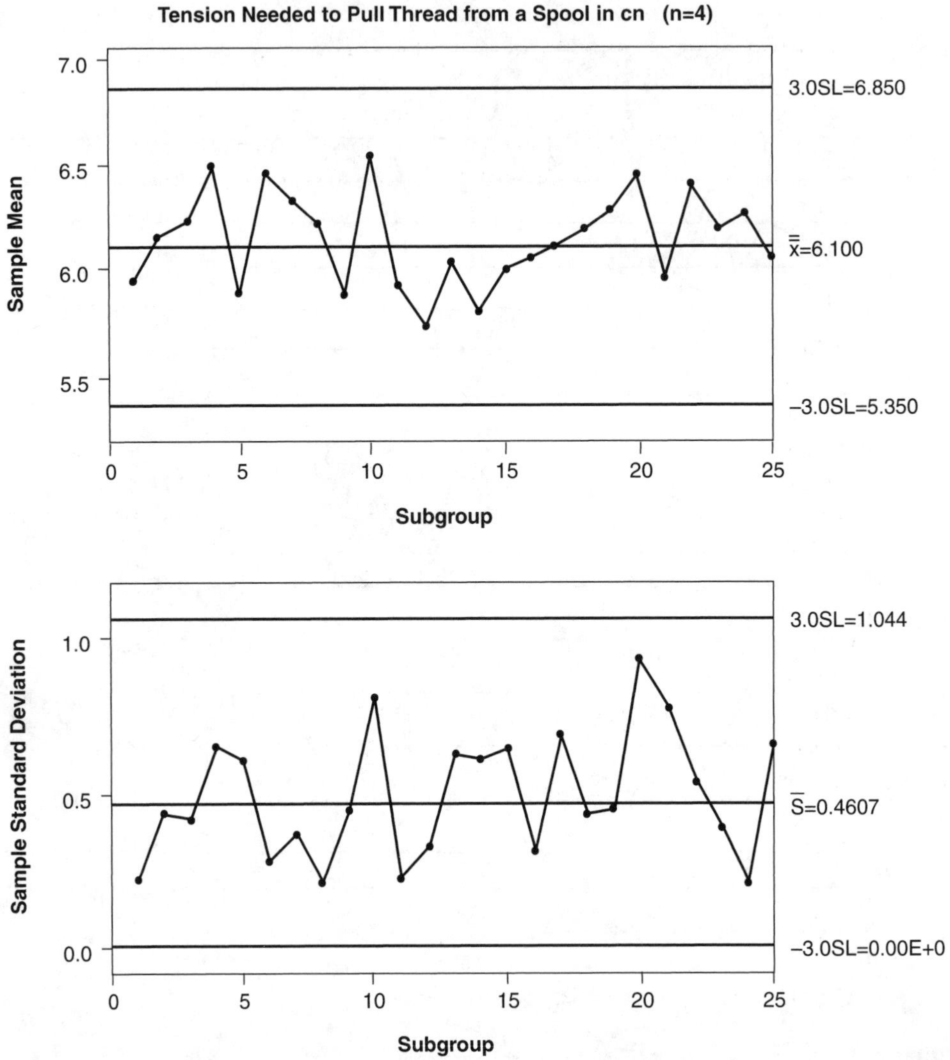

FIGURE 3.19 A Monotone Pattern for Seven Consecutive Points that is Not Really an Out-of-Control Signal.

Seemingly Compelling, but Not Really an Out-of-Control Signal

Figure 3.19 illustrates a pattern that at first glance might suggest that the process has changed or is changing. Notice that there are six points in a row, each of which is higher than its predecessor. Actually, until recently, SPC texts and materials have routinely listed six points in a row, either all of which are rising or all of which are falling, as an out-of-control signal. Point sequences that are all rising or all falling are sometimes called *monotone* patterns. If there are only seven points on a graph, plotted in chronological sampling order, and each of the last six is higher than its predecessor, it would be almost intuitive to take this as a visual indicator that the distribution's location is shifting or has shifted. Such a pattern is very compelling when the rising or falling points are viewed in isolation.

However, Figure 3.19 shows earlier and later sample means. Process location and variation both appear to be stable when all samples are viewed in totality. What has happened here is that the process has been in control and the pattern occurred by chance. Of course, there is a small chance that any specific out-of-control signal can occur when the process is in control. But remember the key to what constitutes a good out-of-control signal. The rising pattern of points on an $\overline{X}$-chart must have not only just a small chance of occurring when the process is in control. One should also expect to see this pattern frequently when the process mean is shifting upward or has shifted upward.

It is this second criterion that shows the weakness of monotone patterns as a tool for knowing when to look for assignable causes. Other routine out-of-control signals mentioned earlier are more sensitive to detecting gradual or sudden shifts. For example, *two-out-of-three-consecutive-points-beyond-2.0-SL* or *four-out-of-five-consecutive-points-beyond-1.0-SL*, are two out-of-control signals that would be expected to register first. Consider Figure 3.18, which depicts a process whose mean did experience a gradual upward shift. Many signals registered after that, but not once did a six-point monotone pattern occur. Basically, even though the process mean was rising gradually, having to wait for a pattern of six subsequent points that are monotone rising without having even one mean be less than its predecessor is a low probability event in comparison to other out-of-control signals. As a result, monotone patterns have fallen out of favor, with good justification.

Zone Charts Have One Out-of-Control Rule

It is one thing for software to be able to recognize when any of the chosen signals has occurred, and another matter for an alert human to do the same. A tool that can prove helpful when an employee monitors a control chart is the zone chart. As an illustrative example, Table 3.1 provides the bolt-length data for the process whose mean is gradually shifting upward from the target value of 40 mm (see Figure 3.18). The pooled standard deviation $S_p = 1.01$ mm is used to estimate the process standard deviation.

Figure 3.20 shows the zone chart for the sample means based on these data. For a brief introduction to zone charts, see Jaehn (1989). Just as for the $\overline{X}$-chart given in Figure 3.18, the interval between the 3-sigma limits is divided into six zones of equal width $\frac{\sigma}{\sqrt{n}} \cong \frac{1.01}{\sqrt{5}} = .452$ mm. A comparison

TABLE 3.1 Bolt Lengths from 25 Samples of Size 5.

x_1	x_2	x_3	x_4	x_5	**Mean**
38.4	40.5	40.7	40.1	38.9	39.73
41.8	40.1	38.6	40.6	39.8	40.18
41.3	40.1	40.7	40.0	39.5	40.32
41.0	37.7	41.5	41.2	40.8	40.42
40.0	41.7	39.8	39.3	38.5	39.86
40.9	41.8	40.2	40.8	40.8	40.90
39.4	40.4	39.0	39.1	41.3	39.84
41.3	40.8	40.8	39.1	38.4	40.11
41.1	40.3	40.4	41.0	41.2	40.79
41.8	42.4	40.6	40.1	39.5	40.89
40.6	39.3	40.9	41.2	41.1	40.63
39.1	40.4	40.7	40.9	43.2	40.84
41.2	39.9	39.5	41.1	42.7	40.87
41.8	41.4	41.3	43.0	41.0	41.70
41.4	40.3	39.7	41.4	40.1	40.58
42.0	42.5	41.3	41.0	42.8	41.91
40.8	41.3	40.2	39.3	40.9	40.50
42.9	41.1	39.5	40.6	42.3	41.28
42.5	41.2	40.7	42.4	41.7	41.68
42.1	40.3	41.3	41.7	40.8	41.23
42.5	40.8	40.8	42.4	40.4	41.39
41.6	42.1	41.4	42.8	41.0	41.77
41.4	41.0	42.1	42.3	42.3	41.81
41.5	40.7	39.1	42.6	41.4	41.06
39.0	40.6	41.3	41.6	41.4	40.77

of the values on the right side of these two figures shows that the zone boundaries are the same, save for roundoff error. On a zone chart, each zone has a score (or weight) attached to it. The zone nearest the center line typically has a score of 0 or 1 attached, and the zones farther from the center line have increasingly larger scores attached—usually 2, 4, and 8. These are given on the left side of Figure 3.20.

We now discuss how to obtain the numbers inside the circles. The value inside a circle represents a *cumulative score* whose value determines whether an out-of-control signal has occurred.

The first sample mean is *plotted* in its appropriate zone by drawing a circle in the middle of the zone and writing the zone score for that zone in-

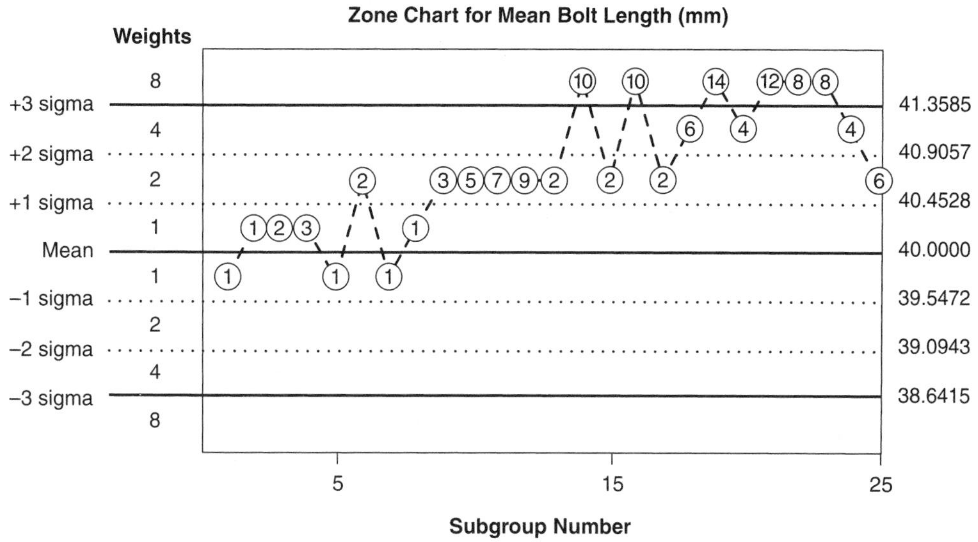

FIGURE 3.20 Zone Chart for Mean Bolt Length (mm).

side the circle. In our case, the first sample mean is 39.73, which falls in the lower 1-zone. Accordingly, a circle is drawn in the middle of that zone and the zone score, 1, is written inside.

With each new sample mean, a circle is plotted in the middle of the proper zone and the zone chart is updated according to the following rules:

- If the new point falls on the same side of the center line as the preceding point, the cumulative score is updated by adding the zone score for the new point to the cumulative score shown inside the preceding circle. The sum, that is, the new cumulative score, is entered inside the new circle. Any point that falls exactly on any of the zone boundaries ±1.0SL, ±2.0SL, and ±3.0SL is considered to have fallen in whichever adjacent zone is nearer to the center line. A point that falls exactly on the center line is considered to be on the same side of the center line as the preceding point, but carries a zone weight of zero.
- If the cumulative score reaches 8 or more, an out-of-control signal has occurred.
- A fresh cumulative score is begun if the new point falls on the opposite side of the center line from the preceding point or on the point immediately following an out-of-control signal. A fresh cumulative score is also started on the first point following a process adjustment designed to recenter the process.

These rules and the sample means from Table 3.1 produce the zone chart of Figure 3.20. The first out-of-control signal registers on Sample 12.

For whatever reason, the process was allowed to operate unchanged, and a total of seven out-of-control signals occurred.

A zone chart is a simple tool for a human operator to use because there is only one rule to remember for what constitutes an out-of-control signal—*Has the cumulative score reached 8 or more?*

A second benefit of a zone chart is that its one rule for an out-of-control signal incorporates a number of signals that are or are similar to standard out-of-control signals. For example, a single point beyond a 3-sigma limit, either by itself or on the tail end of a run of points on the same side of the center line, registers 8 or more. The sum is also 8 or more when, of two or three points in a row on the same side of the center line, two fall beyond 2.0 SL (thus having individual zone scores of at least 4). Similarly, when four or five points are on the same side of the center line and four of them fall beyond 1.0 SL (thus having individual zone scores of at least 2), the cumulative score will be 8 or more. This is also the case when eight points in a row fall on one side of the center line, provided that a zone score of 1 is used for the zones immediately above and below the center line. A reason some people prefer a zone score of 0 to a score of 1 for these two zones is that 8 points in a row near the target or center line, all happening to be on the same side of the center line, does not make as compelling a case for a shift in the mean as, say, two points in a row in the same *4-zone.*

The whole idea of a zone chart is that when a process is centered, points should fall randomly on both sides of the center line. A cumulative score of at least 8 can come from a long run of points on one side of the center line or a short run of points sufficiently far from the center line. Either is evidence of a possible shift in the process mean.

MINITAB Procedure for Zone Charts

The procedure for obtaining a zone chart is similar to that for obtaining an $\overline{X}$-chart. From the **Stat** menu, choose **Control Charts,** then **variables charts for subgroups,** and then **Zone. . . .** A box with windows and buttons will appear. First indicate how the data are stored in the spreadsheet. Suppose the data are in a single column, say column 3, and the subgroup size is 5. Choose the option **All observations for a chart are in one column,** click on the box immediately below, and type c3. Then click on the window labeled **Subgroup size:** and enter 5. If the data are stored in columns so that the values in each row are the values from one sample, select the option **Observations for a subgroup are in one row of columns:,** click on the corresponding box, and enter the columns—for example, c1-c5 indicates that columns 1 through 5 contain the data. If the historical area is known, click on **Zone Options . . . ,** and enter the mean under the **Parameters** tab. If no historical mean is specified, the mean-of-means, $\overline{\overline{X}}$, will be used. There is another window where the historical sigma can be entered, if known. If left unspecified, the pooled standard deviation will be used. Any of $\overline{S}$, $\overline{R}$, or the pooled standard deviation can be used to estimate σ. A click on the tab labeled **Estimate. . .** will produce a window that allows this choice to be made. A click on the tab labeled **weights/reset** allows the user to specify zone weights. Also in this box, there is a window labeled **Reset cumulative score after each signal** that should be clicked in order that a fresh cumulative score is begun after each out-of-control signal. Otherwise, the cumulative score is not reset after reaching 8 or more. There are settings in which a cumulative score larger than 8 should be used as the out-of-control signal. This is discussed later. Also in this box is the **Display** tab, where the user can prescribe whether the chart is to show all points or only the last 25 points. Click **OK.** Click on the **Labels . . .** button in order to give the zone chart a title.

PROBLEMS

1. In Figure 3.21, using the four out-of-control signals discussed at the beginning of the section, identify the number of the sample at which an out-of-control signal occurs (a) on the $\overline{X}$-chart; (b) on the *S*-chart.
2. In Figure 3.22, using the four out-of-control signals discussed at the beginning of the section, identify the number of the sample at which an out-of-control signal occurs (a) on the $\overline{X}$-chart; (b) on the *S*-chart.
3. The sample means in Problem 1, read by rows are:

32.22	33.67	32.11	32.41	33.12	33.49	34.05	33.48	33.37
34.01	33.28	33.74	33.41	32.20	32.77	32.36	33.24	32.30
32.14	32.08	33.18	32.39	33.57	32.64	32.46	32.58	33.51
32.05	32.82	33.26	34.51	32.75	34.37	33.30	33.10	

   Construct a zone chart for sample means using the zone boundaries shown in Figure 3.21. Use zone weights of 1, 2, 4, and 8. When do any out-of-control signals occur?
4. The sample means in Problem 2 are: 22.52, 22.43, 22.12, 21.89, 22.24, 21.43, 21.49, 21.77, 21.45, 21.38, 21.84, 21.35, 21.38, 21.13, and 21.04. Construct a zone chart for sample means using the zone boundaries shown in Figure 3.22. Use zone weights of 1, 2, 4, and 8. When do any out-of-control signals occur?
5. A company produces hollow fiberglass rods for tree trimmers. Rod length has a lower spec limit of 182.05 cm and an upper spec limit of 182.20 cm. The quality manager's standing rule known to all operators is, "When using the cutting equipment, you can always take off a little if a rod is too long, but you can't put it back if the rod is cut too short. So be sure the cutter is set nearer to the upper spec

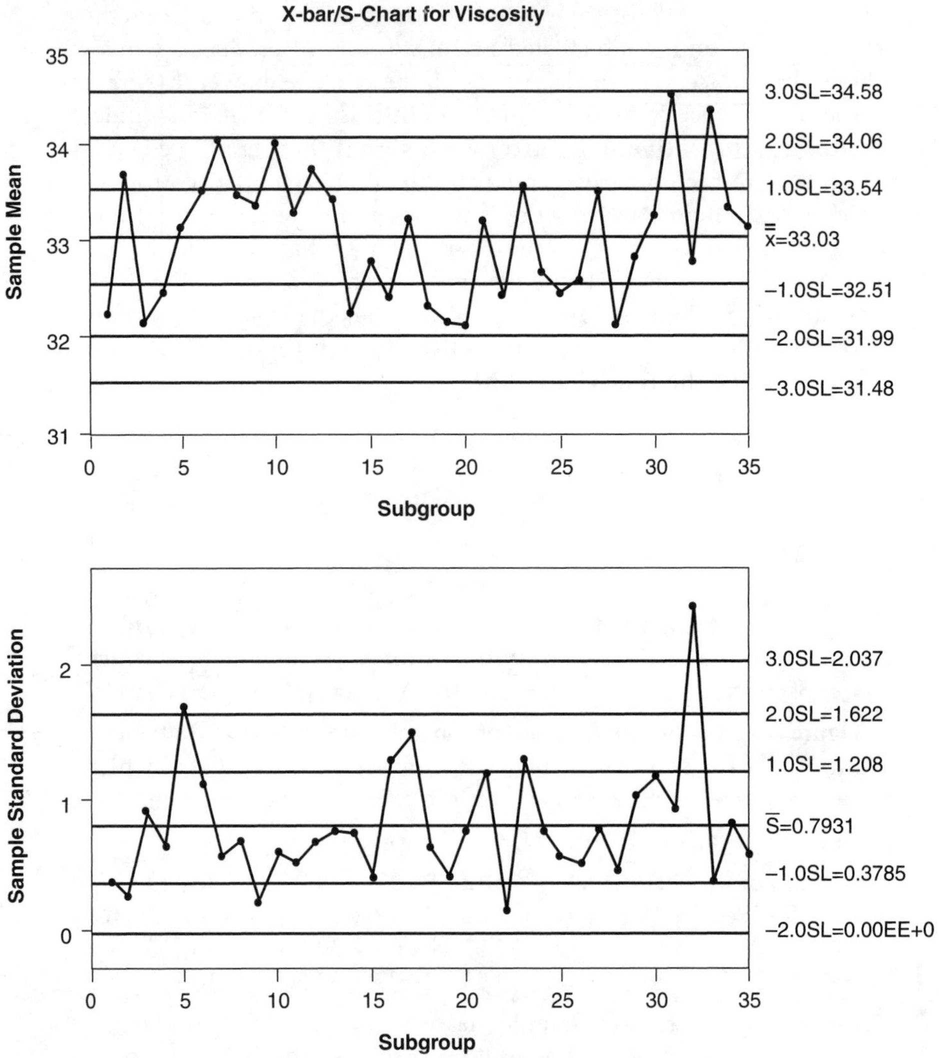

FIGURE 3.21 *X* Bar/*S*-Chart for Viscosity.

than the lower spec. It's easy enough to take off a shim if we have to." Accordingly, the cutter is set at 182.18 cm. Twice a day, a sample of three rods is taken from production and rod length is measured. Figure 3.23 compares sample means to spec limits. All sample means are within specs. Management gets upset with cutting operators because during the assembly process, 17% of the rods are found to have lengths exceeding the upper spec and are returned to have shims taken off.

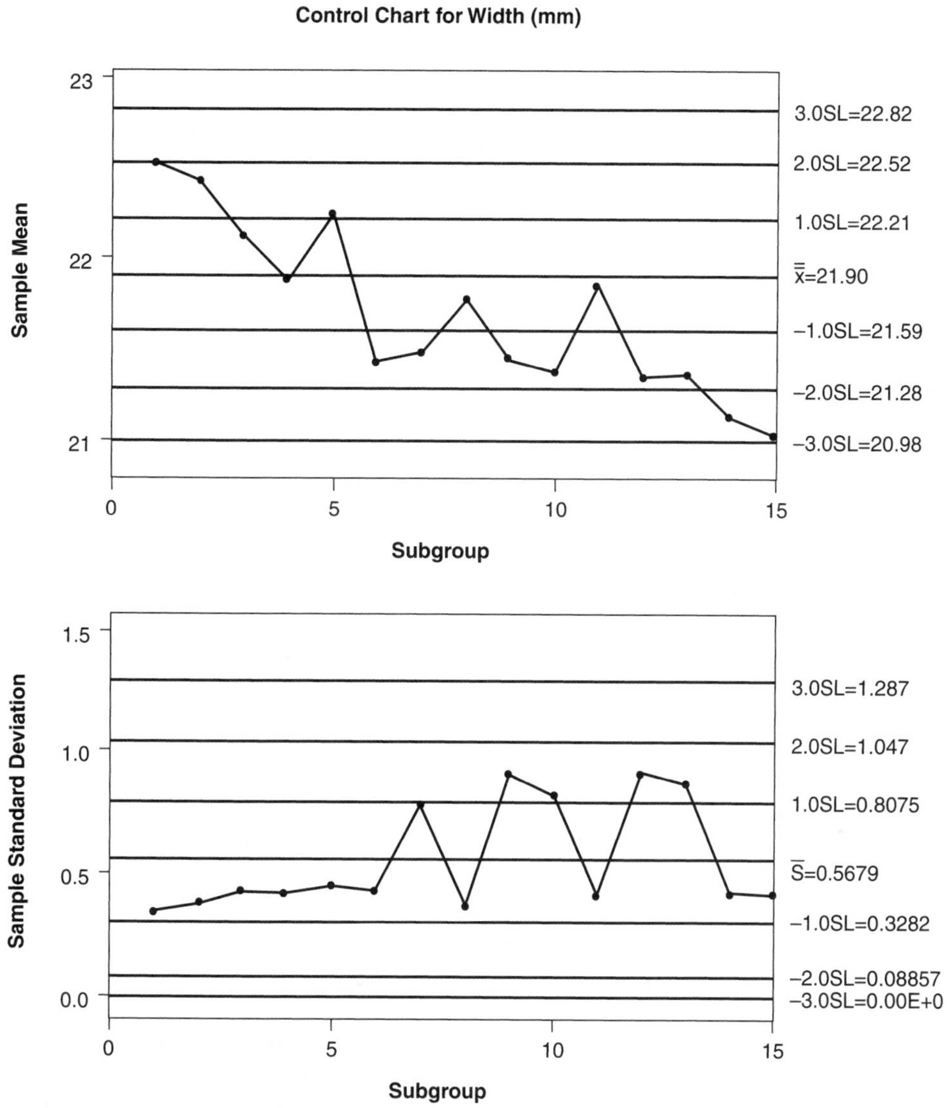

FIGURE 3.22 Control Chart for Width (mm).

a. What is wrong with Figure 3.23 as a quality tool?
b. What is wrong with management's being upset with the cutting operators?
c. Suppose the process standard deviation σ is estimated to be .017 cm. What is the largest practical value at which the cutter can be set so that rod lengths exceeding the upper spec limit will hardly ever occur?

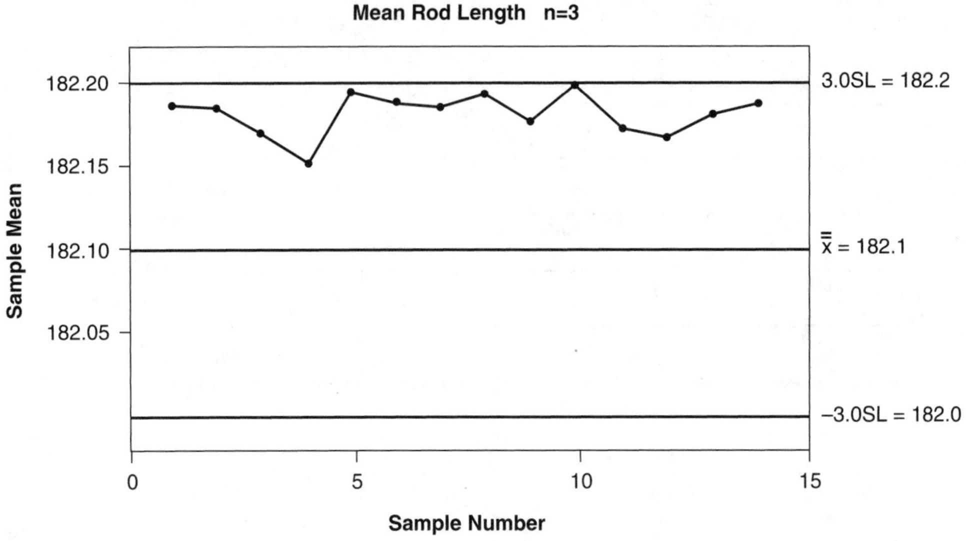

FIGURE 3.23 Mean Road Length (n = 3).

d. If the process mean is maintained at the value determined in part c., will this produce an appreciable proportion of rods that fall below the lower spec?

6. Using MINITAB, obtain a zone chart from the data in Table 3.1 based on a target length of 40.00 mm and historical standard deviation σ = 1.01 mm. Compare to Figure 3.20.

3.3 PROCESS CAPABILITY INDICATORS

Sampling can be done for a variety of purposes. One is to generate control charts used to monitor or control a process. This typically involves small, periodic samples. Another purpose would be to determine whether a shipment of incoming raw materials should be accepted. A typical method for doing this involves selecting a sample of a given number of items from the entire shipment, examining (i.e., measuring or checking) each item in the sample, and following some decision rule for accepting or rejecting the entire shipment based on the data from the one sample. This is called *acceptance sampling.* Some typical acceptance sampling protocols could proceed as follows. Select 30 items at random from the lot of 1,000 and check each one. Then accept the lot only if all 30 items meet specs. A different decision rule could be: If two or more items fail to meet specs, reject the lot; otherwise, accept the lot. A protocol for a different product could begin with measuring the resistance of each item in a sample of 50 items taken from the lot. The decision rule might be to accept the lot if the mean resistance is between 6.00 ± .03 ohm, and reject the lot otherwise. In addition to sampling incoming shipments from a supplier, a company can sample its own manufactured items to help it determine whether a particular shipment should be sent to a certain customer.

The more precision that is desired in estimating either the mean resistance of all the items in the lot or the percentage of items that don't meet specs in a lot, the bigger the sample size will need to be. In other words, the more time and money will be needed. In recent years, acceptance sampling has been recognized as a generally undesirable activity, an activity whose role in the manufacturing process is to be avoided if possible. The rationale comes from statistics and SPC. If control charts show that the process has been in control and the customer is provided the values of some basic descriptive statistics, such as the process mean, process standard deviation, and appropriate capability indices (we discuss some of these indices in this section), there is no need to do acceptance sampling because, as we shall see, the customer now has enough information to know how capable the process is of making items that meet specs. The beauty of this is that the periodic samples that beget the points on control charts typically involve small sample sizes, and these samples would need to be taken anyway in order to know when to make adjustments to keep the process in control. That the customer or manufacturer would not also need to spend the time or money needed for acceptance sampling is a happy residual benefit.

Of course, there are reasons why acceptance sampling sometimes would have to be done anyway. One example would be when, for whatever reason, the regular supplier of raw materials can't meet a customer's urgent need and the customer is forced to consider making do with an unfamiliar vendor who either can't or won't reveal process statistics, or who has process statistics that show that too high a percentage of items produced will not meet the customer's specs. Then either the customer or the vendor will need to do acceptance sampling, or even worse, 100% sampling—that is, examine every item in the shipment to ferret out all items that won't meet specs. In any case, the movement away from acceptance sampling and toward customers wanting to see, before making a buy, statistics on the process that produced the items has been a beneficial paradigm shift in the quality business.

We now discuss some of the capability indices that customers like to see. We begin by recalling the company that made copper wire spools. The process standard deviation was estimated as 3.503 grams, but unfortunately, the spec range for the product was 562.5 ± 5.5 grams. Because the natural tolerance of the process is $6\sigma \cong 6(3.503) = 21.0$ grams, which exceeds $2(5.5) = 11.0$ grams, the process produces a sizeable percentage of spools having weights that don't meet specs. This state of affairs is portrayed at the top of

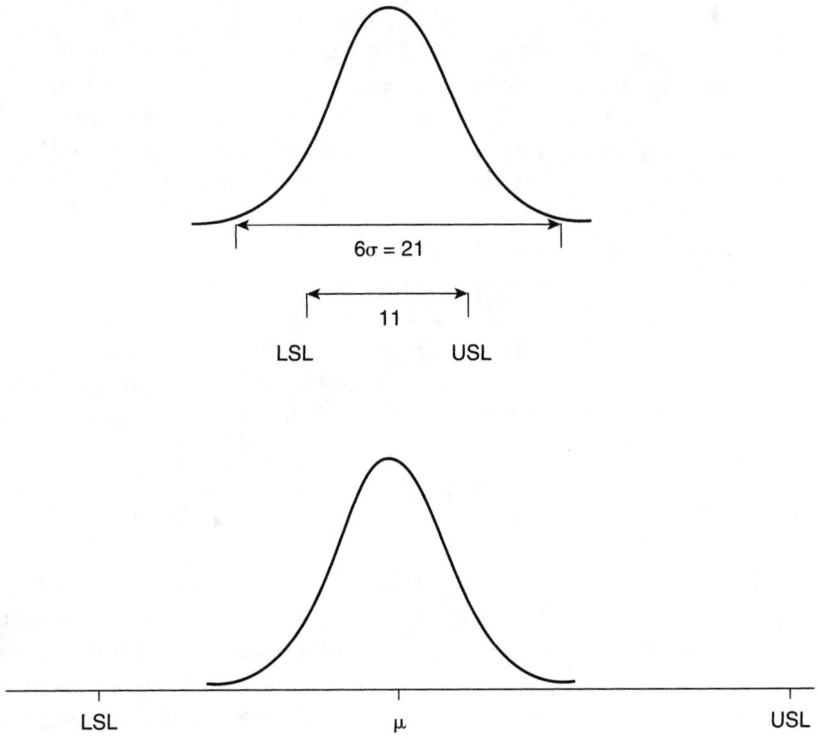

FIGURE 3.24 How the Spec Range and the Natural Tolerance of a Process Compare Provides a Measure of Process Capability.

Figure 3.24. The company would rather see a situation like the one portrayed at the bottom of the figure.

One commonly used index is the *capability potential*, C_p, where

$$C_p = \frac{USL - LSL}{6\sigma}$$

When σ is not known for the process, an estimate based on $\overline{R}$, $\overline{S}$, or the pooled standard deviation is used. The desirable scenario depicted in Figure 3.24 has the distance between spec limits exceeding the process natural tolerance of 6σ. In other words, it is desirable to have the numerator of C_p exceed the denominator, so that $C_p > 1$. The bigger C_p is, the more capable the process is of not producing items that don't meet specs *if* the process mean is between spec limits and far from either limit. This highlights a weakness of C_p. It does not involve the process mean. It would be possible, for example, to have $C_p = 1.4$, which is considered good, but still have virtually 100% of the items produced not meet specs because of where the process is centered.

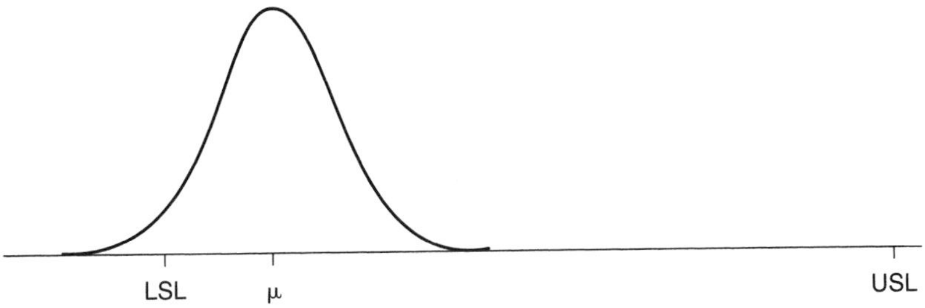

FIGURE 3.25 $C_{pk} < 1$ for this Process, but Could Exceed 1 by Adjusting the Process Mean.

A much more commonly used index that does take into account where the process centered is

$$C_{pk} = min\left\{\frac{\mu - LSL}{3\sigma}, \frac{USL - \mu}{3\sigma}\right\}$$

This measure is sensitive to both how far the mean falls below the upper spec limit and how much the mean exceeds the lower spec limit. By taking C_{pk} to be the minimum of the two fractions in the formula, the only way for C_{pk} to exceed 1 is for both fractions to exceed 1. For example, C_{pk} would be less than 1 for the process depicted in Figure 3.25. Just as σ must be estimated if unknown, we would estimate μ with $\overline{\overline{X}}$ if necessary.

For the company that makes copper wire spools, we illustrate by computing the value of C_{pk} that applies to the period when the process was in control during start-up. Spec limits and descriptive statistics on spool weights are

$LSL = 557.0$ g $\quad$ $USL = 568.0$ g

$\overline{\overline{X}} = 561.16$ g $\quad$ $S_p = 3.503$ g

Then

$$C_{pk} = min\left\{\frac{561.16 - 557}{3(3.503)}, \frac{568 - 561.16}{3(3.503)}\right\} =$$

$$min\left\{\frac{4.16}{10.509}, \frac{6.84}{10.509}\right\} = min\{.396, .651\} = .396$$

Typically, a company would compute C_{pk} in a systematic way, say for each month based on the data from the month, or for every product merge based on data from the merge, and so on. We would like C_{pk} to be at least 1.0 every month, merge, and so on. Since means and standard deviations fluctuate from month to month, it is desirable to have values of C_{pk} average out to more than 1.0. A commonly used rule of thumb is that an average of at least 1.33 is considered good.

For some product characteristics, only one specification limit is appropriate. For example, the amount of pull required to break a meter of fiber may be required only to exceed a lower spec limit. Depending on the product, there might be no reason to have an upper spec limit. When the focus is on how much the mean exceeds the lower spec limit, the appropriate capability index is

$$C_{pL} = \frac{\mu - LSL}{3\sigma}$$

Similarly, when there is only an upper spec limit, the capability index to use is

$$C_{pU} = \frac{USL - \mu}{3\sigma}$$

Inherent in all of these indices is the assumption that virtually all individual measurements fall at least 3σ from a spec limit of interest. This will be the case if the measurements during a period of control come from a normal population. While normal populations are not the only populations with this property, it has become customary to check whether the individual measurements during a period of control could reasonably have come from a normal population. We examine how to do this in MINITAB.

Evaluating Capability Indices with MINITAB

A company is starting a new product run of spools of thread ordered by a customer. Two spools are sampled every half hour and tested to find the minimum tension (in centinewtons) needed to make thread pull from the spool without snagging. Data for the first 26 samples are given in Table 3.2. The accompanying time series plots of means and standard deviations are given in Figure 3.26. The process appears to be in control because the process mean and standard deviation both appear to be stable. The upper and lower specs for tension are LSL = 6.50 cn and USL = 8.50 cn. The mean-of-means and pooled standard deviation during this production run are $\overline{\overline{X}} = 7.236$ cn and $S_p = .352$ cn.

Figure 3.27 is a MINITAB graphic that provides several pieces of information. The menu sequence that obtains the graphic is **Stat,** then **Quality Tools,** then **Capability Analysis (normal). . . .** In the box that appears, the user keys into the appropriate window the location of the data in the usual way (e.g., all in one column). The **Estimate. . .** button allows the user

TABLE 3.2 Sample Data for Minimum Tension.

Sample	1	2	3	4	5	6	7	8	9	10	11	12	13
Obs 1	7.42	6.94	6.82	7.15	7.16	7.22	7.05	7.06	6.65	7.43	7.33	7.54	6.75
Obs 2	7.47	7.18	7.24	7.38	7.13	7.41	7.51	7.85	7.35	7.02	7.45	7.15	7.43

Sample	14	15	16	17	18	19	20	21	22	23	24	25	26
Obs 1	6.86	7.10	6.66	6.99	7.08	6.79	7.38	7.55	7.84	7.31	7.52	6.48	7.60
Obs 2	7.16	7.36	7.52	7.51	7.69	7.54	7.19	7.42	7.48	7.13	7.08	7.23	6.71

to choose the method of estimating sigma, if unknown. Also in the box are windows labeled **Upper Spec:** and **Lower Spec:** in which the user enters one or both of these values.

The first thing to do with this graphic is determine if it is plausible for the data to have come from a normal population. Superimposed on the histogram of the 52 sample values is the best normal curve (solid and labeled ST for *short term* labeled or *within*) that fits the data. This curve is based on the pooled standard deviation and assumes there are no fluctuations in the current process mean. The long term (LT), or *overall*, curve allows for drift in the process mean consistent with the amount of sample-to-sample variation in the existing data. As a rule, the more drift or fluctuation there is in the process mean, the less concentration of area there will be near the process mean under the LT curve, and the fatter the tails will be, in comparison the ST curve. In this case, the curves are very close. In fact, the LT curve even has a slightly higher concentration of area near the process mean than the ST curve, indicating that forcing an allowance for drift with these data does not contribute to projecting the long-term behavior of the process.

In any case, the histogram's shape is consistent with either normal distribution, and one is justified in computing values for the capability indices that describe this process during the time period when the data were collected.

The graphic shows that $C_{pU} = 1.20$. In other words, the upper spec limit is 1.20 times three standard deviations above the mean. Accordingly, it is rare for this process to produce a spool with fiber that exceeds USL—the graphic shows an estimate of how rare. This is the number of *parts per million exceeding USL* (PPM > USL), which is estimated at roughly 164, giving a single-item probability of .000164. However, the ST curve does show substantial area in the left tail to the left of LSL. A projected 18,230 spools in one million can be expected to have fiber that would measure below LSL based on normal distributions. This is 1.823 spools per hundred, for a single-item probability

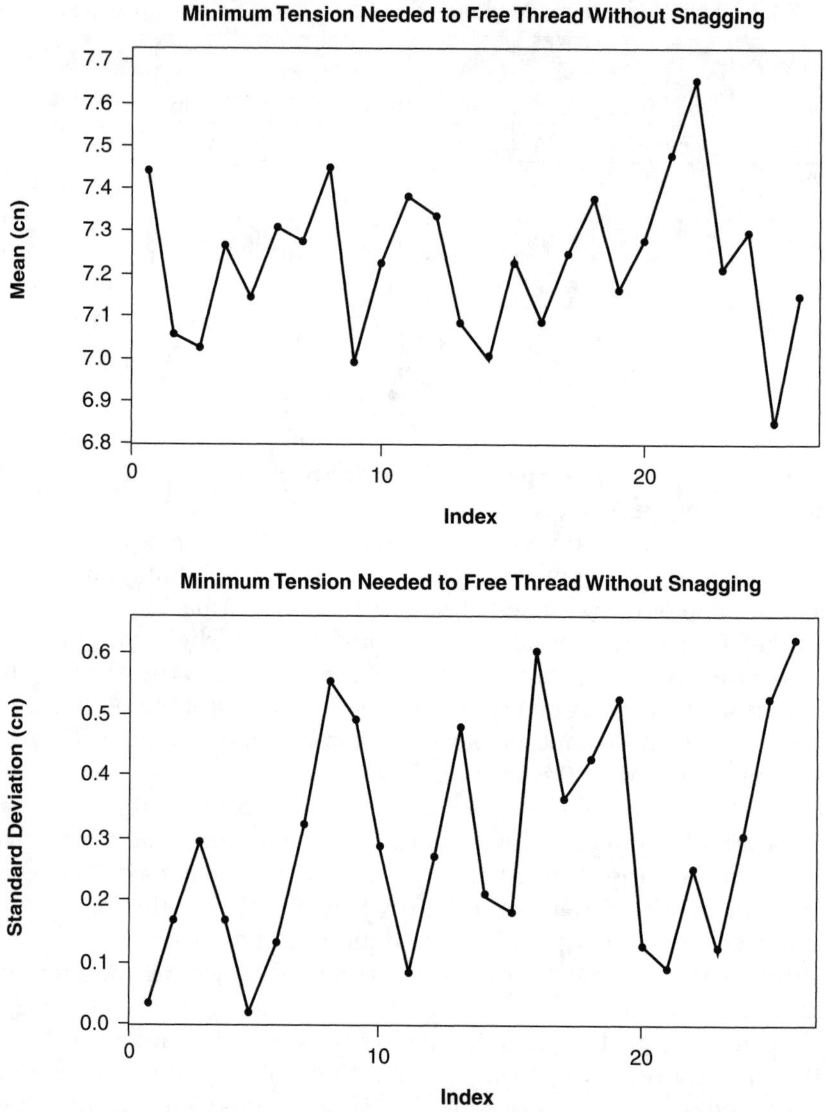

FIGURE 3.26 Means and Standard Deviations for Minimum Tension.

of .01823. Figure 3.27 also shows that the observed rate at which spools fell below the lower spec limit was 19230.77 per million. This comes from the fact that 1 of the 52 values was less than 6.50 cn and $\frac{1}{52} = 0.019230769$.

Since $C_p = .95$, or USL − LSL 0 $\cong .95(6\sigma)$, the proportion of out-of-spec spools will not be negligible, no matter where the process is centered, so long as the process standard deviation doesn't change. Suppose the cost to

Process Data	
USL	8.50000
Target	*
LSL	6.50000
Mean	7.23596
Sample N	52
StDev (ST)	0.351838
StDev (LT)	0.306608

Process Capability Analysis for Minimum Tension May 21-May 27

Potential (ST) Capability	
Cp	0.95
CPU	1.20
CPL	0.70
Cpk	0.70
Cpm	*

Overall (LT) Capability							
Pp	1.09	Observed Performance		Expected ST Performance		Expected LT Performance	
PPU	1.37	PPM<LSL	19230.77	PPM<LSL	18229.87	PPM<LSL	8190.02
PPL	0.80	PPM>USL	0.00	PPM>USL	163.65	PPM>USL	18.73
Ppk	0.80	PPM Total	19230.77	PPM Total	18393.53	PPM Total	8208.74

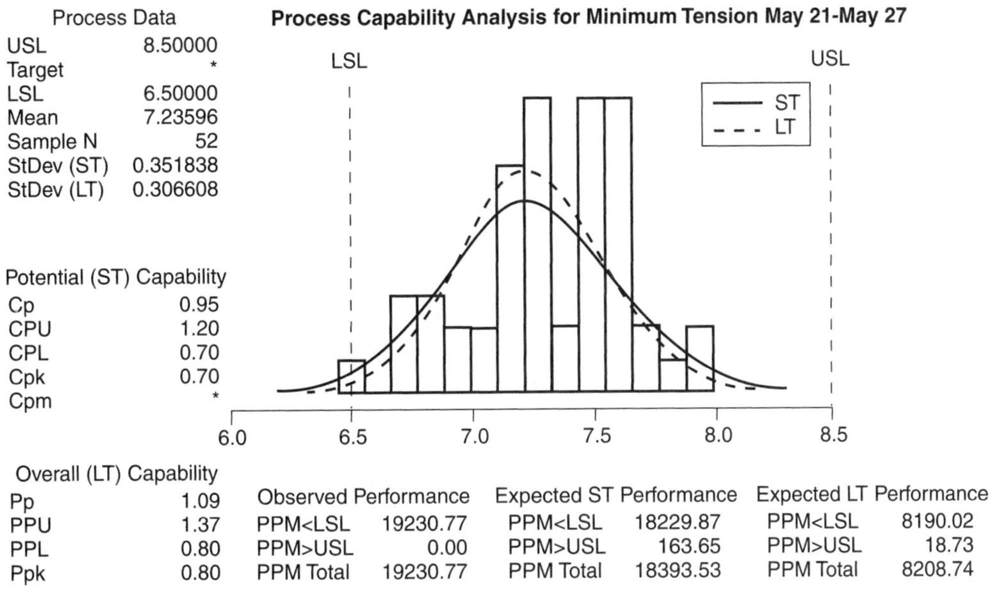

FIGURE 3.27 MINITAB Graphic Gives Process Capability Feedback.

the company of having a spool exceed the upper spec is equal to the cost
when a spool falls below the lower spec. This would be the case if an out-
of-spec spool must be scrapped or would generate a customer claim. Then,
in order to minimize the proportion of out-of-spec spools produced, the immediate goal should be to center the process equally far from either spec
limit—

namely, at $\dfrac{\text{USL} + \text{LSL}}{2} = \dfrac{8.50 + 6.50}{2} = 7.50$ cn. The next management

goal should be to work on reducing σ. This should be a never-ending goal
for any process, but is imperative for this particular process. It also may be
worthwhile to explore whether the spec limits are needlessly tight.

PROBLEMS

1. During the production run to fill a particular buyer's order, $S_p = 1.5$
   and $\overline{\overline{X}} = 8.1$. Also, the lower and upper spec limits for the product are
   3.0 and 12.7 and the target value is 8.0. Also, the sample values had
   an approximate normal distribution.

   a. For this production run, what values should be reported for C_p,
      C_{pL}, C_{pU}, and C_{pk}?

b. Is the target value the value μ would need to equal in order to maximize C_{pk}?

c. Suppose $\overline{\overline{X}} = 7.85$. Find C_{pk}.

2. An efficiency observer records how long it takes each of a sample of 5 customers to receive his or her order at a fast-food restaurant at noon.

   a. Which capability index would be most appropriate for measuring this variable?

   b. After observing for the required one week, the observer found that the average amount of time needed to receive an order was $\overline{\overline{X}} = 153.4$ seconds with $S_p = 46.3$ seconds. If the restaurant chain's upper spec limit is 180 seconds and the observed times have an approximate normal distribution, what value should the observer report for the capability index?

   c. Identify two ways the restaurant could improve the process as measured by this index.

3. For a steel ball used in a certain product, it is imperative that the diameter not be too small. The lower spec limit is 3.70 mm. Suppose the process has been in control for the last 31 samples (subgroup size 4) during which $\overline{\overline{X}} = 3.807$ mm and $\overline{R} = .0577$ mm.

   a. Which capability index would be most appropriate for this variable?

   b. Estimate μ and σ.

   c. What value should be reported for the capability index?

4. a. Can C_{pk} be negative? If yes, what would have to happen?

   b. Based on normal distributions, at least what percentage of production would not meet specs when $C_{pk} = 0$?

5. In Figure 3.27, there is a *shortage* of values in the interval centered at 7.3 cn. Someone might use this to argue that the data come from a bimodal (two-peak) population rather than a normal one.

   a. Give an argument as to why the histogram is nevertheless consistent with the shape of a normal distribution.

   b. Confirm the values given for $\overline{\overline{X}}$ and S_p for the data in Table 3.2.

   c. Confirm the values of C_p, C_{pL}, C_{pU}, and C_{pk} shown in Figure 3.27.

   d. Use a statistical software package to produce the capability analysis given in the figure.

6. In Problem 5, suppose that while it is undesirable to have a spool exceed the upper spec limit, there is no cost in doing so. It is simply understood for this product that the process mean not be *too high*. However, suppose there is a cost when a spool falls below the lower spec limit.

   a. What is the smallest value of the process mean that should be considered acceptable?

Process Data	
USL	7.00000
Target	*
LSL	1.00000
Mean	3.23899
Sample N	160
StDev (ST)	2.52051
StDev (LT)	2.53641

Process Capability Analysis for Distance from Center

Potential (ST) Capability	
Cp	0.40
CPU	0.50
CPL	0.30
Cpk	0.30
Cpm	*

Overall (LT) Capability		Observed Performance		Expected ST Performance		Expected LT Performance	
Pp	0.39	Observed Performance		Expected ST Performance		Expected LT Performance	
PPU	0.49	PPM<LSL	162500.00	PPM<LSL	187187.76	PPM<LSL	188688.37
PPL	0.29	PPM>USL	100000.00	PPM>USL	67828.66	PPM>USL	69062.71
Ppk	0.29	PPM Total	262500.00	PPM Total	255016.42	PPM Total	257751.08

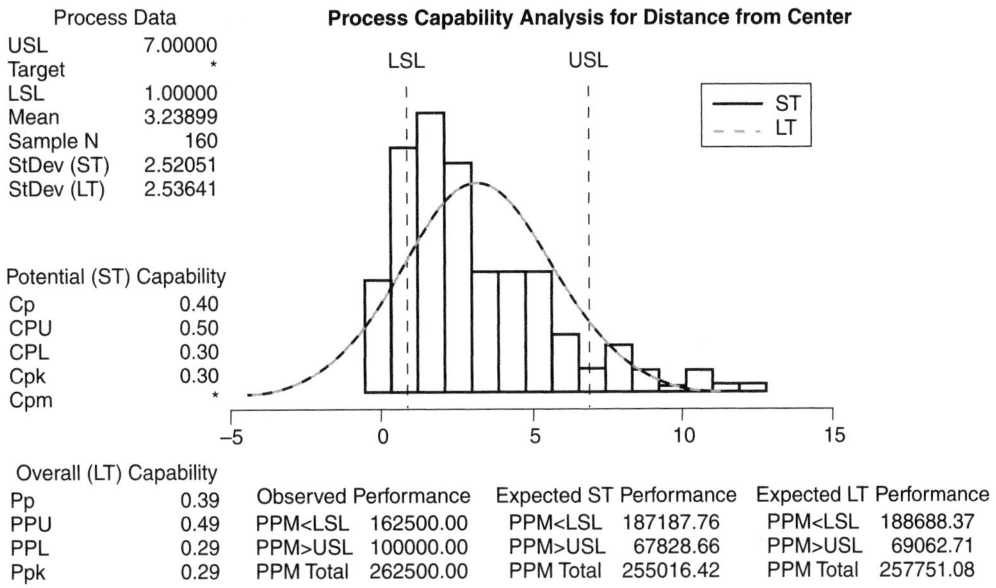

FIGURE 3.28 Capability Analysis for Problem 7.

   b. After centering the process, what should be the next immediate goal?

7. Figure 3.28 is a capability analysis based on 40 samples of size 4 from a process that has been in control.

   a. What is $\overline{\overline{X}}$ for these data?

   b. Estimate σ using the short-term standard deviation.

   c. Is it appropriate to use capability indices for these data? Explain.

8. Figure 3.29 is a capability analysis based on 35 samples of size 5 from a process that has been in control.

   a. What is $\overline{\overline{X}}$ for these data?

   b. Estimate σ using the short-term standard deviation.

   c. Is it appropriate to evaluate capability indices for these data? Why or why not?

   d. What are the values of C_p, C_{pL}, C_{pU}, and C_{pk}?

   e. How many standard deviations above the process mean is the upper spec limit?

   f. How many product units out of one million would be expected to exceed the upper spec limit? Fall below the lower spec limit? Not meet spec limits?

   g. How many product units out of one million would be expected to fall below the lower spec limit if the process were centered at the target value of 59.8? (Hint: Obtain the process capability analysis for these data but enter a process mean of 59.8.)

Process Data	
USL	65.5000
Target	*
LSL	56.5000
Mean	60.9012
Sample N	175
StDev (ST)	1.42363
StDev (LT)	1.41078

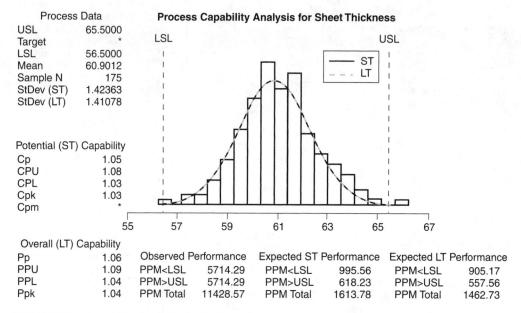

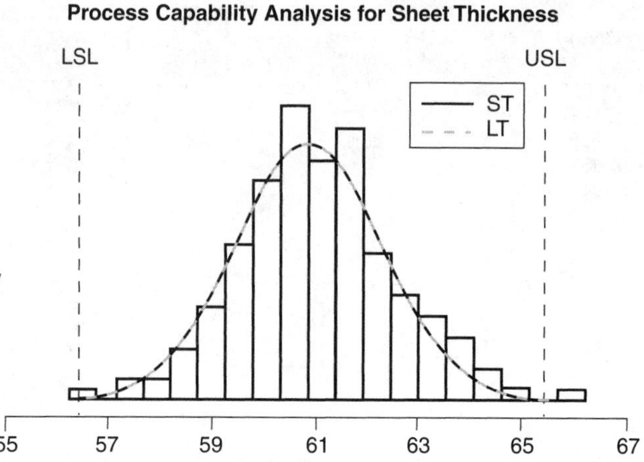

Potential (ST) Capability	
Cp	1.05
CPU	1.08
CPL	1.03
Cpk	1.03
Cpm	*

Overall (LT) Capability	
Pp	1.06
PPU	1.09
PPL	1.04
Ppk	1.04

Observed Performance		Expected ST Performance		Expected LT Performance	
PPM<LSL	5714.29	PPM<LSL	995.56	PPM<LSL	905.17
PPM>USL	5714.29	PPM>USL	618.23	PPM>USL	557.56
PPM Total	11428.57	PPM Total	1613.78	PPM Total	1462.73

FIGURE 3.29 Capability Analysis for Problem 8.

4

Charts

OBJECTIVES

Explain how to set up and use:

- Individuals charts
- *p*-charts
- *np*-charts
- *c*-charts
- *u*-charts
- EWMA charts
- CUSUM charts
- Short-run charts

Give an example of how a zone chart was adapted to deal with dependent samples

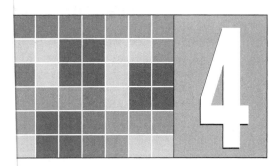

4.1 INDIVIDUALS CHART

In some settings, it is appropriate to have one observation constitute the sample. An example would be monitoring the quality of an incoming chemical. Whenever a tanker truck of it arrives, a small amount is analyzed, producing a single number. Another would be to monitor or control the quality of batches produced in a chemical process, with the quality of each batch adequately portrayed by one number. Other examples would be any production process in which the cost of obtaining a measurement is tremendous, or the time required to measure an item is large, or if obtaining the measurement means a destructive test of the item and items cannot be squandered for whatever reason.

Actually, an individuals chart is just an $\overline{X}$-chart with $n = 1$. The interpretations of the chart are the same. And the assumptions—that measurements of the continuous variable of interest come from normal populations and that samples are independent—are the same.

What is different about an individuals chart is how the process standard deviation is estimated. Because $n = 1$, σ cannot be estimated based on sample ranges or sample standard deviations. A common method is to use *moving ranges*, a topic we introduce in connection with the following example.

The example is not from industry, but illustrates how a control chart could be used to help those who monitor bird populations determine when to suspect that a change may have occurred in the population. The values in Table 4.1 are approximations gleaned from a graph of the annual January estimates of the number of ducks in a region of the northwestern United States. See Perrins, Lebreton, and Hirons (1991).

Figure 4.1, an Excel® time series plot of these data, is indicative of a process with a stable mean. In the spirit of the usual mean-of-means, the estimate of the process mean during this period is the mean of all the individual values:

$$\mu \approx \overline{X} = \frac{19.0 + 22.6 + \cdots + 20.5}{25} = 20.43 \ (\textit{million ducks})$$

TABLE 4.1 Estimates of the Number of Ducks (in Millions) in the Region in January.

1955	19.0	1962	17.5	1968	17.5	1974	18.8
1956	22.6	1963	20.0	1969	17.6	1975	20.5
1957	19.0	1964	23.8	1970	22.3	1976	23.3
1958	22.9	1965	20.0	1971	22.9	1977	21.5
1959	21.8	1966	20.0	1972	22.5	1978	17.4
1960	18.7	1967	21.0	1973	19.9	1979	20.5
1961	19.7						

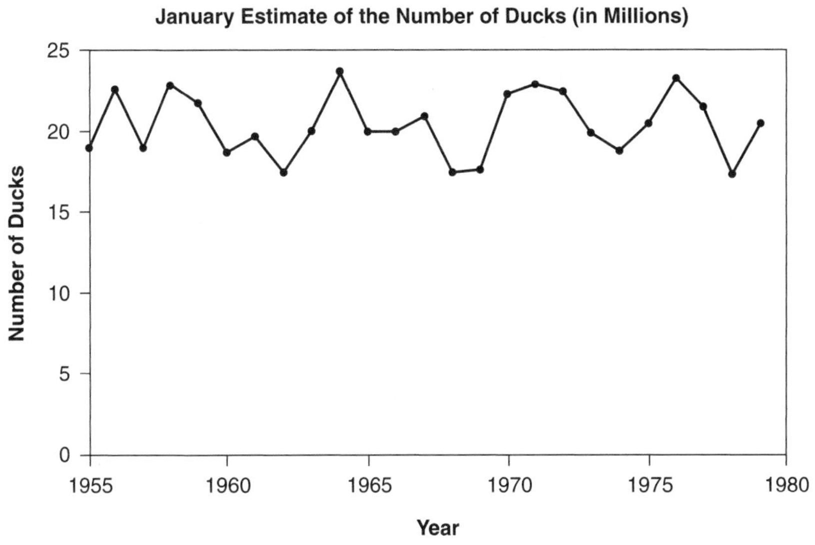

FIGURE 4.1 Time Series Plot of Duck Estimates.

To determine whether variation has been stable, and if so, to estimate σ, we examine moving ranges. First, the *length* of the moving ranges (the number of numbers in each group from which a range is computed) must be decided upon. Let k, an integer satisfying $k \geq 2$, denote the length. Now, acting as if each group of k consecutive individual values constitutes a sample, we compute the range for each group. The moving ranges are, in time order, the range of individual values 1 through k, the range of individual values 2 through $k + 1$, the range of individual values 3 through $k + 2$, and so on.

Table 4.2 shows both the moving ranges of length 2 and the moving ranges of length 3 for the data in Table 4.1. The first moving range of length 2 is 3.6, the range of 19.0 and 22.6. The second moving range is also 3.6, the range of 22.6 and 19.0. The third moving range is 3.9, the range of 19.0 and 22.9. The fourth moving range is 1.1, the range of 22.9 and 21.8, and so on. Because there are 25 individual values, there will be 24 moving ranges of length 2. Similarly, there will be 23 moving ranges of length 3.

Figure 4.2 is a time series plot of the moving ranges of length 2. The ranges appear to be stable in terms of both location and variation, and we can now go forward with estimating σ. It is very common to use moving ranges of length 2. This produces lots of ranges, and averaging these ranges gives a basis for assessing short-term variation in the system or long-term variation, provided the system remains in control. If a larger range length is used, the time series plot will tend to be smoother and any trends more apparent. On the other hand, the larger the value of k, the greater the contribution any location drift present in the sample data will be in estimating the process standard deviation.

TABLE 4.2 Moving Ranges on Estimates of the Number of Ducks (in millions).

Individual value	19.0	22.6	19.0	22.9	21.8	18.7	19.7	17.5	20.0
Moving range, $k=2$	3.6	3.6	3.9	1.1	3.1	1.0	2.2	2.5	3.8
Moving range, $k=3$	3.6	3.9	3.9	4.2	3.1	2.2	2.5	6.3	3.8

Individual value	23.8	20.0	20.0	21.0	17.5	17.6	22.3	22.9	22.5
Moving range, $k=2$	3.8	0.0	1.0	3.5	.1	4.7	.6	.4	2.6
Moving range, $k=3$	3.8	1.0	3.5	3.5	4.8	5.3	.6	3.0	3.7

Individual value	19.9	18.8	20.5	23.3	21.5	17.4	20.5		
Moving range, $k=2$	1.1	1.7	2.8	1.8	4.1	3.1			
Moving range, $k=3$	1.7	4.5	2.8	5.9	4.1				

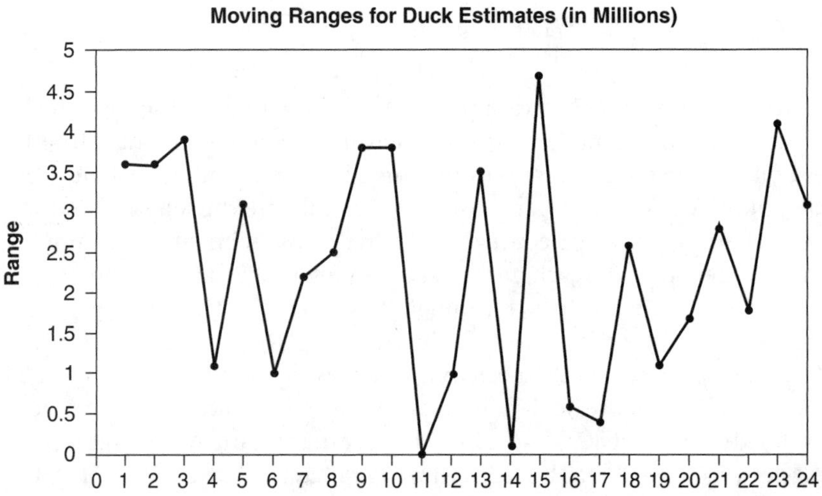

FIGURE 4.2 Moving Ranges for Duck Estimates (in millions).

Let $\overline{R}$ denote the mean of the ranges. The estimate for σ is the same one that is used when the ranges come from independent samples of size k: $\sigma \approx \frac{\overline{R}}{d_2}$. For the 24 ranges of length 2, $\overline{R} = \frac{3.6 + 3.6 + 3.9 + \cdots + 3.1}{24} = \frac{56.1}{24} = 2.34$ and $\sigma \approx \frac{2.34}{1.128} = 2.07$. Once σ is estimated, the rules for 3-sigma limits on the individuals chart are the same as the ones for an $\overline{X}$-chart with a subgroup size of $n = 1$. So for the individuals chart, the center line is at 20.43, and

$$UCL_I = \mu + 3\sigma \approx 20.43 + 3(2.07) = 20.43 + 6.21 = 26.64$$

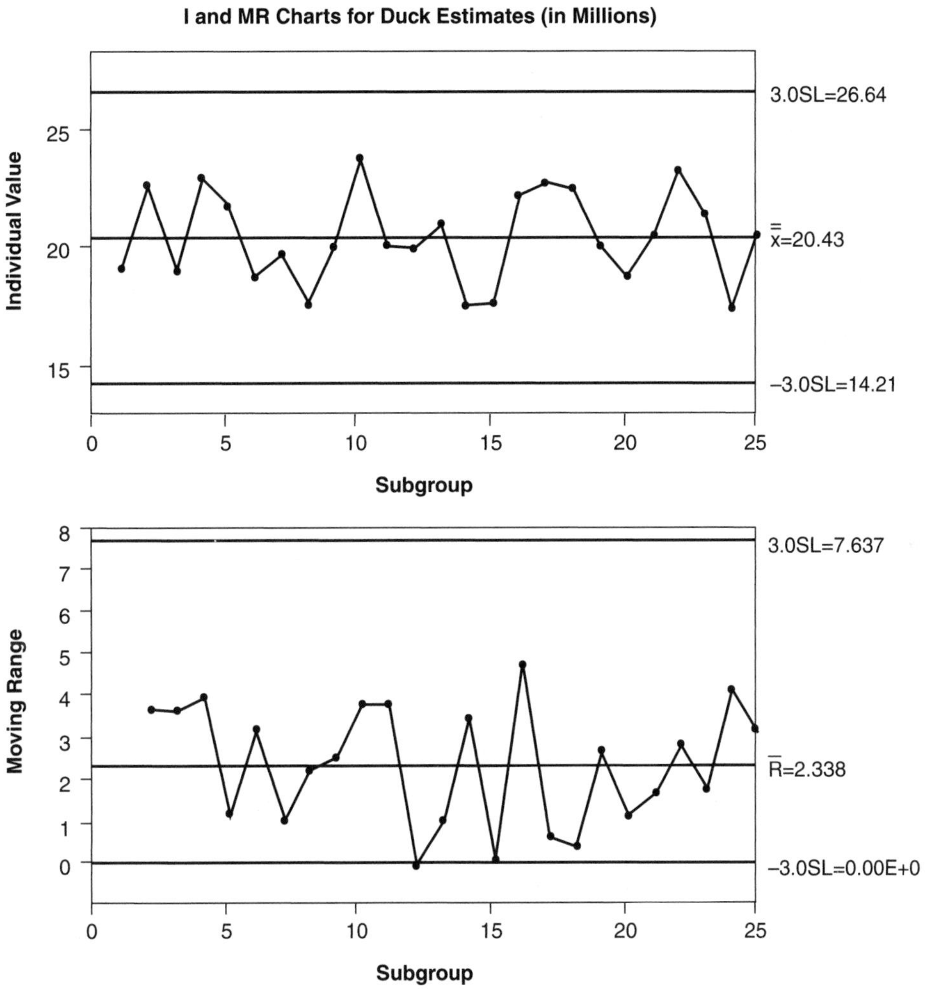

FIGURE 4.3 I- and MR-Charts for Duck Estimates (in millions).

and

$$LCL_I = \mu - 3\sigma \approx 20.43 - 3(2.07) = 20.43 - 6.21 = 14.22$$

Also, with σ estimated, the rules for 3-sigma limits on the moving range chart, which usually accompanies the individuals chart, are the same as the ones for an R-chart that shows ranges of subgroups of size k. Namely, the center line is at $\overline{R}$, while the lower and upper 3-sigma limits are, $LCL_{MR} = D_1\sigma$, and $UCL_{MR} = D_2\sigma$. On the moving range chart for duck estimates, the center line is at $\overline{R} = 2.34$, while $LCL_{MR} = 0(2.07) = 0$ and $UCL_{MR} = 3.69(2.07) = 7.64$. Figure 4.3 shows the individuals chart accompanied by the moving range chart. These are sometimes referred to as I-charts and MR-charts. Control limits just computed agree with those in the figure, save for roundoff error.

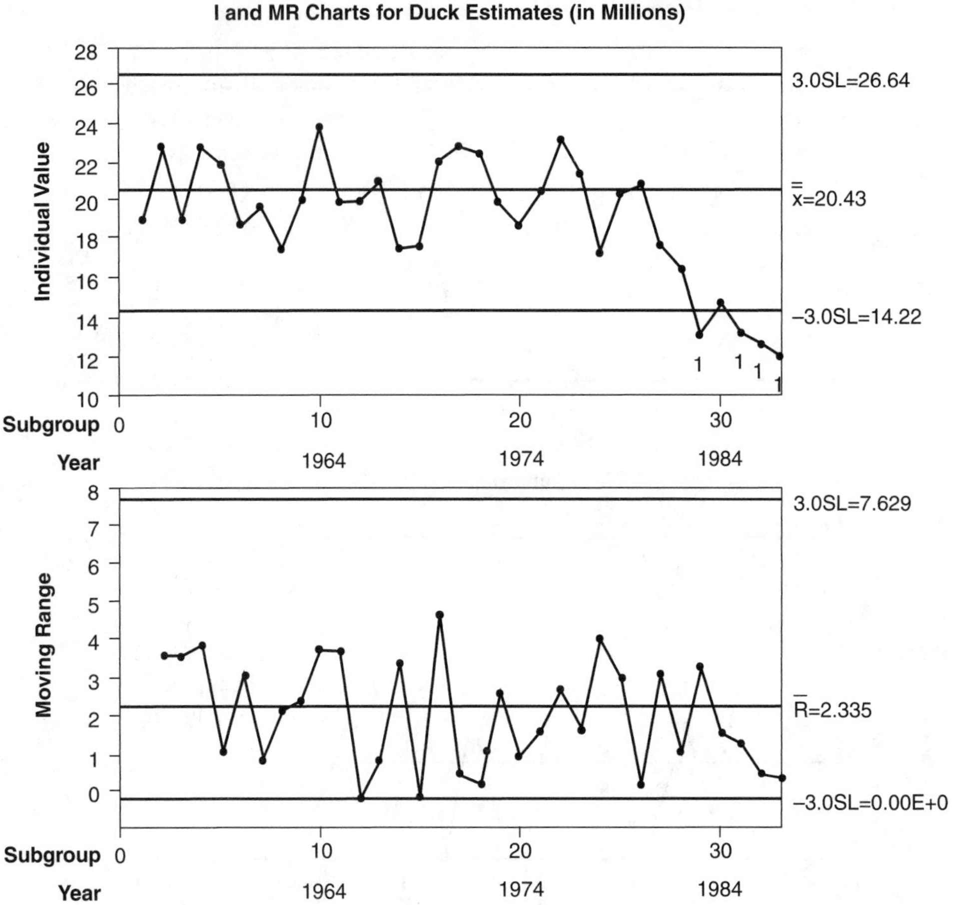

FIGURE 4.4 I- and MR-Charts for Duck Estimates (in millions).

On the basis of control exhibited in duck estimates during the 25-year period from 1955 to 1979, one could have used a historical mean of 20.43 (millions) and a historical standard deviation of 2.07 (millions) on an individuals chart to monitor duck estimates in succeeding years. Any ensuing out-of-control signal would indicate when a potential change in the population has taken place due to an assignable cause.

Duck estimates from 1980 to 1987 were:

1980	1981	1982	1983	1984	1985	1986	1987
20.9	17.7	16.5	13.1	14.8	13.3	12.6	12.0

An individuals chart that plots all data and uses $\mu = 20.43$ and $\sigma = 2.07$ is shown in Figure 4.4. The earliest signal of a potential change in population size occurred in 1983 and there were three reinforcing signals in the next four years.

Obtaining I- and MR-Charts in MINITAB

We will begin by giving first the sequence of steps needed to produce just an individuals chart using the duck population estimates from 1955 through 1979. These individual values (19.0, 22.6, 19.0, and so on) should be entered in some column, say C1. From the menu, choose **Stat,** then **Control Charts,** then **Variables Charts for Individuals** then **Individuals. . . .** A box will appear. In the window in the box labeled **Variables:,** either type C1 or double-click on C1 in the window that identifies columns containing data. If the historical mean and/or historical sigma are known, the value(s) can be entered after clicking on the **I Chart Options. . . .** button and choosing the **Parameters** tab. If sigma is unknown, the MINITAB default is to estimate it based on the average of the moving ranges with a length of 2—namely, $\sigma \cong \overline{R}/d_2$. A user who desires a different estimate of sigma should click on the tab labeled **Estimate.** The user can then identify whether the average of the moving ranges or the median of the moving ranges should be used. In this window, the user also can choose a run length other than 2. The user should then click **OK** to return to the first box. The **Labels. . .** button can be used if a title is desired for the individuals chart. The user should now click **OK** and the chart will appear.

If it is desired that the individuals chart be accompanied by a moving range chart, as in Figure 4.3, the menu sequence should be **Stat,** then **Control Charts,** then **Variables Charts for Individuals,** then **I-MR.** Opportunities to define the historical mean and/or sigma and prescribe how sigma is to be estimated are the same as for the individuals chart.

An Individuals Chart with Correlated Samples

The duck count estimates in Figure 4.1 appear to be high and low at random. This is reinforced by the individuals chart in Figure 4.3, which shows a mix of points above and below the center line, with no long runs on one side of the line. Besides normality, the usual $\overline{X}$-chart or individuals chart presumption is that sample values are independent. Independence allows us to estimate σ for $\overline{X}$-charts based on $\overline{S}$, $\overline{R}$, or S_p and for individuals charts based on moving ranges. Independence along with approximate normality of the population being sampled enable statisticians to evaluate how often the various out-of-control signals/patterns we've discussed will occur by pure chance when there is nothing wrong with the process, and how often they will occur when the process has changed in a given way. For some processes, however, it is known to all those involved with the process that sample values are not independent.

We now consider an example of an individuals chart in a chemical batch process. It is common in such processes for sample values to not be independent when the process is in control. As we shall see for this particular process, a large sample value tends to be followed by another large value and a small value tends to be followed by a small value. In statistical parlance, consecutive sample values are *positively correlated.* It is left to the reader to guess what term would apply when a large value tends to be followed by a small value and a small value by a large value.

We first examine why the individuals chart looks wrong if σ is estimated using moving ranges (as if sample values are independent). We then examine how an individuals chart can be constructed for this setting.

This example is based on a chemical batch process in which many batches of a polyurethane product are made from chemicals drawn from large storage tanks. For proprietary reasons, the product is unspecified and the data have been rescaled. That said, the concentration of "chemical AB" in the completed batch is used as the quality measure of the batch. After each batch has been made, the AB concentration is measured, recorded, and plotted on a chart. The target concentration is 5.65. The quality of any given chemical varies from one tanker truck shipment to the next. Because of this, it is not unusual for AB concentrations to shift location somewhat when batches are made after a new chemical shipment is placed in a storage tank. Figure 4.5, a plot of AB concentrations from the 52 consecutive batches given in Table 4.3, illustrates.

The arrows in Figure 4.5 indicate when newly arrived shipments were put to use. Obvious location shifts make it clear that AB concentrations in consecutive batches are positively correlated. Nevertheless, the company must accept these shifts as inherent in the process, and this is how the company expects the graph to look when the process is centered at 5.65. However, by

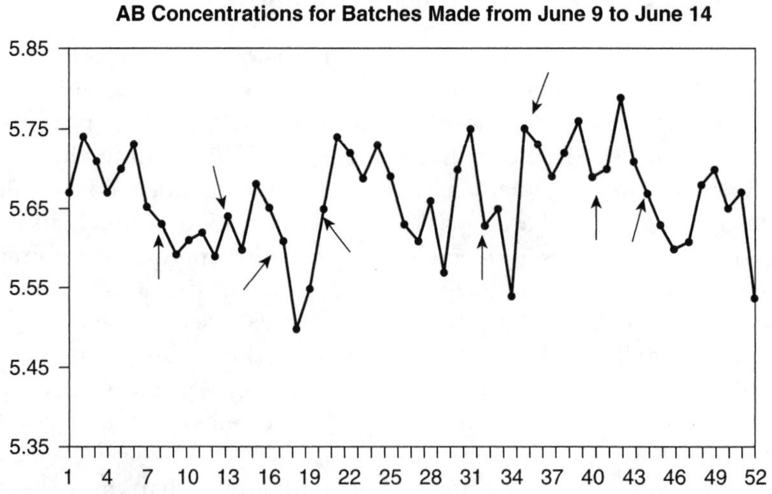

FIGURE 4.5 AB Concentrations for Batches Made from June 9 to June 14.

TABLE 4.3 AB Concentration Data.

obs	AB	obs	AB	obs	AB	obs	AB	obs	AB
1	5.67	11	5.62	21	5.74	31	5.75	41	5.7
2	5.74	12	5.59	22	5.72	32	5.63	42	5.79
3	5.71	13	5.64	23	5.69	33	5.65	43	5.71
4	5.67	14	5.6	24	5.73	34	5.54	44	5.67
5	5.7	15	5.68	25	5.69	35	5.75	45	5.63
6	5.73	16	5.65	26	5.63	36	5.73	46	5.6
7	5.65	17	5.61	27	5.61	37	5.69	47	5.61
8	5.63	18	5.5	28	5.66	38	5.72	48	5.68
9	5.59	19	5.55	29	5.57	39	5.76	49	5.7
10	5.61	20	5.65	30	5.7	40	5.69	50	5.65
								51	5.67
								52	5.54

TABLE 4.4 Moving Ranges of Length 2 from Seven Samples.

obs	AB	MR
1	5.67	
2	5.74	0.07
3	5.71	0.03
4	5.67	0.04
5	5.70	0.03
6	5.73	0.03
7	5.65	0.08

using moving ranges of length 2, we proceed to estimate σ as if batches were independent. The first few moving ranges are illustrated in Table 4.4.

The average of the 51 ranges is $\bar{R} = .054314$, so that

$$\sigma \approx \frac{\bar{R}}{d_2} = \frac{.054314}{1.128} = .04815$$

Then $3\sigma \approx .1445$ and UCL $=$ Target $+ 3\sigma \approx 5.65 + .1445 = 5.7945$, while LCL $=$ Target $-3\sigma \approx 5.65 - .1445 = 5.5055$.

The individuals chart with run length 2 appears in Figure 4.6. Note the high percentage of points that would obviously lie outside the 1SL zones. This is not what we picture for a process viewed as in control. The reason for the high percentage is that, of the ranges averaged to obtain an estimate of σ, most do not involve the shifts that result from new chemical shipments.

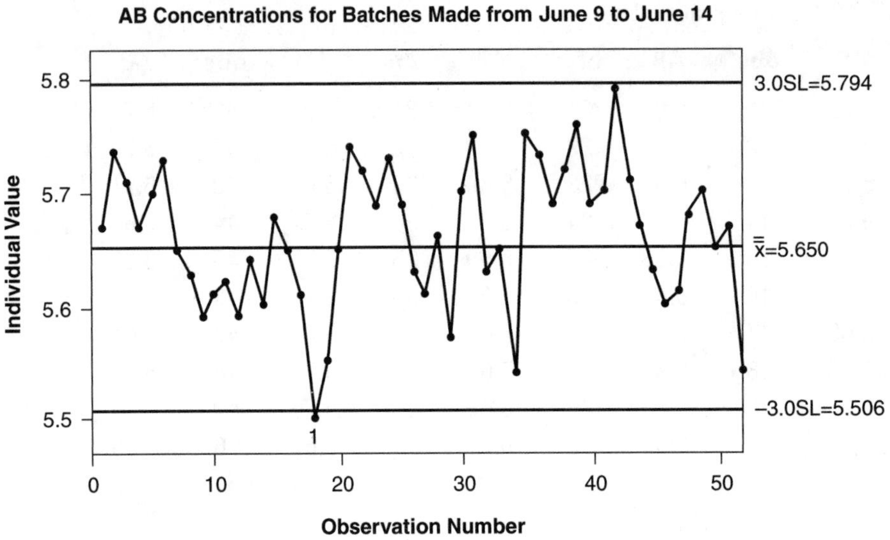

FIGURE 4.6 AB Concentrations for Batches Made from June 9 to June 14.

We are underestimating what a proper σ should be, causing control limits to be too tight.

Since the plotted points are viewed as what "in control" should look like for this process, an alternative would be to estimate an appropriate σ by using S, the sample standard deviation of the 52 AB concentrations (S = .0629). Recall when sampling from a population that S is routinely used to estimate σ when the sample size is large, where a rule of thumb for *large* is $n \geq 30$. Of course here, Figure 4.5 shows that there have been 9 chemical shipments, so that samples of varying sizes come from 9 populations. However, management feels that the mix of upward and downward shifts is representative and the chart is representative of what they consider control to mean for their process. The resulting I-chart based on σ = .0629 is shown in Figure 4.7. At least visually, the point fallout seems to be more in line with the 68%, 95%, and 99.7% values of the Empirical Rule (see Exercise 4).

The company decides to set up an individual's chart for July with the historical σ defined to be .0629 and to review each month's data at the end of the month to determine if σ should be changed. The chart for July appears in Figure 4.8.

An AB concentration of 5.36 on July 15 was an out-of-control signal. It was discovered that through laboratory test error, a shipment of chemical which did not meet specs was accepted and put to use on July 13. Since 5.36 was an acceptable concentration and the cost of disposing of the rest of the shipment was prohibitive, it was decided to use the balance of the shipment.

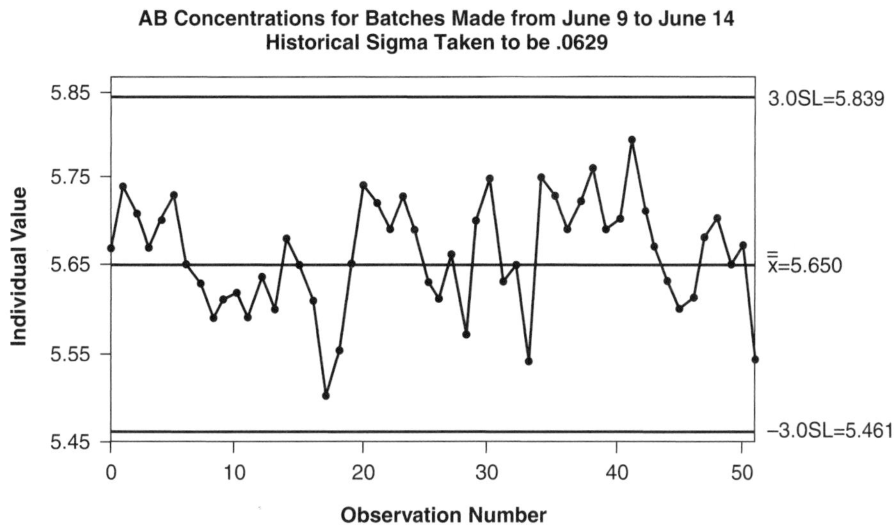

FIGURE 4.7 AB Concentrations for Batches Made from June 9 to June 14. Historical Sigma Taken to be .0629.

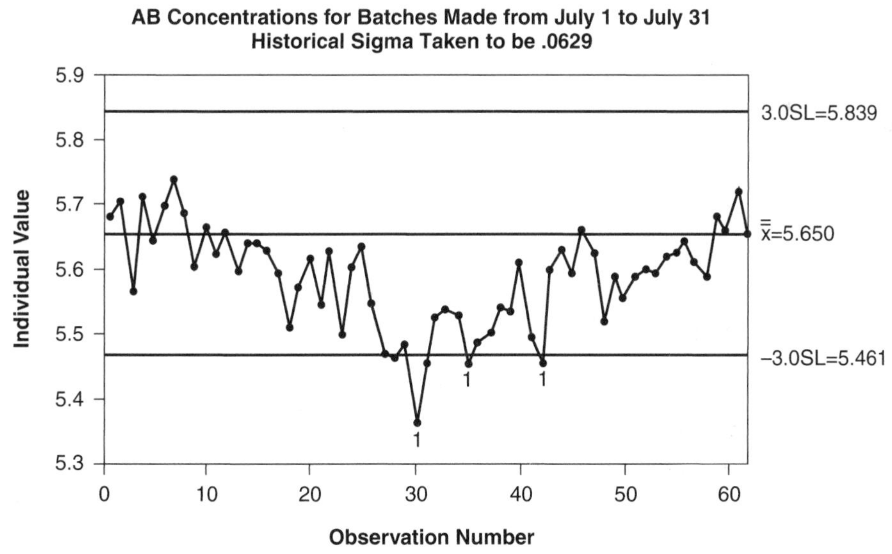

FIGURE 4.8 AB Concentrations for Batches Made from July 1 to July 31. Historical Sigma Taken to be .0629.

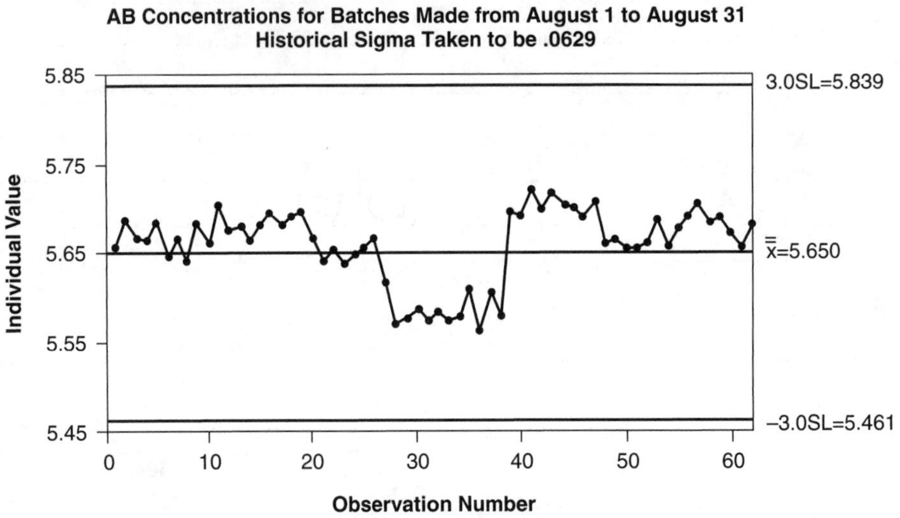

FIGURE 4.9 AB Concentrations for Batches Made from August 1 to August 31. Historical Sigma Taken to be .0629.

A new shipment of the chemical was put to use on July 20. Values of $\overline{X}$ and S for July were:

July Data:	**Mean**	**S**
$n = 62$ (all batches)	5.59	.0776
$n = 50$ (12 batches deleted)	5.61	.0609

With all data connected to the bad shipment deleted, the standard deviation was .0609. It was decided to continue to use .0629 as the historical sigma.

On August 1, a more precise instrument for measuring the amount of chemical C that goes into each batch was put to use. The chart for August appears in Figure 4.9. The descriptive statistics for August were:

August Data:	**Mean**	**S**
$n = 62$	5.66	.0418

The process change—that is, equipment change—clearly reduced the amount of variation in AB concentration other than that attributable to the arrival of new chemical shipments. It is decided to use a historical sigma of .042 for September. With the reduction in process variation, it is easier to tell from the I-chart when chemical shipments have arrived and obvious that differences between chemical shipments of the same chemical contributes more to variation in AB concentrations than other sources of variation combined. Accordingly, management explores if specs for incoming shipments of key chemicals can be tightened.

Chart updated by									
8 5.84 or more									
4 5.78 to 5.83									
2 5.72 to 5.77									
1 5.66 to 5.71									
0 5.65									
1 5.59 to 5.65									
2 5.53 to 5.58									
4 5.47 to 5.52									
8 5.46 or less									
AB Concentration									
Date									
Batch No.									
Event Log									

FIGURE 4.10 Individuals Chart for AB Concentrations.

Logging in Events

Control charts should be accompanied by an event log in order to record when changes occur that could affect quality (chemical shipment arrival, subassembly arrival, new machinery, machinery adjustment, operator change, vendor change, procedure change, etc.). Without such notations, an obvious location shift seen on a control chart can be an unsolved and unremedied mystery. Charts updated by hand typically are designed with room for notations. When control charts are computerized, a process event file is common. An example of a control chart with event log appears in Figure 4.10.

PROBLEMS

1. Explain why any method of estimating the process standard deviation that involves sample ranges or standard deviations cannot be used when the subgroup size is 1.

TABLE 4.5 Ozone Level Data.

1/1/98	2/1/98	3/1/98	4/1/98	5/1/98	6/1/98
309	287	253	284	336	304
7/1/98	8/1/98	9/1/98	10/1/98	11/2/98	12/1/98
317	292	294	285	280	259
1/3/99	2/1/99	3/1/99	4/1/99	5/1/99	6/1/99
258	263	292	283	367	328
7/1/99	8/1/99	9/1/99	10/1/99	11/1/99	12/1/99
306	305	295	285	269	344
1/1/00	2/1/00	3/1/00	4/1/00	5/1/00	6/1/00
287	276	274	305	335	315
7/1/00	8/1/00	9/1/00	10/1/00	11/1/00	12/1/00
308	306	293	282	269	268
1/1/01	2/2/01	3/2/01	4/1/01	5/1/01	
318	255	261	335	341	

2. Refer to the data in Table 4.2.
   a. How many moving ranges of length 4 are there?
   b. Find the values of these ranges, compute $\overline{R}$.
   c. Estimate σ and compare to the estimate of σ based on moving ranges of length 2. These should be comparable when samples are independent and the process is stable.

3. In Problem 2, moving ranges of length 4 were used to estimate σ.
   a. Is the value of $\overline{R}$ in Problem 2 comparable to 2.34, the value of $\overline{R}$ for moving ranges of length 2?
   b. Which should one expect to be larger, $\overline{R}$ based on moving ranges of length 2 or $\overline{R}$ based on moving ranges of length 4?

4. Suppose total column ozone amounts at a particular latitude and longitude, in Dobson units, are as given in Table 4.5. Actual values at a latitude and longitude of interest may be obtained from a NASA Web site (see Ozone-TOMS [2003]).
   a. Make a time series plot of these data. Do the measurements appear to be independent?
   b. In spite of the answer to part a., suppose it is decided to interpret the time series plot as stable and to maintain an individuals chart. Estimate σ based on a moving range of length 2. Determine 3-sigma control limits.

c. Obtain an individuals chart. Choose as out-of-control signals any point beyond the 3-sigma limits, 2-out-of-3 beyond the 2-sigma limits, 4-out-of-5 beyond the 1-sigma limits, and 9-in-a-row on one side of the center line. Describe the many out-of-control signals that arise from these so-called "stable" data.

5. Each morning one spool of Spandex fiber designed for pantyhose is taken from the machine that made it and the denier (weight in grams of 9,000 unstretched meters of fiber) is measured. For the last 30 mornings, the deniers have been:

   19.8, 20.4, 20.2, 19.6, 20.5, 19.1, 20.4, 19.5, 19.3, 21.1, 19.8, 20.6, 19.6, 19.6, 20.3, 20.5, 19.9, 20.1, 19.7, 19.6, 19.8, 20.0, 20.2, 20.6, 19.5, 19.5, 19.9, 19.9, 20.3, 20.3.

   a. Obtain a time series plot of the data. Is there apparent control?
   b. Estimate the mean and standard deviation (using moving ranges of length 2) for the denier of fiber produced by this machine.
   c. Obtain 3-sigma control limits assuming that the target denier is 20.

6. Suppose the data in Problem 5 are used to establish a historical value for σ for the next month. Also suppose the deniers for the first five days of the next month are 19.3, 18.9, 19.3, 19.0, and 19.0.

   a. Do any points fall outside the control limits?
   b. Obtain an I-and MR-chart showing these data from the first five days of the month. Identify any standard out-of-control signals other than point(s) outside the 3-sigma limits.

4.2 *p*-CHARTS AND *np*-CHARTS

p-Charts

Sometimes instead of the interest being on a characteristic whose measurement is on a continuous scale, the interest is only on whether an item has a particular attribute. An item either has an attribute or it does not—an engine either will start or it won't, a part will fit or it won't, an item either meets specs or it does not. The attribute may be associated with a characteristic that has an underlying and natural continuous scale—for example, resistance in ohms. However, if having resistance less than 12.0 ohms equates to an item not meeting specs, then management may want a method to track/monitor either the *proportion* or *number* of parts that doesn't meet specs in addition to charting the average resistance. A part that does not meet specs is often referred to as *nonconforming* or *defective.*

A control chart for the proportion of nonconforming items inspected is called a *p*-chart.

TABLE 4.6 April Data.

Number Inspected	108	118	118	128	121	135	103	132	109	113	120	118	108	134	122	118	122	105	126	132
Number Needing Repair	10	7	7	11	11	16	7	9	8	10	5	13	19	6	6	3	9	10	15	13

EXAMPLE 1

In recent days a company has noticed a large number of warranty repairs due to coolant loss on the air conditioner units it manufactures. While studying the cost of different options that would fix the problem, the company began doing 100% inspection of its daily production in order to fix the problem in the units needing it. The results of this effort for the last 20 days of April appear as shown in Table 4.6.

For this process, it is natural to track the proportion of nonconforming items on a per-day basis. To see what a chart might look like, we need the three reference lines:

$$\text{center line: } \bar{p} = \frac{\text{total number of nonconforming items}}{\text{total number of items inspected}}$$

$$UCL_p = \bar{p} + 3\sqrt{\frac{\bar{p}(1-\bar{p})}{n}} \qquad LCL_p = \bar{p} - 3\sqrt{\frac{\bar{p}(1-\bar{p})}{n}}$$

In this setting, we let p denote the population proportion of nonconforming items that the process can produce. The parameter p generally is unknown and is estimated by $\bar{p}$. From the data of Table 4.6, our initial estimate of p is

$$\bar{p} = \frac{10 + 7 + 7 + \cdots + 13}{108 + 118 + \cdots + 132} = \frac{195}{2{,}390} = .082$$

The formulas for the control limits make it clear that the control limits valid on a particular day depend not just on $\bar{p}$, but also on the number of items inspected that day, n. For the control limits to graph as horizontal lines, the number of items inspected would have to be the same every day, which is not the case here.

To determine if the process has been stable, an informal p-chart is examined based on all the data from Table 4.6. See Figure 4.11. The process appears to be in control, except on April 18. A check of the activity log shows that the large percentage of the defectives that day coincides with the replacement of a spot welder that de-

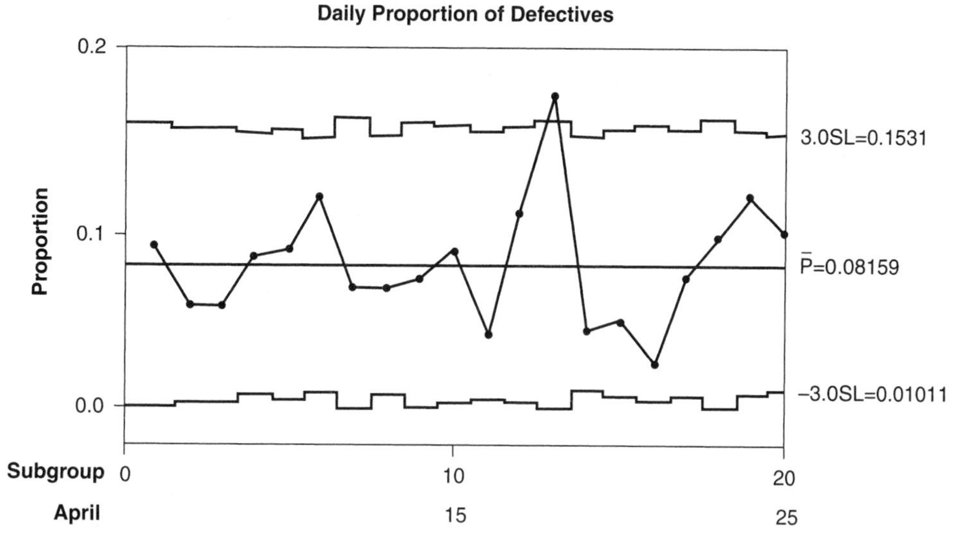

FIGURE 4.11 Daily Proportion of Defectives.

veloped an electrical short. This almost certainly created most of the defectives produced that day. The log shows no other event that would require the data for any day to be viewed as unrepresentative of the process. Management decides to maintain a *p*-chart to monitor the proportion of nonconforming air conditioners produced daily. In order to establish the value of $\bar{p}$ to use for the next month, the data from April 18 are deleted as unrepresentative. This gives

$$\bar{p} = \frac{195 - 19}{2{,}390 - 108} = .077$$

While the current value of $\bar{p}$ locates the center line on the *p*-chart, the value of $\bar{p}$ is never to be viewed as a target or goal. No one should *want* to produce 7.7% defectives. In fact, the goal should be 0% defectives. The way to interpret the value 7.7%, the value chosen for the center line for the next month, is in terms of process capability. Management believes that the process, as it now stands, produces nonconforming air conditioners at a rate of around 7.7% and that is what the process currently is capable of doing.

Control limits on *p*-charts are based on the Central Limit Theorem, which provides that certain variables should be approximately normally distributed when the sample size n is large enough. A typical rule of thumb is to assume that each plotted point represents a value taken from an approximate normal distribution if both

$$n\bar{p} \geq 5 \text{ and } n(1 - \bar{p}) \geq 5$$

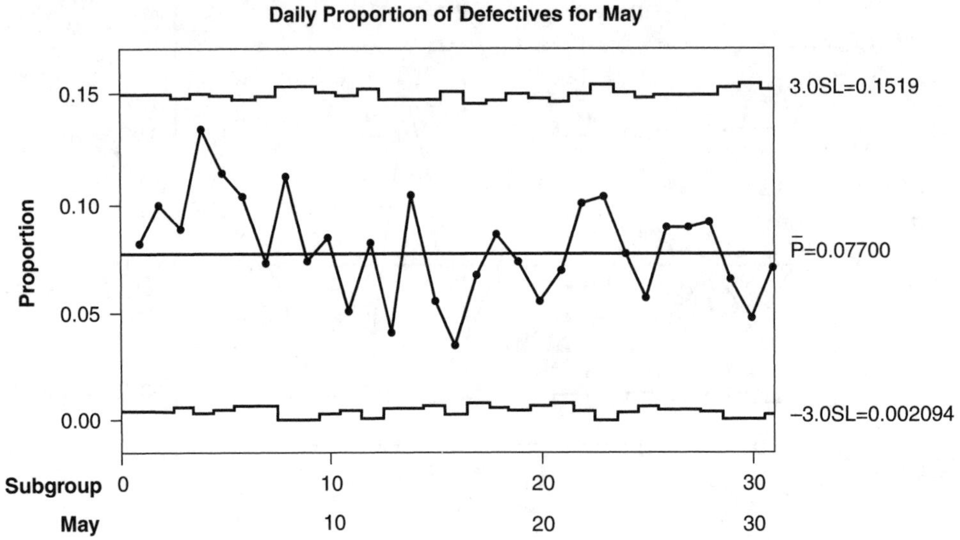

FIGURE 4.12 Daily Proportion of Defectives for May.

Since n, the number of items inspected daily, exceeds 100 and $\bar{p} = .077$, then $n\bar{p} \geq 5$. Likewise, $n(1 - \bar{p}) \geq 5$. So the proportion of nonconforming items produced in a day may be taken to be approximately normally distributed, and the control limits will be set at the appropriate "3-sigma" from $\bar{p}$, namely $3\sqrt{\frac{\bar{p}(1 - \bar{p})}{n}}$ from $\bar{p}$.

The p-chart for proportion nonconforming in May with center line defined at .077 is given in Figure 4.12. The chart reflects control. The actual computed value of $\bar{p}$ based on the data from May is .079. Management decides that for June, the center line will be set at a *historical* $\bar{p}$ of .078. It also would be defensible to continue to use a *historical* $\bar{p}$ of .077, in the spirit of viewing .079 as within routine fluctuation and not conceding that the process has a higher nonconforming rate than before without more than just the data from May.

Obtaining *p*-Charts in MINITAB

The number of nonconforming items from each sample should be entered in chronological order into a MINITAB column, say C2. For the data of Table 4.6, the entries beginning at the top of C2 would be 10, 7, 7, and so on. When n varies from sample to sample, the sample sizes should be entered into a second column. For the data of Table 4.6, the first few entries in C1, say, would be 108, 118, 118, and so on. From the menu, choose **Stat,** then **Control Charts,** then **Attributes Charts,** then **P. . . .** A box appears. In the window in the box labeled **Variables,** enter C2 or double-click on C2 in the window that identifies columns containing data. Click on the box labeled **Subgroup sizes:** and enter either the sample size, if all sample sizes are the same, or identify which column contains the sample sizes, C1 in this case. If there is a prescribed value for $\bar{p}$, click on the button labeled **P Chart Options,** and enter this (historical) proportioned in the window under the **Parameters** tab. If no value is entered here, $\bar{p}$ will be computed based on the data. The usual control chart features such as **Labels. . .** are available as well. Finally, click **OK.**

np-Charts

The setting is the same as the one for p-charts. Namely, items are inspected, and each item either meets specs or is nonconforming. If the interest is on plotting the actual *number* of nonconforming items in each sample rather than on the proportion of such items, these values can be plotted on an np-chart. Such a chart is easiest to interpret when the number of items in each inspected group remains the same, as we shall see. As for p-charts, n should be large enough that $n\bar{p} \geq 5$ and $n(1 - \bar{p}) \geq 5$. If these inequalities are met, there is justification in assuming that the number of nonconforming items in a sample has an approximate normal distribution, which in turn allows appropriate "3-sigma" limits for the number of nonconforming items to be used.

We illustrate by modifying the air conditioner example with the assumption that the manufacturer routinely inspects 100 units per day and that for the month of April, nonconforming unit data appears as in Table 4.7.

The three reference lines on an np-chart are:

center line: $n\bar{p}$

control limits: $n\bar{p} \pm 3\sqrt{n\bar{p}(1 - \bar{p})}$

TABLE 4.7 April Nonconforming Air Conditioner Data.

Date						4/1	4/2	4/3	4/4	4/5	4/6
Number of air conditioners not meeting specs						10	5	7	8	8	4
4/7	4/8	4/9	4/10	4/11	4/12	4/13	4/14	4/15	4/16	4/17	4/18
9	10	5	11	5	6	9	8	9	6	9	13
4/19	4/20	4/21	4/22	4/23	4/24	4/25	4/26	4/27	4/28	4/29	4/30
3	8	9	8	5	10	6	4	11	6	7	7

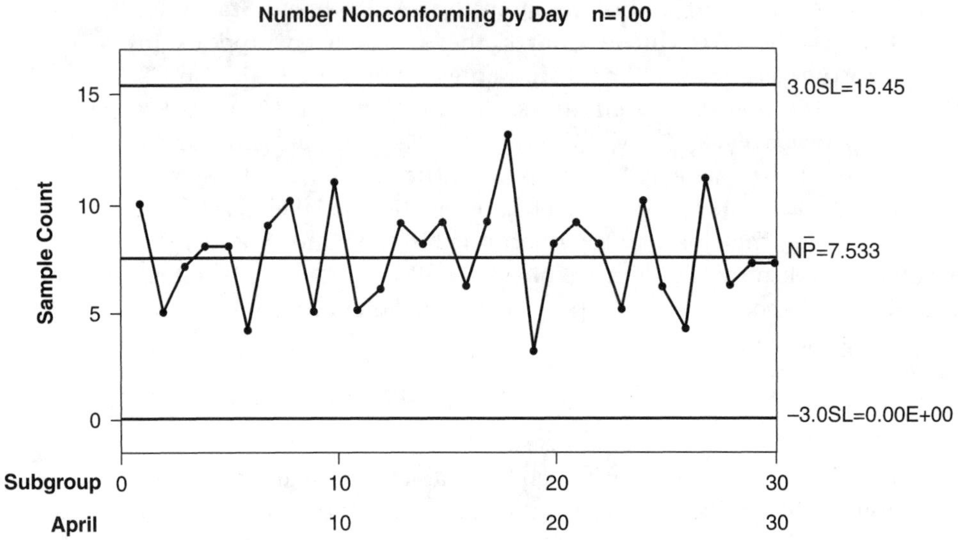

FIGURE 4.13 *np*-Chart.

For these data the initial estimate of p is $\bar{p} = \dfrac{10 + 5 + 7 + \cdots + 7}{30(100)} = .07533.$

Therefore, in each inspected group we would expect to see $np = 100\ (.07533) = 7.533$ nonconforming units. An informal *np*-chart for these data, Figure 4.13, is examined first to determine if the process is stable.

The process nonconforming rate p appears to be constant during the month. Also there are no out-of-control points, which would mean a check of event logs to determine if any data should be deleted for the purposes of estimating p. Accordingly, management will use a historical $\bar{p}$ of .07533 on its *np*-chart for May.

Obtaining *np*-Charts in MINITAB

The number of nonconforming items from each sample should be entered in chronological order into a MINITAB column, say C2. For the data of Table 4.7, the entries beginning at the top of C2 would be 10, 7, 7, and so on. From the menu, choose **Stat,** then **Control Charts,** then **Attributes Charts,** then **NP. . . .** A box will appear. In the window in the box labeled **Variables:,** enter C2 or double-click on C2 in the window that identifies columns containing data. Since the sample size in our example is always 100, next click on the box labeled **Subgroup sizes:** and enter 100 as the

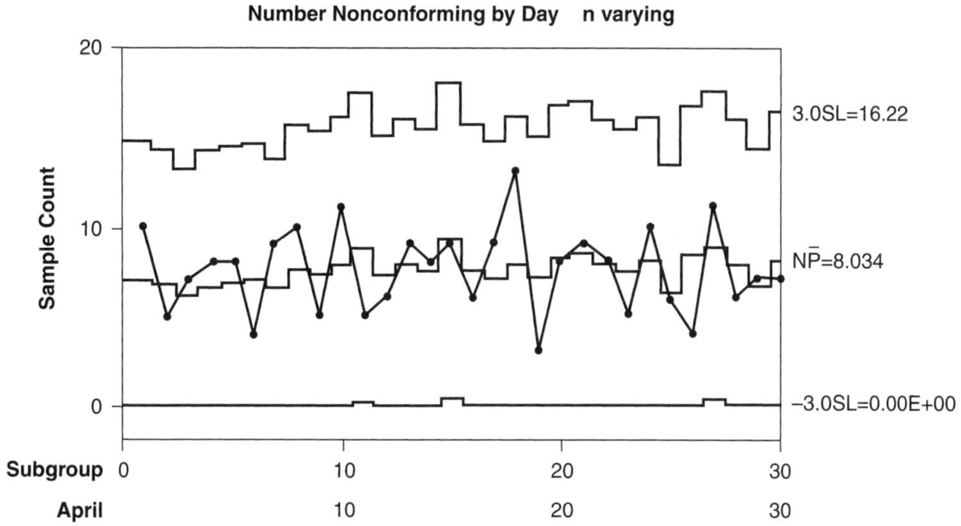

FIGURE 4.14 *np*-Chart with Varying *n* Values Can Be Difficult to Interpret.

value to be used for *n*. Next, click on the button **NP Chart Options. . .**, and under the parameters tab, enter the historical $\bar{p}$, 07533, in the window. If no value is entered here, $\bar{p}$ will be computed based on the data. The usual control chart features such as **Labels** are available as well. Finally click **OK** to obtain the *np*-chart shown in Figure 4.13.

It is possible to allow for different sample sizes and enter these in a second MINITAB column as was discussed is setting up *p*-charts. Figure 4.14 shows what an *np*-chart could look like with varying values of *n*. Because the center line value $n\bar{p}$ varies from sample to sample, the center *line* is really more of a center zigzag and the chart is more difficult to interpret visually, particularly when combinations of successive points are adopted as out-of-control signals—for example, 8 points in a row on the same side of the center line. If the sample size varies, use of a *p*-chart would avoid this problem.

Exact Probability Limits

Recall that control limits on *np*-charts are based on the Central Limit Theorem, which provides that certain variables should be approximately normally distributed when the sample size *n* is large enough. Having both $n\bar{p} \geq 5$ and $n(1 - \bar{p}) \geq 5$ was given as a rule of thumb for determining when *n* is large enough. Even when *n* is "large enough," the number of nonconforming items in a sample has only an approximate normal distribution. In fact, the number of nonconforming items in a sample, being a discrete variable, could at best be only approximately normal. This means the expectation that 3 points in 1,000 on an *np*-chart will not fall between $n\bar{p} \pm 3\sqrt{n\bar{p}(1 - \bar{p})}$ when the process is in control may be slightly off the mark. If n is large enough, and for a particular combination of n and p, the actual expectation were closer to, say, 5 points in 1,000 or 2 points in 1,000, it is easy to argue from a practical perspective that the 3-sigma limits $n\bar{p} \pm 3\sqrt{n\bar{p}(1 - \bar{p})}$ should be used anyway. This is because the formula is simple, and *3 points in 1,000* will be close to right. Besides, *3 points in 1,000* comes from the Empirical Rule's three standard deviations from the mean, which, while universally adopted in statistical circles, is arbitrary. Why not use, say, 2.326 standard deviations from the mean, so that 10 points in 1,000 would be expected to not fall between the control limits when the process is in control?

Nevertheless, in the spirit of completeness, we observe that some people prefer using control limits that are determined by a second paradigm. These so-called *exact limits* or *probability limits* give a better understanding of how often a point would be expected to fall beyond a control limit when the process is in control. While the choice is arbitrary, typically, this probability is required to be at most .005.

The probability that an individual point from an in-control process will fall beyond a given probability limit is as close to .005 as possible, but not more than .005.

We illustrate by finding probability control limits using the data of Table 4.7. Recall that the subgroup size was 100, that our estimate of p was $\bar{p} = .07533$ based on the data in the table, and that a *np*-chart for May would be maintained using .07533 as the historical $\bar{p}$. As before, the value for the center line is $n\bar{p} = 7.533$.

When a production process produces items in such a way that each item has probability p of being nonconforming, independent of whether other items are nonconforming, the actual probability distribution for the number of nonconforming items in a sample of n items is well known to statisticians. It is called a *binomial distribution*. One need know only the values of n and p in order to compute the probability that a given number of nonconforming items will show up in a sample. For $n = 1, 2, \ldots, 12$ or $n = 15$, 20 or 25 and selected values of p, cumulative probabilities for binomial distributions may be found in Table V in the Appendix. Statistics software packages typically provide more flexibility by allowing the user to input a value for n, a value for p (any value between 0 and 1), and then generate probabilities for that particular binomial distribution. We illustrate with a table generated by MINITAB.

Table 4.8 is a portion of the table of cumulative probabilities for the binomial distribution with $n = 100$ and $p = .07533$. In the table, k denotes

TABLE 4.8 Cumulative Binomial Probabilities for n = 100 and p = .07533

k	$P(X \leq k)$
0	0.000396
1	0.003630
2	0.016669
3	0.051369
4	0.119922
5	0.2227149
6	0.365461
7	0.516772
8	0.660071
9	0.779407
10	0.867877
11	0.926846
12	0.962476
13	0.982125
14	0.992072
15	0.996718
16	0.998729
17	0.999538
18	0.999842
19	0.999949
20	0.999985

the number of nonconforming items possible in a sample. In this example, k could equal any integer from 0 to 100. X denotes the number of nonconforming items there are in the next sample, a number that is unknown before the sample is obtained. Finally, $P(X \leq k)$ means, literally, *the probability that the number of nonconforming items in the next sample is less than or equal to k*. For example, the probability that the next sample of 100 items has 3 or fewer nonconforming items is .0514, whereas the probability of 12 or fewer nonconforming items is .9625. Notice that the probabilities increase with k (Why?). And since we expect, theoretically, 100(.07533) = 7.533 nonconforming items in a sample, it is comforting to see that the probability of having this many nonconforming items or fewer in 100 (as a practical matter, this means 7 or fewer such items in 100) is .5168, which is close to $\frac{1}{2}$. Which is what we should expect from a distribution that is approximately normal.

We first use Table 4.8 to find the lower probability limit, denoted as LPL. This means we want to choose a value for LPL so that $P(X \le \text{LPL})$, the probability of having LPL or fewer nonconforming items in 100, is as close as possible to .005 without exceeding .005. This suggests that we use LPL = 1. Since over time it is easy to forget whether *less than one* or *one or less* is the out-of-control signal, it is customary to adjust the integer up or down by .5, as needed, to obtain the official probability limit. In this case, LPL = 1.5 and there is no doubt about whether one nonconforming item in a subgroup constitutes an out-of-control signal.

As a comparison, the lower 3-sigma limit formula $n\bar{p} - 3\sqrt{n\bar{p}(1-\bar{p})}$ produces $-3.0SL = 100(.07533) - 3\sqrt{100(.07533)(1-.07533)} = -1.49$, which MINITAB routinely redefines to be 0 since the number of nonconforming items in a subgroup cannot be negative.

Finding the upper probability limit, UPL, is a bit trickier. Here we want to choose a value for UPL so that the probability of having UPL or more nonconforming items in 100 (i.e., $P(X \ge \text{UPL})$) is as close to .005 as possible without exceeding .005. This means that the probability of having *fewer than* UPL nonconforming items is as close to .995 as possible, but not less than .995. Or equivalently, the probability of having UPL −1 or fewer nonconforming items is as close to .995 as possible, but not less than .995. In Table 4.8, the cumulative probability closest to .995, but not less, is $.996718 = P(X \le 15)$. So UPL − 1 = 15 and UPL = 16. That is, 16 nonconforming items or more constitutes an out-of-control signal. Officially, we take UPL = 15.5 as the upper probability limit for the np-chart.

In this case, the exact upper limit agrees with the result obtained from the formula for the upper 3-sigma limit (see Figure 4.13):

$$3.0SL = n\bar{p} + 3\sqrt{n\bar{p}(1-\bar{p})} =$$

$$100(.07533) + 3\sqrt{100(.07533)(1-.07533)} = 15.45$$

Probability limits allow the user to know exactly what the probability is of having a point not fall between the control limits when the process is in control for given values on n and p. Note that this probability usually is not 2(.005) = .01. Actually, the procedure for determining exact limits guarantees that the actual probability will be no greater than .01. For a given n and p, one expects the actual probability to be less than .01. For our process with $n = 100$ and $p = .07533$, so long as the process does not change, the probability that the next subgroup of 100 items will produce an out-of-control signal is

$$
\begin{aligned}
P(X \le 1.5) + P(X \ge 15.5) &= P(X \le 1) + P(X \ge 16) \\
&= P(X \le 1) + [1 - P(X \le 15)] \\
&= .0036 + [1 - .9967] \\
&= .0069.
\end{aligned}
$$

Using MINITAB to Find Cumulative Probabilities for Binomial Distributions

For cumulative binomial distributions not found in Table V, MINITAB will provide the cumulative probabilities corresponding to the values of k that the user enters into a column. Since the number of defectives in a sample could be, at least theoretically, any integer from 0 to n, all of these integers could be entered into any column, say C1. There is a quick and simple procedure for entering such patterned data. We need not produce cumulative probabilities for all possible values of k, however, because we know that the exact probability limits will be in the vicinity of the 3-sigma limits. In our example, Figure 4.13 shows that -3.0SL $= 0$ and 3.0SL $= 15.45$. So, one could enter into C1 the integers from 0 to 20, as was done to produce Table 4.8.

From the menu, choose **Calc,** then **Make Patterned Data,** then **Simple Set of Numbers . . . .** A box will appear. In the window labeled **Store patterned data in:,** enter C1. In the window labeled **From first value:,** enter 0. In the window labeled **To last value:,** enter 20. The integers we are entering are to be consecutive, so the increment from one column entry to the next is 1, which should be entered in the window labeled **In steps of:.** Finally, the pattern we want for the numbers in C1 is that each value appears once in the list and is not repeated and the whole list appears once and is not repeated. There are two windows used to specify the number of times each value is listed and the number of times the whole list is to appear. Be sure a 1 is entered in each window. Then click **OK.** The integers from 0 to 20 will appear in C1.

To obtain the column of cumulative probabilities, go to the menu, choose **Calc,** then **Probability Distributions,** then **Binomial . . . .** A window will appear. Click the window labeled **Cumulative probability.** In the window labeled **Number of trials:,** enter the value of n. In our example, n is 100. In the window labeled **Probability of success:,** enter the assumed value of p. In our example, this is .07533. In the window labeled **Input column:,** enter C1. In the window labeled **Optional storage:,** enter C2, say, as the column into which the cumulative probabilities will be placed. Click **OK.**

TABLE 4.9 Data for Problem 2.

Orders	Orders in wrong boxes
400	23
398	17
407	25
421	15
372	17
386	20
434	19
414	24
418	20

PROBLEMS

1. A company makes, boxes, and ships a product to typically 400 or more customers each day. Since getting an order to the right customer is important, the company decides to monitor the daily proportion of orders put in boxes with the wrong shipping labels. After collecting data for 21 days, management found that 535 orders were put in boxes with the wrong labels out of a total of 8613 orders. A plot of the daily proportion of orders in boxes with the wrong labels indicates apparent control.
   a. Find a value for the center line for a p-chart.
   b. Find 3-sigma limits for the p-chart on a day when 411 orders are boxed.

2. In Problem 1, the company decides to implement a p-chart for the proportion of orders placed in boxes with the wrong labels, using .062 as the historical mean nonconforming rate. The company also has made changes in an effort to reduce p.
   a. Using the data for the next nine days given in Table 4.9, create a p-chart.
   b. Do any points fall outside the 3-sigma limits?
   c. Are there other indications that p has been reduced? (Hint: Don't be limited to a point outside the control limits as the only out-of-control signal.)

3. Upon receipt of an order, a company retrieves the ordered item from a warehouse. A record is kept of whether the forklift operator finds the

item in the location it is supposed to be. A p-chart is to be maintained on nonconforming retrievals. Data for the next five work-weeks are:

Orders sought:	45	36	49	58	46	52	46	37	41	45	46	28
Nonconforming:	3	2	4	5	5	5	2	1	4	3	5	3
56	47	45	45	40	39	51	46	33	47	51	40	46
5	3	4	5	3	5	4	3	3	1	6	4	1

a. Find a value for the center line for a p-chart.

b. Find 3-sigma limits for the p-chart on a day when 46 orders are sought (which occurred on the last day).

4. Create a p-chart for the data in Problem 3. Do not prescribe a historical $\bar{p}$, which should confirm your answers in Problem 3. Does the process appear to reflect control?

5. a. In Problem 1, are the guidelines $n\bar{p} \geq 5$ and $n(1 - \bar{p}) \geq 5$ met on a typical day?

   b. Are the guidelines met for the data in Problem 3? From a practical perspective, one can use the p-chart with the understanding that the 99.7% figure associated with the 3-sigma limits from the Empirical Rule will be somewhat off the mark. Another approach would be to plot each point based on data from two days rather than one day.

6. A company maintains a p-chart on customer claims based on the number of orders shipped each week. Suppose the historical $\bar{p}$ for customer claims is 4.8% and that for the next four weeks, the pertinent data are

Number of orders shipped	1,113	1,089	1,152	1,253
Number of customer claims	62	53	30	39

Do customer claims appear to be in a state of control? If not, when did any out-of-control signals occur?

7. Upon receipt of an order, a company retrieves the ordered item from a warehouse. Each day, 500 of the retrievals are checked to determine if the manufacturing history (shift and time of day produced) stamped on the box is consistent with the contents in the box. An np-chart has been maintained on nonconformities and the historical nonconformity rate has been 2.4%.

   a. Find the center line value and values for the 3-sigma control limits.

   b. If nonconformity counts for the next five days are 10, 11, 19, 20 and 24, does the process still appear to be in control? If not, when do any out-of-control signals occur?

8. In Problem 7, find probability control limits based on historical data and compare to the 3-sigma control limits.
9. At a company's maintenance division, a worker who checks out a tool is supposed to return it by the end of the worker's shift. Each week 60 tools from the workroom are selected at random by means of their bar codes and someone checks to determine how many of the 60 cannot be accounted for (e.g., the tool is not logged as having been checked out recently and is not at its position on the storage rack). It is decided that missing tools are an important enough issue to warrant establishing an *np*-chart. Twenty consecutive days of missing tool counts are as follows (by rows):

8	9	8	10	4	5	1	4	8	8
7	4	4	6	5	8	6	5	9	6

   a. Estimate p, the proportion of missing tools, under the assumption of control.
   b. Find values for the center line and control limits of an *np*-chart.
   c. Check your values by obtaining an *np*-chart for these data without prescribing a historical $\bar{p}$.
   d. Does the missing tool process appear to be in control?

10. Find probability control limits for an *np*-chart if (a) the sample size is 12 and the historical p is .5; (b) the sample size is 25 and the historical p is .21.

4.3 *C*-CHARTS

As with *np*-charts, a control chart for the number of items not conforming to specs is needed. Suppose, however, that each sample presents lots of opportunities to produce nonconformities, but the probability that any single opportunity produces a nonconformity is extremely small. This is the setting in which *c*-charts are used. All one needs to have in order to set up a *c*-chart is a good estimate of μ_C, the average number of nonconforming items per sample for the process.

Before examining the specifics of *c*-charts, we consider some *lots-of-opportunities/small-probability* settings in which it is natural to use *c*-charts. We also examine why some of them make it difficult to use an *np*-chart instead.

EXAMPLE 1

Suppose 1,100 spot welds are inspected daily. Recently the mean rate of nonconforming welds per day has been 3.9, which also serves as our estimate of μ_C for this process. The value of 1,100 for n is large. Also, our estimate of p, the proportion of nonconforming welds lately, is $\bar{p} = \frac{3.9}{1{,}100} = .0035$, which is small. Regardless of whether n

and p are large or small, we can say that the number of nonconforming welds in a sample has a binomial distribution, justifying the use of an np-chart. Note that $n\bar{p}$, the expected number of nonconforming welds per day based on historical data, is $1{,}100\left(\frac{3.9}{1{,}100}\right) = 3.9$ Since it is not the case that $n\bar{p} \geq 5$, which is part of our rule of thumb for being able to treat the distribution as approximately normal, the 3-sigma limit formula $n\bar{p} \pm 3\sqrt{n\bar{p}(1 - \bar{p})}$ should not be used. Probability limits should be used for the np-chart. Or a c-chart could be used with 3.9 as the estimated value of μ_C.

If an np-chart could always be used instead of a c-chart, there would be no need to learn about c-charts. The key here is that when n is extremely large and p extremely small, a c-chart is easy to apply. Also, there are occasions when an np-chart may not be easy to apply.

EXAMPLE 2

A sample consists of 5,000 meters of insulated copper wire. A sample is taken twice a day. The nonconformity of interest is insulation holes. The mean rate has been 1.6 such holes per 5,000 meters. That is, the estimate of μ_C for this process is 1.6. One could try to contrive a way to view the number of holes in 5,000 meters as having a binomial distribution. Suppose one views the 5,000 meters of wire as consisting of 5,000 one-meter segments. Each segment either is conforming (has no holes) or it isn't. Almost always, the total number of nonconforming segments should equal the total number of holes on the whole length of wire. But not always. Why? Suppose a discrepancy in the two counts is viewed as occurring so rarely as to be ignored. The number of holes in a sample could be taken to have a binomial distribution. The sample size would be $n = 5{,}000$. The estimate of p, the proportion of nonconforming one-meter segments for the process (i.e., the probability that any given segment is nonconforming), would be $\bar{p} = \frac{1.6}{5{,}000} = .00032$. Of course, the expected number of nonconforming segments per 5,000 meter sample would be $n\bar{p} = 5{,}000(.00032) = 1.6$.

Instead, one could view the 5,000 meters of wire as consisting of a sequence of 500,000 one-centimeter segments. Then $n = 500{,}000$ and our estimate of the proportion of nonconforming one-centimeter segments would be $\bar{p} = \frac{1.6}{500{,}000} = .0000032$. What remains the same is that $n\bar{p} = 1.6$, which is the estimate of μ_C, the average number of holes in one sample.

Unlike the faulty weld example where the choice of n was obvious, using an *np*-chart to track the number of holes in the insulation means deciding upon a sample size that is artificial, arbitrary, and a needless construct. It also would mean being willing to ignore the rare event that the number of insulation holes may be different than the number of nonconforming segments. The key observation in working with the number of holes in a long length of wire is that there are lots of *places* on the wire where a hole can occur and the probability that any single *place* has a hole is extremely small. One need not try to define how big a *place* is. But a *c*-chart can be used knowing only the fact that $\mu_C \approx 1.6$.

A *c*-chart also would be the natural control chart to use when at regular time intervals the number of nonconformities that occurred during the just completed time interval is recorded, provided nonconformities are rarely occurring. Tracking the number of murders that occur in Montana each year would be such a setting (see "How not to Think—An Example" in Section 1.1).

EXAMPLE 3

A *c*-chart is a natural tool for monitoring the number of accidents per month at an industrial plant because it involves counting the number of nonconformities that occur during each of a succession of equally long, more or less, time intervals. Suppose the plant has averaged .36 accidents/month over the course of 25 months and counts have been stable during that time. Consider one month as being partitioned into many time intervals of equal length. One minute, say. Then a month provides lots of opportunities for nonconformities—accidents—and in each short time interval there is a very small probability that an accident will occur. Therefore, a *c*-chart based on the estimate $\mu_C = .36$ is natural to use for tracking future monthly accident counts.

We now examine how *c*-charts are constructed. For a concrete frame of reference, assume that at regular time intervals, a record is made of the number of nonconformities that have occurred during the just-completed time interval. Also assume that nonconformities are rare, so that the *lots-of-opportunities/small-probability* model pertains.

Let C denote the number of nonconformities that occur during a time interval. Under some technical but very reasonable mathematical assumptions (that are discussed in Section 5.2), C will have a Poisson distribution. To compute Poisson probabilities one needs to know only μ_C, the average value of C for the process. While we will not use the formula for computing Poisson probabilities in this section, we give it for sake of completeness:

When C has a Poisson distribution and the expected number of nonconformities is μ_C, then the probability of exactly k nonconformities is $\dfrac{\mu^k{}_C e^{-\mu_C}}{k!}$ for $k = 0, 1, 2, \ldots$.

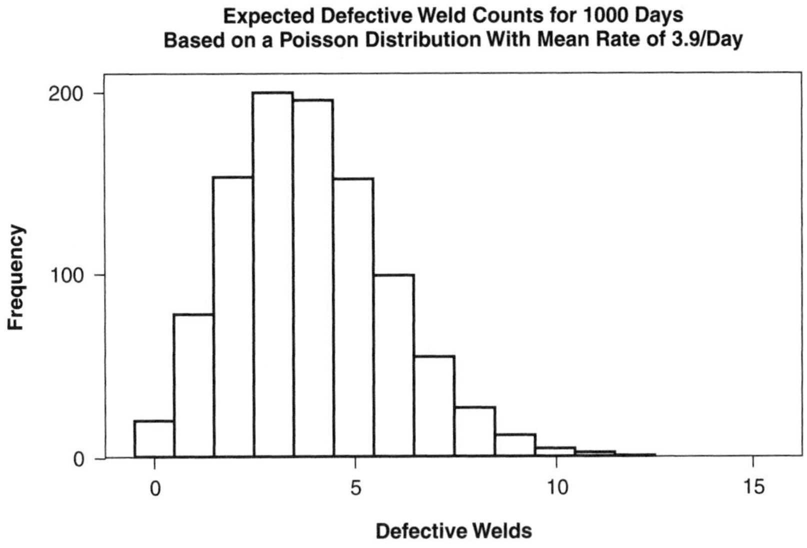

FIGURE 4.15 Poisson Distribution.

Of course, the mathematical assumptions behind the scenes present an idealized view of nature, and the Poisson distribution model is just that—a model. For example, even though 1,100 welds are inspected each day, the Poisson probability formula just shown accords a positive, but submicroscopic, probability of there being 1,235 defective welds in a day. Nevertheless, as a practical model, a Poisson distribution does a remarkably good job of describing the occurrence of nonconformities. As with binomial distributions, we will use either selected tables of cumulative Poisson probabilities (see Table VI in the Appendix) or tables generated by MINITAB.

Poisson distributions are not symmetric. They are always skewed to the right. Figure 4.15 illustrates this fact about Poisson distributions by showing how many days various defective weld counts would be expected to occur over the course of 1,000 days for a process that generates an average of 3.9 defective welds per day. The most common defective weld count, 3, should occur on approximately 20% of the days (from 200/1,000 or from using $k = 3$ in the just-cited formula). A close second is four defective welds per day. A defective weld count of 0 is less likely than a count of 8 but more likely than a count of 9.

Suppose when counting nonconformities that the number of items inspected, or the length of material inspected, or area of material inspected, or amount of material examined, or the amount of elapsed time since the last recording, or whatever, is the same for every sample. Also suppose that each sample provides lots of opportunities for nonconformities but that the probability of a nonconformity on a single opportunity is small. Let μ_C denote the average number of nonconformities per sample for the process.

μ_C may be estimated by

$$\overline{C} = \frac{\text{number of nonconformities in all samples combined}}{\text{number of samples}}$$

The three reference lines on a *c*-chart are:

center line: μ_C

3-sigma limits: $\mu_C \pm 3\sqrt{\mu_C}$

When μ_C is unknown, its estimate, $\overline{C}$, should be used in the reference line formulas.

For a process that is in control, the expected "three-samples-in-a-thousand outside the control limits" rule, tied to normal distributions, may be slightly off the mark, particularly for small values of μ_C. A table of cumulative Poisson probabilities may be used to determine whether exact probability limits are preferable for the situation.

EXAMPLE 4

Book availability was considered important at a college library, but no systematic availability measurement plan was in place. In an effort to develop a plan, a statistician in the mathematics department was asked to help. In a discussion of library processes, it was observed that every book had its own identifying bar code. Bar codes had a sequential numeric pattern that made a monthly sampling plan easy (see Norton, Seaman, and Sprankle, 1996). By means of a statistics software package, a random sample of 50 bar codes was obtained each month, and the corresponding books sought. A book was considered to be *accounted for* when it was on the shelf in the right spot, or recently checked out, or known to be in Rebinding, and so on. Books that could not be accounted for were identified as *missing.* Library management decided to monitor a *c*-chart for the number of missing books.

Some assignable causes of missing books revealed themselves, usually weeks after the affected book was declared missing. Some examples were interlibrary loan sending books to other libraries without recording the event, a shelver who developed casual work habits prior to a job change, and an insufficient number of temporary shelvers hired to cover peak activity months. Also, data revealed that an old belief was a misconception, namely that availability was at its worst at the end of the fall semester and again at the end of the spring semester. One perceived cause for this belief was that some students who put off library projects until the end of the semester could be re-

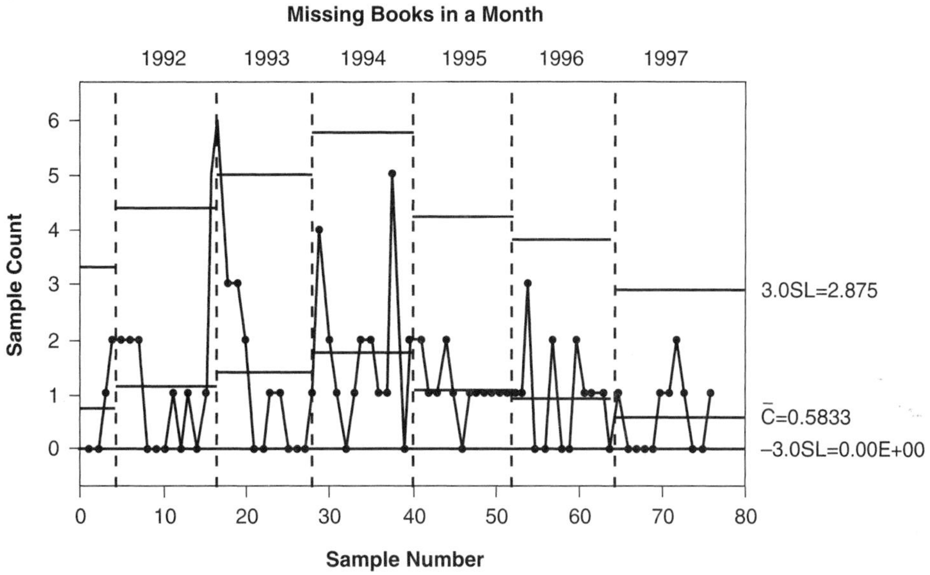

FIGURE 4.16 Missing Book c-Chart.

lied upon to hide needed books under wrong call numbers so as not to be inconvenienced by others who might need the books. However, data from the first year made it clear that high missing book counts occurred throughout the period from December to March, then returned to a calm when students went home at the end of April (see Figure 4.16, which compares *c*-charts by year). Samples from the next few successive years showed this peak to be a predictable annual event.

It took roughly three years to discover and fix some of the big-ticket assignable causes. One fix, for example, was that the amount of shelf reading was increased (walkthroughs to examine the stacks for books shelved in the wrong location). By the beginning of 1995, the annual peak had been virtually eliminated and the plot appeared relatively stable through the end of 1997. For the last 12 months, the missing book rate was down to less than .6 missing books out of 50.

A *c*-chart was maintained for monitoring purposes, and the library included an availability report every year in its annual report. Figure 4.16 presents the historical data in a way that allows comparisons across years.

Suppose at the end of 1997 someone had wanted to obtain an estimate of μ_C, the average number of missing books per sample for this process, and to use the estimate to set up a *c*-chart for the immediate future. Traditional guidelines are that a plot of recent data should reflect apparent stability for at least 25 to 30 samples before estimating a parameter like μ_C, setting up control limits, and so on. It would have been reasonable, then, to use data from, say, the last

TABLE 4.10 Cumulative Probabilities for a Poisson Distribution with Mean $\mu_C = .5833$.

0	0.55805
1	0.88357
2	0.97850
3	0.99696
4	0.99965
5	0.99997
6	1.00000
7	1.00000
8	1.00000
9	1.00000

25 months. It also would have been defensible to use the 36 samples from the years 1995 through 1997 to estimate μ_C. Including 1995 data would give a more conservative estimate of μ_C because of the rarity of 0-counts that year. It also would be defensible to use the data beginning with April, 1994 (thereby omitting any data connected to the first quarter peak of 1994) and excluding the 5-count from October, 1994. The 5-count had an assignable cause—the shelver who went through the motions prior to a job change. Sometimes in the workplace, judgments have to be made and reasonable people can make different decisions. The key in deciding how far back in time to go with data in order to estimate μ_C is that the plot of recent points appear stable—that is, points appear to fall randomly on both sides of a horizontal line.

Nevertheless, for reporting purposes, it is not uncommon to see graphics like the one in Figure 4.16 that depicts a succession of charts for which the control limits and center line are based on a small number of points. To illustrate the computation of the 3-sigma limits and center line for the 12 months of 1997, observe from Figure 4.16 that the monthly counts were 1, 0, 0, 0, 0, 1, 1, 2, 1, 0, 0, and 1. The value that locates the center line is, then,

$$\overline{C} = \frac{1+0+0+0+0+1+1+2+1+0+0+1}{12} = .5833$$

Also, $3.0SL = .5833 + 3\sqrt{.5833} = 2.875$. The negative result of the computation $-3.0SL = .5833 - 3\sqrt{.5833} = -1.708$ forces us to say either that a lower 3-sigma limit does not exist or that, officially, $-3.0SL = 0$ and that any smaller count, an impossibility, is an out-of-control signal.

Table 4.10, which shows cumulative probabilities for a Poisson distribution with mean $\mu_C = .5833$, can be used to determine exact probability limits. Since the smallest cumulative probability exceeds .005, there is no lower probability limit. The smallest cumulative probability greater than or equal to .995 is .99696, which is the probability of 3 or fewer missing books. Hence 4 or more missing books constitutes an out-of-control signal, and the upper probability limit is defined to be 3.5. Cumulative distributions for selected Poisson distributions appear in Table VI in the Appendix.

Note that a missing book count of 3 is interpreted differently depending on whether a 3-sigma limit or a probability limit is used. Table 4.10 shows that the probability of 3 or more missing books—that is 1 minus the probability of 2 or fewer missing books—is $1 - .9785 = .0215$, a value most would consider far too large for the probability of an out-of-control signal. Probability limits are preferable here. Note also that the library could have used an *np*-chart instead of a *c*-chart to monitor missing books.

Obtaining *c*-Charts in MINITAB

The number of nonconforming items in each sample should be entered in chronological order into a MINITAB column, say C2. For the missing book data, the entries beginning at the top of C2 would be 0, 0, 1, 2, 2, and so on. From the menu, choose **Stat,** then **Control Charts,** then **Attributes Charts,** then **C . . . .** A box will appear. In the window in the box labeled **Variables:,** enter C2 or double-click on C2 in the window that identifies columns containing data. If a value is known for μ_C, click on the window labeled **C Chart Options,** choose the **Parameters** tab, and enter the historical areas. If no value is entered here, μ_C will be computed based on the data. The usual control chart features such as **Labels . . .** are available in the box as well. Finally click **OK** to obtain the *c*-chart.

If there is a natural grouping variable, as *year* was in the library example, and one wants a succession of charts, one chart per group as in Figure 4.16, values of the group variable need to be recorded in a second MINITAB column (see Table 4.11). Under **C Chart Options. . .** click on the tab labeled **stages,** follow the instructions, and click **OK.**

Obtaining Cumulative Poisson Probabilities in MINITAB

If Table VI in the Appendix does not have the cumulative distribution for the desired value of μ_C, MINITAB will provide the cumulative probabilities that go with the values of k that the user enters into a column. Using a Poisson model, the number of defectives in a sample could be, at least theoretically, any nonnegative integer. However, we also know that the exact probability

TABLE 4.11 Monthly Missing Book Counts.

C2	C3
Missing	Year
0	1991
0	1991
1	1991
2	1991
2	1992
2	1992
2	1992
0	1992
0	1992
0	1992
1	1992
0	1992
1	1992
0	1992

limits will be in the vicinity of 3-sigma limits. In our missing book example, the 3-sigma limits for 1997 were $-3.0SL = 0$ and $3.0SL = 2.875$. So, one could enter into C1 the integers from 0 to 9, say, as was done to produce Table 4.10.

Either enter the integers from 0 to 9 into the first 10 cells of C1 or, from the menu, choose **Calc,** then **Make Patterned Data,** then **Simple Set of Numbers . . . .** A box will appear. In the window labeled **Store patterned data in:,** enter C1. In the window labeled **From first value:,** enter 0. In the window labeled **To last value:,** enter 9. The integers to be entered are consecutive, so the increment from one column entry to the next is 1, which should be entered in the window labeled **In steps of:.** Finally, the pattern we want for the numbers in C1 is that each value appears once in the list and is not repeated and the whole list appears once and is not repeated. There are two windows used to specify the number of times each value is listed and the number of times the whole list is to appear. Be sure a 1 is entered in each window. Then click **OK.** The integers from 0 to 9 will appear in C1.

To obtain the column of cumulative probabilities, go to the menu, choose **Calc,** then **Probability Distributions,** then **Poisson . . . .** A box will appear. Click the window labeled **Cumulative probability.** In the window labeled **Mean:,** enter the value of μ_C or its estimate $\overline{C}$. In our example, $\overline{C}$ is .5833. In the window labeled **Input column:,** enter C1. In the window labeled **Optional storage:,** enter C2, say, as the column into which the cumulative probabilities will be placed. Click **OK.**

PROBLEMS

1. a. Find values for the center line and 3-sigma limits for an *np*-chart based on the 1996 missing book data of Figure 4.16. Compare these to the center line and 3-sigma limits for a *c*-chart based on these data.
   b. Find probability limits for both the *np*-chart and *c*-chart. Compare these values.

2. A sample consists of 5,000 meters of insulated copper wire. The mean historical rate for insulation holes has been 1.6 per 5,000 meters.
   a. Find values for the center line and 3-sigma limits on a *c*-chart.
   b. If hole counts for the next few samples are 1, 0, 5, 1, 1, 2, 6, 5, 6, 4 and 5, does the process appear to be in control?
   c. Obtain the appropriate *c*-chart.

3. Suppose that the sample size is always the same and that the number of nonconformities per sample recently have been:

   3 6 7 5 8 4 4 3 7 8 7 10 11 7 4 6 4 8
   9 3 4 8 6 8 13 11 8 10 6 6 4 9 9 3 6 16
   8 4 7 10.

   Suppose the historical mean rate of nonconformities has been 7.0 per sample.
   a. Find control limits for this process based on the historical mean.
   b. Does the process appear to be in control?
   c. Suppose an assignable cause investigation reveals that for the last 5 samples, nonconformities were determined by a faulty viscometer. Estimate the current process mean.
   d. Should the historical mean be redefined?

4. In Problem 3, find exact limits (i.e., probability limits) for a *c*-chart based on a historical nonconformity rate of 7.0 per sample.

5. Recall that in the example of "how not to think" in Section 1.1, Montana experienced 23 murders in 1989 and 30 murders the following year. This 30% increase was cause for concern to those who didn't know better than to compare one sample value to the previous one. One way to explore whether a change of 7 is unusual is to consider whether a one-year count of 30 murders is exceptionally

large under the assumption that the historical mean murder rate is 23 per year.

a. Compare 30 to the 3-sigma limits for a *c*-chart.
b. A contrasting percentage increase was in Los Angeles, which experienced 877 and 983 in the same two years, a 12% increase. Compare 983 to the upper 3-sigma limit on a *c*-chart with 877 as the historical mean. Which of Montana or Los Angeles suggests a search for an assignable cause?

6. In Problem 5, use cumulative Poisson probabilities to find the probability that 30 or more murders will occur in a year, assuming a stable process with a historical mean rate of 23 per year.
7. Large sheets of plate glass are made by pouring a hot liquid onto a tray and letting the liquid cool. Once a day a completed sheet is taken from production and the number of surface flaws counted. Counts for the last 27 days (in chronological order by row) are:

6	3	4	3	4	5	5	2	3	9	7	2	4	7
4	4	3	3	4	7	2	7	3	3	4	5	3	

a. Make a time series plot of the data. Does the process appear to be in control?
b. Assuming that no assignable causes could be found to explain unusual points, estimate μ_C for this process.
c. Find the center line and 3-sigma limits for this process.
d. Find exact probability limits for this process.
e. Plot the data given on a *c*-chart.
f. Suppose the manufacturer contends that its historical mean number of surface flaws is 2.5 per sheet. Respond to the contention.

8. Suppose in Problem 7 that there are two kinds of flaws: imbedded particles and wave distortions (which do not involve particles). The manufacturer has been maintaining two control charts, one for imbedded particles and one for wave distortions. Through process improvements, the occurrence of each type of flaw has become so rare that management decides to maintain just one *c*-chart for the total (combined) number of flaws on a sheet. The historical mean rate for imbedded particles is .6 per sheet and for wave distortions is 1.3 per sheet. The number of imbedded particles and the number of wave distortions on a sheet are independent. What should be the historical mean for this new chart?
9. Suppose the number of imperfections in the paint job of a new car occur at a historical rate of 2.3/car. If a sample consists of 5 cars and a *c*-chart is maintained on the total number of paint job flaws in a sample, what value should be used for the historical mean on the *c*-chart?

10. Voters in Helena, Montana, a city of population 26,000, passed a broad indoor smoking ban in June, 2002 (see newspaper article: "Heart attacks in Montana town drop after indoor smoking ban"). The ban lasted for six months, ended by a legal challenge. Did cleaning the air in bars and restaurants reduce heart attacks? For four years prior to the ban, heart attacks had averaged about 7/month. During the six months the ban was in place, heart attacks averaged 3/month. Assume that there are no heavy or slow months for heart attacks, since the article says that these figures have been "adjusted" for seasonality. Even though actual monthly counts in the six-month span are not given, argue how a *c*-chart that is updated every six months could be used to help answer the question. Heart attacks returned to earlier levels when the ban ended, according to the article. (Information from the article is used with permission of the Associated Press.)

4.4 *u*-CHARTS

As with *c*-charts, suppose the focus is on the number of items not conforming to specs, that each sample presents lots of opportunities to produce nonconformities, and the probability that any single opportunity produces a nonconformity is extremely small. However, suppose additionally that the number of items inspected, or the length, area, or amount of material examined, and so on, is not the same for every sample. This could happen because daily production varies, work schedules for inspectors vary, or simply because life is not completely predictable. This is the setting in which *u*-charts are used. A *u*-chart is updated by plotting for the most recent sample the value of

$$U = \frac{\text{Number of nonconformities in the sample}}{\text{Number of units in the sample}}$$

EXAMPLE 1

A company manufacturing insulated copper wire examines a sample of each day's production, looking for holes in the insulation. Typically, anywhere from 3,000 to 8,000 meters will be examined on any given day. The company uses 5,000 meters as a convenient length "unit," and summarizes daily inspection results in terms of the average number of holes found per 5,000 meters. The data in Table 4.12 represents 30 days of inspection results. It is values from the last column that will be plotted on the *u*-chart.

TABLE 4.12 Inspection Results.

No. Meters	Ins. Holes	Unit Equiv.	Ins. Holes/5000 m
5695	1	1.1390	0.87796
4471	0	0.8942	0.00000
3990	2	0.7980	2.50627
6362	0	1.2724	0.00000
6147	0	1.2294	0.00000
4304	2	0.8608	2.32342
5224	2	1.0448	1.91424
4085	1	0.8170	1.22399
5615	0	1.1230	0.00000
4674	1	0.9348	1.06975
5589	4	1.1178	3.57846
6056	4	1.2112	3.30251
5106	1	1.0212	0.97924
4526	1	0.9052	1.10473
6049	0	1.2098	0.00000
4103	0	0.8206	0.00000
4392	1	0.8784	1.13843
3648	4	0.7296	5.48246
6902	3	1.3804	2.17328
3187	1	0.6374	1.56887
4957	2	0.9914	2.01735
3546	0	0.7092	0.00000
6501	0	1.3002	0.00000
3621	1	0.7242	1.38083
4977	2	0.9954	2.00924
6676	1	1.3352	0.74895
3996	0	0.7992	0.00000
7289	0	1.4578	0.00000
4790	1	0.9580	1.04384
4354	3	0.8708	3.44511
TOTAL	38	30.166	

All that is needed to set up a *u*-chart is a good estimate of μ_U, the average number of nonconforming items per unit for the process. Using recent data, the natural estimate for this parameter is

$$\overline{U} = \frac{\text{Total number of nonconformities}}{\text{Total number of units inspected}}$$

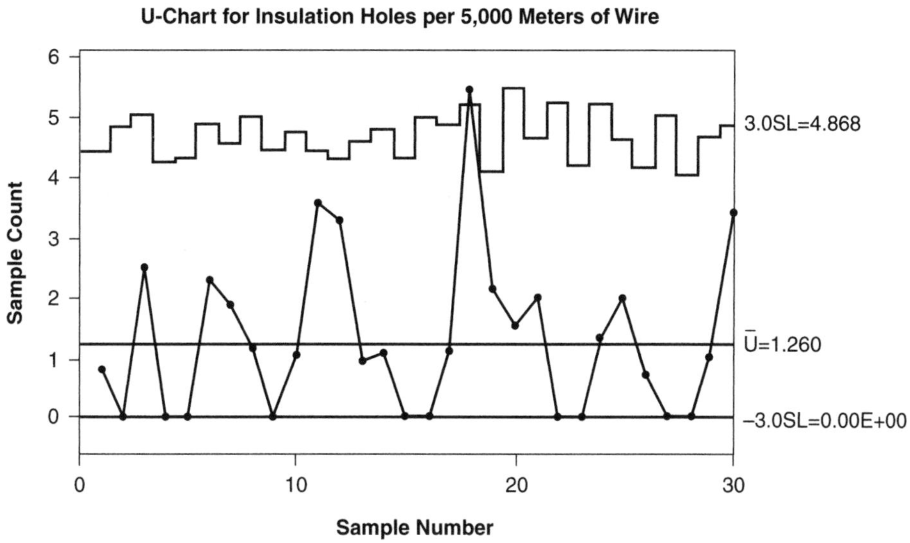

FIGURE 4.17 *U*-Chart for Insulation Holes per 5,000 Meters of Wire.

For the data of Table 4.12, we have $\overline{U} = \dfrac{38}{30.166} = 1.26.$

The three reference lines on a *u*-chart are

center line: μ_U

3-sigma limits: $\mu_U \pm 3\sqrt{\dfrac{\mu_U}{n}}$

where n is the number of units in the sample. If μ_U is unknown, it may approximated by $\overline{U}$. Note that if all the samples are the same size, then each sample constitutes $n = 1$ unit, and the formulas for the center line and 3-sigma limits are the same as for a *c*-chart. When a *u*-chart is used because n varies from sample to sample, the 3-sigma lines are really 3-sigma zigzags.

For the data of Table 4.12, the center line is located at 1.26 while the 3-sigma limits for the last sample are $1.26 \pm 3\sqrt{\dfrac{1.26}{.8708}} = 1.26 \pm 3.61.$ The upper 3-sigma limit for the last sample is 4.87. The lower 3-sigma limit is taken to either not exist or to be defined as 0.

A *u*-chart for the data in Table 4.12 appears in Figure 4.17. The process appears to be in control except for sample 18. The average number of holes per 5,000 meters for this sample was 5.48, which exceeds the upper 3-sigma limit. In order to use these historical data as a basis for estimating μ_U, management should check event logs to determine whether this sample should be excluded from the computation of $\overline{U}$ because an assignable cause explanation for the unusual value of U can be found.

Obtaining *u*-Charts in MINITAB

The number of nonconforming items in each sample should be entered in chronological order into a MINITAB column, say C2. For the data of Table 4.12, the entries beginning at the top of C2 would be 1, 0, 2, 0, 0, 2, 2, and so on. The corresponding number of units in each sample need to be entered in a second column, say C3. The entries beginning at the top of C3 would be 1.1390, .8942, .7980, and so on.

From the menu, choose **Stat,** then **Control Charts,** then **Attributes Charts,** then **U . . . .** A box will appear. In the window in the box labeled **Variables:,** enter C2 or double-click on C2 in the window that identifies columns containing data. In the window labeled **Subgroup sizes:,** enter C3 or double-click on C3 in the window that identifies columns containing data. If a value is known for μ_U, click on μ **Chart Options. . .,** and under the **Parameters** tab, enter the mean. If no value is entered here, μ_U will be computed based on the data. The usual control chart features such as **Labels . . .** and **Estimate** are available as well. Finally, click **OK** to obtain the *u*-chart.

PROBLEMS

1. A restaurant monitors the number of meals served per day as well as the number of complaints about the quality of the meal or the service. The quality indicator of interest is the number of complaints per 100 customers served. During the 40-day period summarized in Table 4.13, there were a total of 225 complaints and 20,444 customers served. A *u*-chart is examined with tentative reference lines based on these data in order to determine if the restaurant process exhibits control.
   a. What value locates the center line of the *u*-chart?
   b. What are the 3-sigma limits on a day when 649 customers are served (as occurred on the last day)?
   c. What should be the value of the first plotted point on the *u*-chart?
   d. Obtain the tentative *u*-chart for these data. Does the process exhibit control? Confirm your computations in parts a. and b. by comparing to the values shown on the *u*-chart.

TABLE 4.13 Customer Complaint Data.

Date	Customers	Complaints	Unit
9/11	428	3	4.28
9/12	443	1	4.43
9/13	492	9	4.92
9/14	472	8	4.72
9/15	600	14	6.00
9/16	691	5	6.91
9/17	409	1	4.09
9/18	466	3	4.66
9/19	508	11	5.08
9/20	459	7	4.59
9/21	470	5	4.70
9/22	647	4	6.47
9/23	715	9	7.15
9/24	392	4	3.92
9/25	480	6	4.80
9/26	485	8	4.85
9/27	462	5	4.62
9/28	487	7	4.87
9/29	677	10	6.77
9/30	670	7	6.70
10/1	431	5	4.31
10/2	446	3	4.46
10/3	459	6	4.59
10/4	512	2	5.12
10/5	419	4	4.19
10/6	673	6	6.73
10/7	632	8	6.32
10/8	502	4	5.02
10/9	423	5	4.23
10/10	452	8	4.52
10/11	456	7	4.56
10/12	411	2	4.11
10/13	637	7	6.37
10/14	656	9	6.56
10/15	462	3	4.62
10/16	438	1	4.38
10/17	437	4	4.37
10/18	444	2	4.44
10/19	451	6	4.51
10/20	649	6	6.49

TABLE 4.14 Holes in Varying Lengths of Insulated Wire.

Holes	13	16	20	19	17	14	15	18	14	14
Meters (10,000)	1.48	1.11	1.55	1.71	1.06	1.32	1.09	1.35	0.93	1.28
Holes	15	24	34	21	18	15	16	12	12	11
Meters (10,000)	1.51	1.57	1.43	1.62	1.46	1.06	1.07	1.10	1.10	1.04
Holes	11	16	17	16	16	14	13	24	20	13
Meters (10,000)	1.34	1.18	1.12	1.10	1.60	1.16	1.28	1.55	1.24	1.47

2. A sample of insulated copper wire from each machine is taken twice a day. The quality indicator of interest is the number of holes per 10,000 meters. For one machine during the 15-day period summarized in Table 4.14, there were altogether 498 holes out of 38.88 units produced (388,800 meters). A *u*-chart is examined to determine if the machine exhibits control.
   a. What value locates the center line of the *u*-chart?
   b. Obtain a *u*-chart with tentative reference lines based on the data in Table 4.14. Does the process appear to exhibit control?
   c. The event log indicates reveals an assignable cause responsible for the high hole count on the sample 13. Recompute an appropriate center line for a second *u*-chart.
   d. Obtain the second *u*-chart with tentative reference lines based on these data with sample 13 deleted. Does the process appear to exhibit control?
   e. What are the 3-sigma limits on a day when 1.47 units are produced (as occurred on the last sample)?
   f. What should be the value of the first plotted point on the *u*-chart? Confirm your computations in parts c. and e. by comparing to the values shown on the second *u*-chart.
3. In Section 4.3, it was observed that a *c*-chart is a natural tool for monitoring the number of accidents per month at an industrial plant because it involves counting the number of nonconformities that occur during each of a succession of equally long, more or less, time intervals. Of course, months are not equally long and a *u*-chart takes this fact into account. Suppose that, historically, accidents have averaged about 3.2 per month. Suppose a 30-day month is taken as the basic unit and 3.2 accidents/month as the historical mean accident rate.
   a. Determine a value for the center line and 3-sigma limits for each of 28-, 30-, and 31-day months.

TABLE 4.15 Accident Count Data.

September	October	November	December	January	February	March	April
4	7	2	2	3	3	4	4

b. Obtain a *u*-chart for the next 8 months if the accident counts are as given in Table 4.15.

c. Does the chart reflect control?

4. Fabric samples are taken from production each day and the number of imperfections of a particular kind counted. The number of square meters of fabric varies from day to day. Suppose 100 square meters is the convenient *unit* and that for a particular kind of fabric produced, the historical mean rate of imperfections has been 12.9/unit. On which of these four days is the number of imperfections most unusual?

Day	Number of Square Meters Inspected	Number of Imperfections in Sample
1	283	54
2	158	32
3	235	48
4	117	26

4.5 A CASE STUDY ON DESIGNING A CONTROL CHART FOR CORRELATED SAMPLES

For some processes it is obvious to all familiar with the process that samples are not independent. This section is a case study about a chemical batch process with positively correlated consecutive samples and how an out-of-control signal was designed for it. It is based on the article by Norton (2001), (portions of which are reprinted by courtesy of Marcel Dekker, Inc.)

The company manufactures a polyurethane liquid. One variable of interest is the concentration of a key chemical in each completed batch of a component used to create the liquid. From a confidentiality perspective, the key chemical is referred to as AB. Also from a confidentiality perspective, the data have been rescaled. That said, having the concentration of AB as close as possible to 5.65 helps ensure the ultimate quality of the product.

The chemicals used in making a batch are pumped through pipes from large storage tanks and mixed according to a chemical recipe. After the resulting chemical reaction, a sample of the completed batch is taken to a laboratory where an operator adds some facilitating chemicals to the sample and measures AB concentration.

A number of factors contribute to batch-to-batch variation in AB concentration. To some extent, each raw chemical used in making a batch affects AB, and the amount of each raw chemical mixed into the batch is never perfectly equal to the amount called for by the recipe. Also, the purity of each chemical affects AB concentration. So do the consistency of the individual doing the lab analysis, variation between the people who do the analyses, and between-lot variation in the chemicals added to a sample that enable AB to be measured. If these variable inputs were the only sources combining to make AB vary from batch to batch, the distribution for AB concentration would have a nice mound shape and could be monitored by an individuals chart.

However, making repeated batches from chemicals stored in large tanks creates an added twist that is common at chemical plants. When the contents of a storage tank are low, an order for the raw chemical is placed. When a tanker truck of the chemical arrives, a sample of the contents is tested. If specifications for the chemical are met, the contents of the tanker truck are transferred into the storage tank. For any given raw chemical, even though specs are met, there is shipment-to-shipment variation. This means that a minor location shift in the AB distribution is common after a chemical shipment arrives. Also, barring assignable cause problems, the center of the AB distribution will not shift again until a new shipment of one of the chemicals is put to use. Such shifts are viewed as inherent in the process—that is, they contribute to common cause variation.

Generally, AB distribution shifts are minor. Also, it is not unusual to have a small number of batches (e.g., three or four) made between the arrival of different chemical shipments, not enough batches by themselves to obtain a reliable estimate of the current process mean. However, one unusual shift, or a cumulative sequence of smaller shifts in the same direction, could warrant adjusting the chemical recipe in order to recenter the process closer to 5.65.

AB concentration in future batches can be raised (lowered) intentionally at any time be reducing (increasing) the amount of the raw chemical CD in the recipe for one batch. System limitations require that changes in the batch amount of CD must always be in integral multiples of 1 kg. Therefore, the current process mean could be slightly off-target, and this would be acceptable because a 1-kg CD adjustment would overcorrect.

Some major equipment/process changes were made. With the changes, it was anticipated that the process standard deviation of .0633 might change, as could the impact on AB of a 1-kg change in CD. We now examine how a zone chart for individuals was implemented to control the new process, why the standard out-of-control signal for zone charts needed to be modified, and what modification was done.

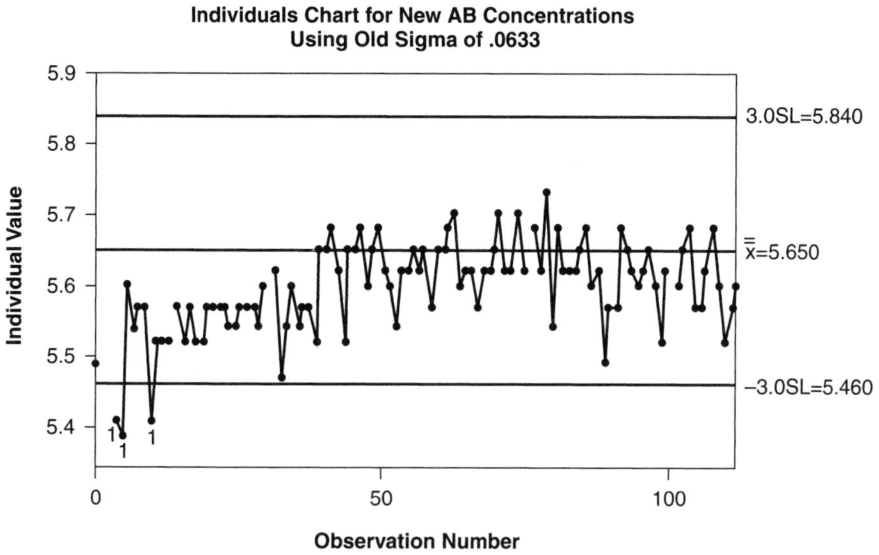

FIGURE 4.18 Individuals Chart for New AB Concentrations Using Old Sigma of .0633.

TABLE 4.16 AB Concentrations by Recipe CD Level.

CD = 998 kg	*	*	*	*								
CD = 995 kg	5.49	*	*	5.41	5.39	5.60	5.54	5.57	5.57	5.41	5.52	5.52
	5.52	*	5.57	5.52	5.57	5.52	5.52	5.57	5.57	5.57	5.57	5.54
	5.54	5.57	5.57	5.57	5.54	5.60	*					
CD = 993 kg	5.62	5.47	5.54	5.60	5.54	5.57	5.57					
CD = 991 kg	5.52	5.65	5.65	5.68	5.62	5.52	5.65	5.65	5.68	5.60	5.65	5.68
	5.62	5.60	5.54	5.62	5.62	5.65	5.62	5.65	5.57	5.65	5.65	5.68
	5.70	5.60	5.62	5.62	5.57	5.62	5.62	5.65	5.70	5.62	5.62	5.70
	5.62	*	5.68	5.62	5.73	5.54	5.68	5.62	5.62	5.62	5.65	5.68
	5.60	5.62	5.49	5.57	5.57	5.68	5.65	5.62	5.60	5.62	5.65	5.60
	5.52	5.62	*	5.60	5.65	5.68	5.57	5.57	5.62	5.68	5.60	5.52
	5.57	5.60										

Over the course of 116 batches made after the process changes were completed, the amount of CD in the recipe had been lowered three times in an effort to raise AB concentrations toward the target of 5.65. See Figure 4.18 and Table 4.16. Because of problems with the start-up, which produced some unusual AB concentrations, 8 of the first 35 batches are omitted from the figure and table, including all 4 batches made using 998 kg of CD, indicated

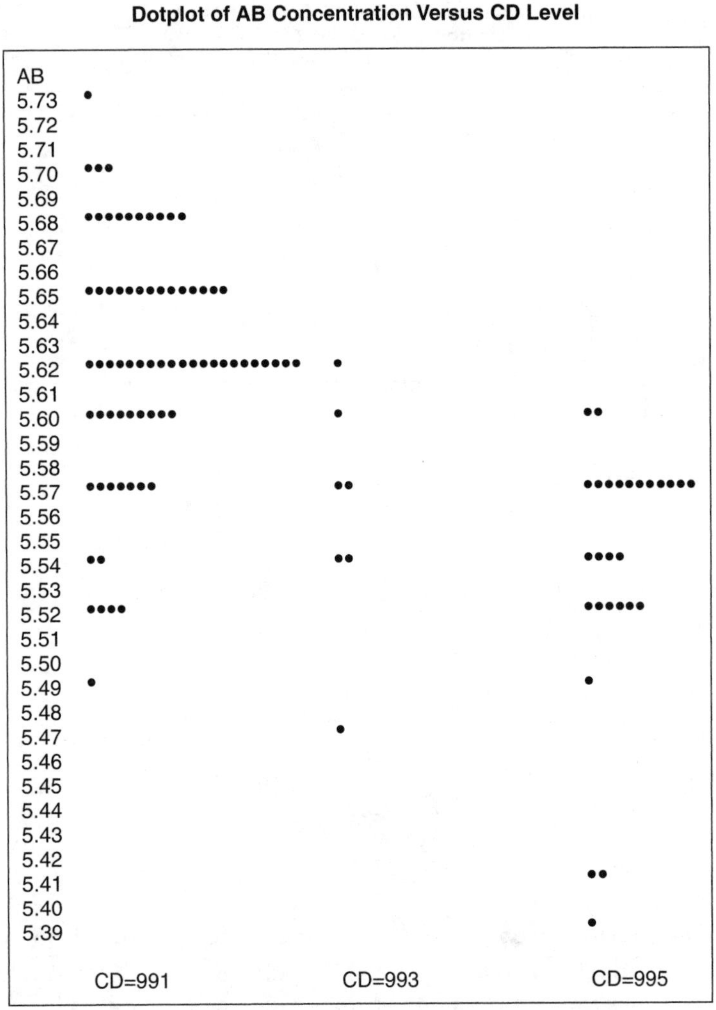

FIGURE 4.19 Dotplot of AB Concentration Versus CD Level.

by asterisks (*). Although the data are no longer available, it can be said that those data exhibited wild swings from the AB target and were clearly associated with getting the kinks out of a new manufacturing process and with people learning to operate the process. Also, two other batches made using 991 kg of CD are omitted because of assignable cause problems. The two AB measurements were not in line with those of the other batches at that time due to operator error in doing the laboratory analyses.

Figure 4.19 is a dotplot of AB concentration against CD level. There are gaps in the range of AB concentrations (e.g., 5.65 and 5.68 are possible, but 5.66 and 5.67 are not). These gaps are due to the rescaling that was done on the original data at the company's request. In disguising the original data

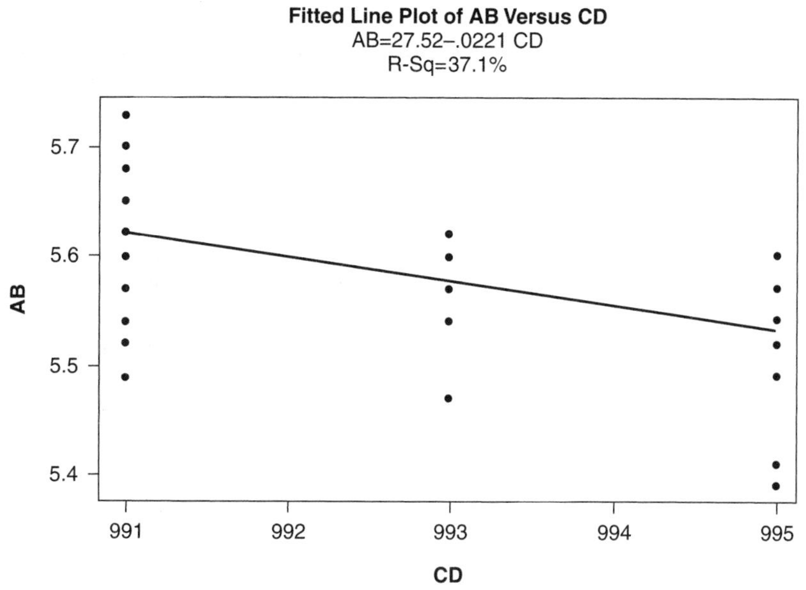

FIGURE 4.20 Fitted Line Plot of AB Versus CD.

value X by the reporting the value $Y = aX + b$, it is the factor a that creates the gaps. The original data were known to two decimal places and there were no gaps.

Figure 4.19 shows not only that AB concentration decreases as CD increases, but also is suggestive that there is an approximate linear relationship between AB and CD. The linear equation that best describes the relationship can be obtained from MINITAB. The graph of this equation is called the *least squares line*, which can be obtained in a graphic that also shows the equation (see Figure 4.20).

It is beyond the purpose of this book to explore the details or theory behind least squares lines. However, we do show how to obtain Figure 4.20, which also gives the equation of the least squares line. In either Figure 4.19 or 4.20, AB concentration is the dependent (i.e., *response*) variable and CD level is the independent (i.e., *predictor*) variable. Suppose the CD values are listed in C1 and the corresponding AB concentrations are listed in C2. From the menu, choose **Stat,** then **Regression,** then **Fitted Line Plot. . . .** A box will appear. In the window labeled **Response (Y):,** enter C2. In the window labeled **Predictor (X):,** enter C1. In the window labeled **Type of Regression Model,** click on **Linear.** If a title is desired for the plot, the user should click on the **Options. . .** button. Finally, click **OK.**

The fitted line plot shows that the equation of the least squares line is

$$AB = 27.52 - .0221\ CD$$

The most important observation here is that the slope of the line is −.0221. So while the mound shapes of Figure 4.19 show that AB concentration cannot be predicted exactly from CD, one *can* expect AB to drop by approximately .0221, on average, for each kg of CD added to the recipe.

The usual out-of-control signal on a zone chart occurs when a succession of points on the same side of the target value has a cumulative total that reaches eight or more. This is an appropriate signal when samples are independent. Here, however, small shifts in AB concentration that result from chemical shipment arrivals are viewed as a source of common-cause variation. It follows that a batch with a high (low) AB concentration will tend to be followed by a batch with another high (low) AB concentration when the process is in control. That is, AB values for consecutive batches are positively correlated, not independent, when this process is in control. Therefore, one expects runs on one side of the target to last longer when the process is in control than they would for a *textbook* process with independent batches. Consequently, running totals reach eight more frequently when this process is in control, and thus having the cumulative total reach eight may not mean much.

To show how an appropriate out-of-control signal was designed, we focus on the 72 batches made when the amount of CD called for by the batch recipe was 991 kg. During the period in which these batches were made, AB concentrations, though low, were stable in terms of both location and variation, and no CD adjustments were made. The reason no CD adjustments were made, despite the fact—obvious from Figure 4.18—that AB concentrations were low, was that AB concentrations were meeting specs, and it was considered important to get the new process established at making acceptable product. This reinforces the concept that there is a difference between meeting specs and having a process that is on target and in control.

Since the sample size is 1, the type of control chart to consider using is either a zone chart with $n = 1$ or an individuals chart. When $n = 1$, the usual tool for estimating the process standard deviation is moving ranges. To see why σ should not be estimated by using moving ranges for this process, consider Figure 4.21, a zone chart for which the historical mean and sigma are left unspecified. The length used for the moving ranges is 2. The figure shows that the average AB concentration of the 72 batches was 5.622 and that the cloud of points is aligned horizontally—that is, is stable. However, what is wrong with the picture is that the 68%/95%/99.7% approximations from the Empirical Rule are too far off the mark. Approximately 32% of the points should fall beyond one standard deviation from the mean, whereas 28 (or 39%) of the 72 points do so. Approximately 5% of the points should fall beyond two standard deviations from the mean, whereas 6 (or 8.3%) of the 72 points do so.

The reason the 3-sigma limits are too tight is that moving ranges underestimate a proper σ for the process. Moving ranges work when samples

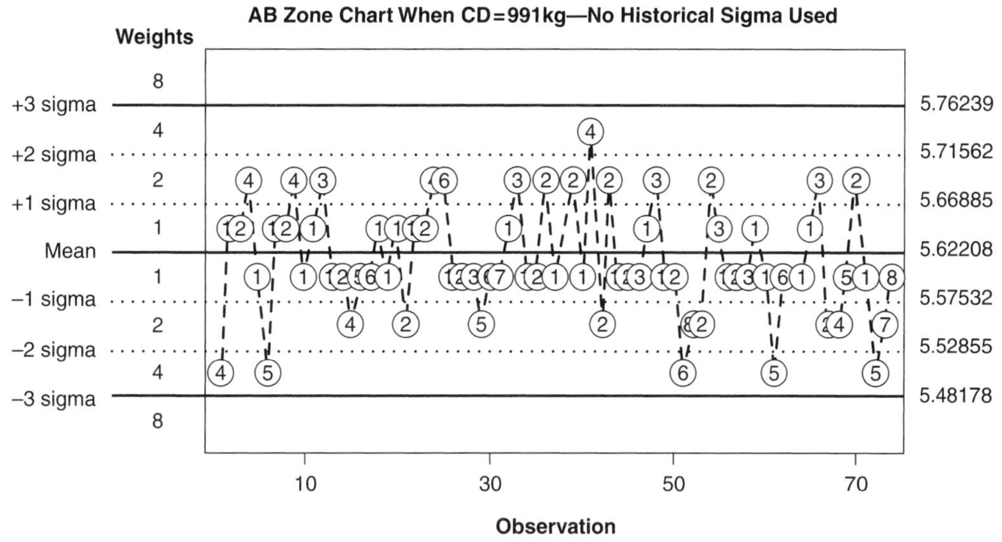

FIGURE 4.21 AB Zone Chart in Which CD = 991 kg—No Historical Sigma Used.

TABLE 4.17 Standard Deviation of AB Concentrations.

CD called for (kg)	995	993	991
Standard deviation	.0547	.0488	.0490
n	27	7	72

are independent. Consider a shipment that produces a large or moderate shift in the AB distribution. Only the one range that involves the last AB concentration before the shift and the first AB concentration after the shift will tend to be large. The ranges immediately preceding the large range and the ranges that immediately follow it should be smaller comparatively. In computing $\overline{R}$, the many small ranges will swamp the few large ranges, resulting in an underestimate of σ. With the larger percentage of points outside the 1-zones, it is easier for cumulative weight totals to reach 8. It already was forseen that a cumulative total of 8 might not mean much, and estimating σ in this way would only make matters worse.

Management avoided the problem by adopting the perspective that since the fallout of points in Figure 4.21 reflected control, albeit off-center, σ should be estimated by using the sample standard deviation of the 72 values.

Further, while 72 values is usually more than enough for estimating σ, it is possible to use all the data from Table 4.16 by pooling the sample standard deviations from each of the three CD recipe amounts. Table 4.17 shows the needed information.

When the sample sizes are the same for all samples, the pooled standard deviation, S_p, is the square root of the mean of the sample variances (see Section 2.2). When the sample sizes are not all the same, S_p is the square root of a weighted mean of the sample variances. The weight accorded each variance is one less than its sample size. This gives

$$\sigma \approx S_p = \sqrt{\frac{26(.0547)^2 + 6(.0488)^2 + 71(.0490)^2}{26 + 6 + 71}} = .0505$$

It was decided to use this value as the standard deviation of the new process on a trial basis.

To arrive at a picture of what a zone chart for this process might look like when centered perfectly on target, Figure 4.22 plots the 72 "modified AB concentrations." The modified values were obtained by adding 5.65 − 5.622 = .028 to each actual AB concentration. In this way, the modified AB values would be centered on target. The value .0505 was used for the historical sigma. In the first 36 batches, one can almost *see* some of the larger shifts that occurred when new chemical shipments were put to use. Examples are the downward shift that occurred at the batch 13, the upward shift that occurred at batch 22 or 23, the downward shift at batch 26.

Figure 4.22 also makes it possible to estimate how often an out-of-control signal would be expected to occur when the process is on target and in control. The cumulative count reached 8 or more twice, the second time being on the batch 72. Therefore, the estimated number of samples per out-of-control signal is 36, which introduces a concept not yet discussed.

The expected number of samples needed until an out-of-control signal occurs is called the *average run length* (ARL).

Consider tossing a fair die repeatedly. The die is not programmed to cycle through its six faces in six tosses. So it does not always take precisely six tosses to get the next 5. However, it can be shown that six-tosses-to-get-a-5 is the average behavior of the die. More generally it can be shown that if the probability of getting a 5 is p, then the average number of tosses needed to obtain a 5 is $1/p$.

Consider a process with independent samples that is being tracked on an $\overline{X}$-chart and suppose the process is in control. Also suppose that a point outside the 3-sigma limits is the only kind of out-of-control signal management uses. Recall that the probability a sample produces this out-of-control signal is .0027. Then just as six-tosses-to-get-a-5 is the average behavior of the die, the ARL for the $\overline{X}$-chart is $\dfrac{1}{.0027} = 370.37$. When a process is in control, the ARL should be large. Upwards from several hundred is common.

An estimated ARL of 36 is far too small for a process that is in control. False signals would lead to too much time being spent seeking assignable

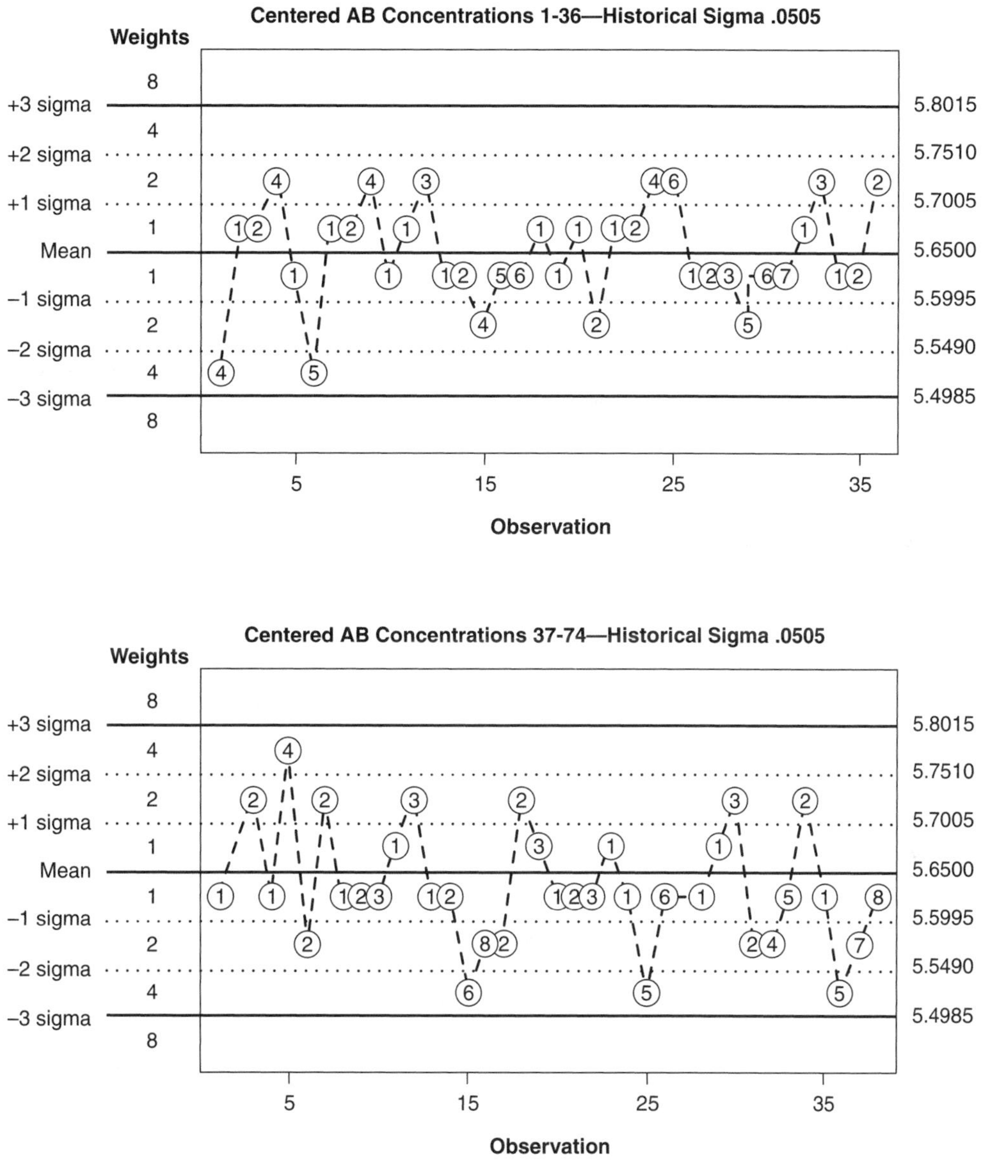

FIGURE 4.22 Zone Chart for AB Concentrations, Values Have Been Adjusted to Illustrate a Process that is On Target.

causes. Management considered using two back-to-back cumulative totals of 8 or more on the same side of the target as the out-of-control signal for the zone chart. This would incorporate a standard out-of-control signal (cumulative total of 8 or more) for a process with independent samples into a practical out-of-control signal for this particular process. There were no two-back-to-back signals during the period when the recipe called for 991 kg of CD.

The next issue in deciding whether a two-back-to-back rule would make a satisfactory out-of-control signal was to obtain an estimate of the ARL when management would want to get a quick signal. Recall that when the recipe called for 991 kg of CD, there was no urgent concern, even though the AB distribution was centered low, the 72 batches having an average AB concentration of 5.622. However, management had seen enough to warrant making a 2-kg CD change after just seven batches had been made using 993 kg of CD. Because each additional kilogram of CD may be expected to reduce AB concentration by .0221, one can estimate that the AB distribution would have been centered at approximately $5.622 - 2(.0221) \approx 5.578$ when 993 kg of CD was used in the recipe.

For illustrative purposes, we take 5.578 as a boundary that separates process means for which management wants a very small ARL from process means that don't need such small ARLs. We now estimate the ARL when the process mean is 5.578.

Again, we adjust the 72 AB concentrations for the batches made with 991 kg of CD, this time by subtracting $2(.0221) \approx .044$. The zone chart with these adjusted concentrations is shown in Figure 4.23, which depicts what a zone chart could look like when the process is centered at 5.578. There are 6 two-back-to-back signals. It took six batches to produce the first such signal, nine batches after that to produce the second signal, and so on.

From these values, the estimate of the ARL when the process mean is 5.578 is

$$\frac{6 + 9 + 31 + 6 + 8 + 9}{6} = 11.5$$

When management can identify a value μ_r that separates those process means that call for very small ARLs from those process means that don't, it is common to use an out-of-control signal for which the ARL is 10 or smaller when the process mean is μ_r. In this example, having an ARL near 11.5 when $\mu = 5.578$ is deemed to be satisfactory.

Almost always after this process has an out-of-control signal, the amount of CD in batches is adjusted. An example of an occasion when an out-of-control signal might not prompt a CD adjustment is when AB measurements seem to have shifted at the same time that the analysis lab began using a new shipment of a chemical it mixes with the batch sample in order to measure AB concentration. Managers and operators are reluctant to make prompt process changes when no new chemical shipment or other event that would shift actual AB concentrations has occurred and an unreliable AB lab measurement is the prime suspect. Data integrity and logged-in process changes and events are considered in making a judgment as to whether an out-of-control signal should be ignored or delayed. Such occasions are rare.

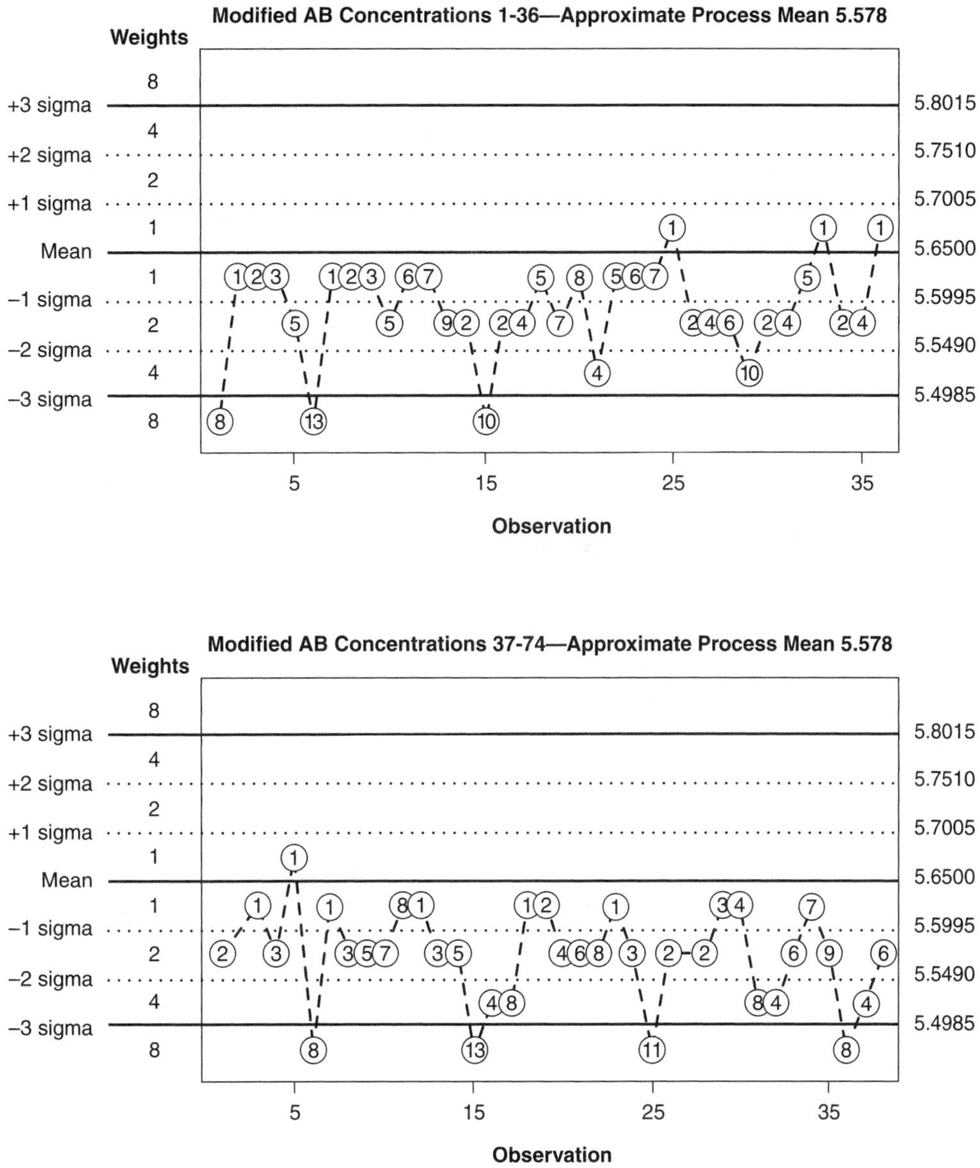

FIGURE 4.23 Zone Chart for AB Concentrations, Values Have Been Adjusted to Illustrate a Process in Need of Immediate Attention.

The rule adopted for how much of a CD adjustment to make following an out-of-control signal made use of the fact that a 1-kg CD adjustment up or down would translate into an approximate .0221 down or up change in AB concentration. The process operator on duty would average those AB

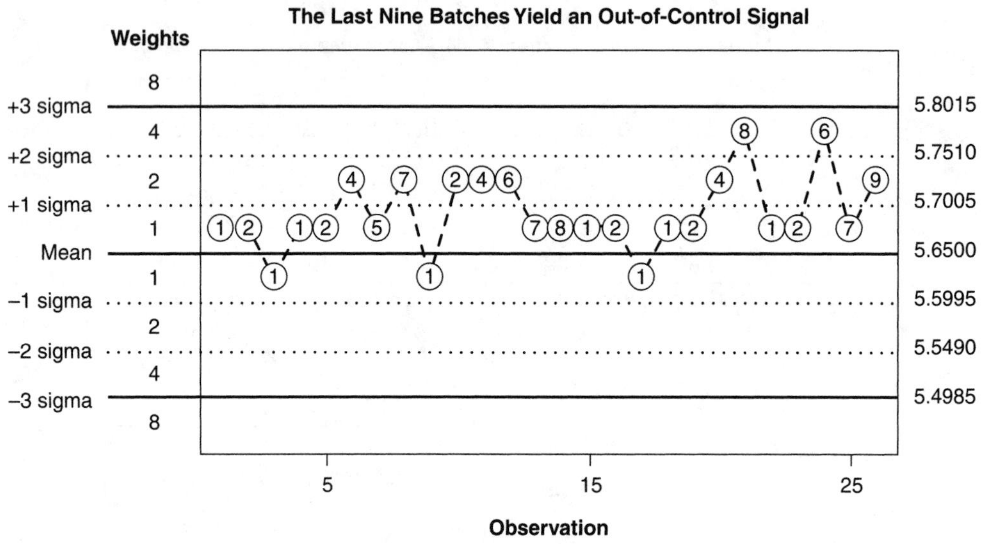

FIGURE 4.24 Zone Chart in Which the Last Nine Batches Yield an Out-of-Control Signal.

concentrations which generated the two back-to-back cumulative totals of 8 or more. The amount of CD adjustment (in kg) would be made by evaluating

$$\text{CD adjustment} = \frac{\text{AB average} - 5.65}{.0221}$$

and rounding to the nearest integer.

To illustrate how the adjustment works, suppose that 992 kg of CD has been called for since the last change in the recipe, and that the most recent 26 AB concentrations have been

> 5.68, 5.66, 5.60, 5.68, 5.68, 5.71, 5.68, 5.71, 5.63, 5.71, 5.71, 5.74, 5.66, 5.66, 5.68, 5.68, 5.63, 5.68, 5.68, 5.71, 5.76, 5.68, 5.68, 5.76, 5.68, and 5.74.

Suppose an operator has just plotted the last point on the zone chart in Figure 4.24. The last nine points provide a two-back-to-back signal. The average AB concentration for these nine batches is

$$\text{AB average} = \frac{5.68 + 5.68 + 5.71 + 5.76 + 5.68 + 5.68 + 5.76 + 5.68 + 5.74}{9} = 5.708$$

Then CD adjustment $= \dfrac{5.708 - 5.65}{.0221} = 2.62$ kg. In order to lower the AB concentration in future batches, the operator should change the recipe so that it calls for 995 kg of CD.

PROBLEMS

1. Using the *fitted line plot* feature of MINITAB, find the equation of the line that best fits the three points (3,5), (7,6), (15,8). For each point, substitute the first coordinate for X in the equation, compute the value of Y, and compare to the second coordinate of the point.
2. In the manufacture of a certain glass insulator, the air temperature at which the insulator cools affects the length of the insulator. Twenty insulators are produced at four temperatures. Temperature (T) is in degrees Celsius, length (L) in centimeters.

T	L	T	L	T	L	T	L
18	1.58	20	1.61	22	1.67	24	1.73
18	1.59	20	1.65	22	1.68	24	1.74
18	1.60	20	1.61	22	1.68	24	1.70
18	1.60	20	1.63	22	1.67	24	1.69
18	1.63					24	1.67
18	1.66					24	1.68

   a. Use the *fitted line plot* feature of MINITAB to find the equation of the least squares line.
   b. Based on the point fallout in the plot, does an approximate linear relationship between L and T seem reasonable?
   c. Suppose the lengths at 20 degrees had been 1.61, 1.60, 1.61, and 1.59, the lengths at 22 degrees had been 1.62, 1.63, 1.63, and 1.62, and the other 12 lengths were as given initially. Obtain the fitted line plot. Does a linear equation appear to be an appropriate model?

3. The number of murders in Charleston County, South Carolina for the years 1991 through 1999 are given below (from "Murder rate increases 30% in tri-county," *The Post and Courier*, Charleston, SC, January 16, 2000).

1991	1992	1993	1994	1995	1996	1997	1998	1999
48	33	35	42	24	32	35	16	23

   a. Use the *fitted line plot* feature of MINITAB to find the equation of the least squares line.
   b. Based on the point fallout in the plot, does an approximate linear relationship between annual murder count and year seem reasonable for this time period?
   c. Interpret any practical significance that may be accorded to the slope of the line.

d. Assuming that the approximate linear relationship holds into the immediate future, how many murders should be *predicted* for the year 2001?
e. Is it reasonable to use the least squares line to predict the number of homicides there will be in 2007?
f. The murder count in 1998 was 16, while in 1999 it was 23, a 44% increase. Is this a meaningful statistic?

4. In Problem 2, the sample standard deviation of length for each of the four temperature settings is as follows.

T	n	S
18	6	.02966
20	4	.01915
22	4	.00577
24	6	.02787

a. Confirm the values in this table.
b. As temperature increases, the standard deviations do not show an obvious increasing pattern, or a decreasing one, which suggests that assuming a common sigma is reasonable. Estimate σ by using the pooled standard deviation.
c. To obtain insulators that on average are closest to the target length, it is decided to operate the process at 23 degrees. Determine the center line and 3-sigma limits for an individuals chart.

5. The following below represent a sequence of AB concentrations for the process discussed in this section, for which the estimate of σ was .0505. The times when chemical shipment arrivals occur are noted.

5.63, 5.64, 5.67, 5.63, New Chemical D, 5.72, 5.64, 5.67, 5.70, 5.64, New Chemical B, 5.70, 5.57, 5.57, New Chemical C, 5.61, 5.66, 5.54, New Chemical A, 5.69, 5.71, 5.67, New Chemical D, 5.61, 5.60

a. Find the sample standard deviation of all 20 values.
b. Use the pooled standard deviation to estimate what σ equals during a period when there are no chemical shipments.
c. Explain why one should not be surprised that the standard deviation in part a. exceeds the standard deviation in part b.

6. Suppose a process is in control. Based on normal distributions, the probability that a sample mean will fall at least 2-sigma from the historical mean on an $\overline{X}$-chart (i.e., at least $2\sigma/\sqrt{n}$ from the mean) is .0456.

a. What is the average number of samples needed until a sample mean falls at least 2-sigma from the process mean?
b. In Figure 4.22, how many points fell at least 2-sigma from the mean? How many samples did it take altogether to produce this

number of points at least 2-sigma from the mean? For this process, estimate the average number of samples needed to produce such a point.

7. Suppose for the process discussed in this section that it is decided that *the* out-of-control signal for the process will be three back-to-back cumulative totals of 8 or more on the same side of the center line. How many such signals are there when the process is centered at 5.578 (see Figure 4.23)? How many samples did it take altogether to produce this number of signals? For this process, estimate the ARL when the process is centered at 5.578.

8. Recall for the product discussed in this section that the target value for AB concentration is 5.65 and $\sigma \approx .0505$. Also recall a property of zone charts—if a point falls on the center line, the point is considered to be on the same side of the line as the preceding point but contributes a zone weight of zero toward any cumulative count.
   a. If the values that follow are the AB concentrations of 20 consecutive batches and, somehow, no CD adjustments were made, what is the first time an adjustment should have been made, based on the two-back-to-back definition of an out-of-control signal?
   b. What kind of CD adjustment should have been made in order to recenter the process?

   5.62 5.67 5.62 5.69 5.55 5.59 5.55 5.68 5.66 5.71 5.62 5.59 5.56 5.71 5.74 5.78 5.80 5.80 5.74 5.78

9. Recall for the product discussed in this section that the target value for AB concentration is 5.65 and $\sigma \approx .0505$. Also recall a property of zone charts—if a point falls on the center line, the point is considered to be on the same side of the line as the preceding point but contributes a zone weight of zero toward any cumulative count.
   a. If the values that follow are the AB concentrations of 20 consecutive batches and, somehow, no CD adjustments were made, what is the first time an adjustment should have been made, based on the two-back-to-back definition of an out-of-control signal?
   b. What kind of CD adjustment should have been made in order to recenter the process?
   c. Obtain a MINITAB zone chart for these data. How does MINITAB appear to deal with plotting values that equal the target value when the two zones nearest the center line carry a zone weight of 1?

   5.62 5.67 5.65 5.65 5.55 5.67 5.55 5.53 5.51 5.56 5.47 5.44 5.41 5.56 5.44 5.48 5.50 5.50 5.44 5.48

10. Recall for the product discussed in this section that the target value for AB concentration is 5.65 and $\sigma \approx .0505$. The two-back-to-back signal described in this section is roughly equivalent to defining an out-of-control signal to occur when the cumulative count reaches 16 or more. Confirm that no such signals occur for the on-target process shown in Figure 4.22 and that eight such signals occur for the process centered at 5.578 (Figure 4.23), giving an estimated ARL of 9. Suppose an ARL of 9 is considered too large when the process is so centered. If an out-of-control signal is defined to occur when the cumulative count reaches 14 or more, confirm that Figure 4.23 shows 10 such signals, for an estimated ARL of 7.2. Use these data to define a cumulative count value for which the estimated ARL is less than 7 when the process is centered at 5.578 but still more than 72 when the process is on target.

4.6 EWMA CHARTS

Exponentially weighted moving average charts, or EWMA charts, are helpful for detecting small shifts in the process mean. When a small shift in the process mean occurs, an EWMA chart tends to produce an out-of-control signal more quickly than an $\overline{X}$-chart will. This increased sensitivity to small shifts is gained by giving up some sensitivity to larger shifts. For a manufacturing process in which a major shift is not a major worry and the main concern is minor drifts in location or keeping the process on target, an EWMA chart can be a helpful tool. The same can be said of another type of chart, the cumulative sum (or CUSUM chart). In this section we consider the EWMA chart, which is closer in spirit to Shewhart-type charts.

The underlying setting is the same as the one in which an $\overline{X}$-chart is used. Samples all have the same sample size, n, which is allowed to be 1. Samples are independent. The population being sampled is normal. We also assume that the process mean has a target value μ_T. There need not be a target value, but it will be insightful to begin by assuming there is one. Let $\overline{X}_1, \overline{X}_2, \overline{X}_3, \ldots$ denote the sample means. The values plotted on an EWMA chart, denoted $A_1, A_2, A_3, \ldots$, are called *exponentially weighted moving averages.* These latter values are obtained from the sample means according to the rules $A_0 = \mu_T$ and

$$A_t = \alpha\overline{X}_t + (1 - \alpha)A_{t-1}, \text{ for } t = 1,2,3, \ldots$$

where α is a weight between 0 and 1 chosen in advance.

Note that A_t is the weighted average of the most recent sample mean, $\overline{X}_t$, and the previous weighted average A_{t-1}, and hence will be between the two. If $\alpha = 1$, then 100% of the weight is accorded to the most recent sample mean, in which case $A_t = \overline{X}_t$ and the EWMA chart is just an $\overline{X}$-chart. If $\alpha = 0$, then no weight is accorded the most recent sample mean. In this case, $A_t = \mu_T$ for all values of t, and there is no point to the chart.

TABLE 4.18 Sample Data.

Sample	X_1	X_2	X_3	X_4	$\overline{X}_i$
1	402	404	405	409	405
2	404	401	409	410	406
3	401	394	403	398	399

For a process with target mean 400, we illustrate how the first three values of A_t would be computed based on the sample data from Table 4.18. We use $\alpha = .2$, which is MINITAB's default weight.

Since $\mu_T = A_0 = 400$, then

$$A_1 = .2\overline{X}_1 + .8A_0 = .2(405) + .8(400) = 401$$

$$A_2 = .2\overline{X}_2 + .8A_1 = .2(406) + .8(401) = 402$$

and

$$A_3 = .2\overline{X}_3 + .8A_2 = .2(399) + .8(402) = 401.4$$

By examining the iterative formula for A_t in a different way, it becomes apparent that A_t is a weighted average of $\overline{X}_t$ and all preceding sample means (and μ_T):

$$\begin{aligned}
A_1 &= \alpha\overline{X}_1 + (1-\alpha)\mu_T \\
A_2 &= \alpha\overline{X}_2 + (1-\alpha)A_1 \\
&= \alpha\overline{X}_2 + (1-\alpha)(\alpha\overline{X}_1 + (1-\alpha)\mu_T) \\
&= \alpha\overline{X}_2 + (1-\alpha)\alpha\overline{X}_1 + (1-\alpha)^2\mu_T \\
A_3 &= \alpha\overline{X}_3 + (1-\alpha)A_2 \\
&= \alpha\overline{X}_3 + (1-\alpha)(\alpha\overline{X}_2 + (1-\alpha)\alpha\overline{X}_1 + (1-\alpha)^2\,\mu_T) \\
&= \alpha\overline{X}_3 + (1-\alpha)\alpha\overline{X}_2 + (1-\alpha)^2\alpha\overline{X}_1 + (1-\alpha)^3\,\mu_T.
\end{aligned}$$

Similarly, A_4 is a weighted average of $\overline{X}_4$, $\overline{X}_3$, $\overline{X}_2$, $\overline{X}_1$ and μ_T, and so on. Note, however, that the older the sample mean is, the less weight is accorded to it.

The three reference lines on an EWMA chart are:

center line: Target value μ_T

3-sigma limits:* $\mu_T \pm 3\sqrt{\dfrac{\alpha}{2-\alpha}}\dfrac{\sigma}{\sqrt{n}}$

*Technically, the 3-sigma limits for A_t depend on t. However, as t increases, the limits approach the values given in the formula, and after several samples will approximately equal the values given in the formula. As usual, if unknown, the process standard deviation can be approximated in any usual way (e.g., pooled standard deviation).

TABLE 4.19 Sample Data from a Process With a Mean That Shifted Upward, Beginning With The Nineteenth Sample.

Sample	X_1	X_2	X_3	X_4
1	45.4	47.4	46.8	42.2
2	43.2	48.9	50.2	42.4
3	43.4	50.3	40.0	42.1
4	45.7	45.1	48.9	43.6
5	45.2	49.0	43.9	44.6
6	43.7	46.1	46.7	41.7
7	48.5	43.4	45.1	43.3
8	39.5	46.2	47.3	44.8
9	39.4	46.9	45.2	46.7
10	43.6	45.7	51.5	41.3
11	41.1	50.5	48.7	48.9
12	42.9	43.3	39.5	48.1
13	45.5	42.9	42.5	42.9
14	42.5	42.6	44.2	47.3
15	49.2	47.4	45.1	42.8
16	45.1	45.2	48.3	51.0
17	43.7	39.9	43.1	46.5
18	46.0	45.3	42.7	44.8
19	42.8	40.3	47.6	52.2
20	45.0	46.6	48.2	45.5
21	51.1	41.5	50.0	42.4
22	49.9	47.8	50.0	50.0
23	53.1	48.8	41.0	41.6
24	47.0	47.7	50.7	50.5
25	48.0	45.0	44.8	48.0
26	49.4	44.3	50.0	49.6
27	45.1	44.7	46.9	47.5
28	45.0	48.4	46.1	47.3
29	47.6	46.0	47.1	41.4
30	41.5	44.2	49.8	48.4

We now compare how the EWMA chart and $\overline{X}$-chart respond to a shift in the process mean. The first 18 samples in Table 4.19 contain data generated from a hypothetical process with target value 45.0 and known sigma of 3.0. Initially, the process is on target, but beginning with sample 19, the

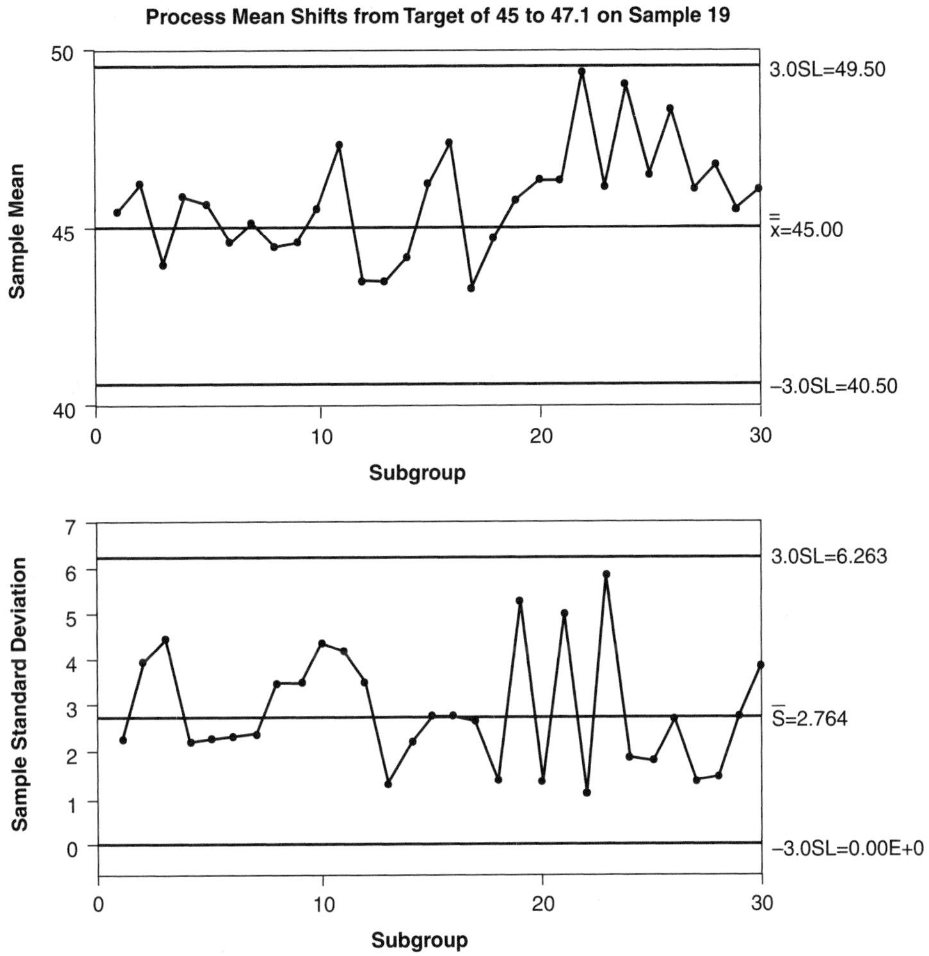

FIGURE 4.25 How an $\overline{X}$-chart Reacts when the Process Mean Shifts from Target of 45 to 47.1 on Sample 19.

process mean shifts to 47.1. Figure 4.25 shows the $\overline{X}$-chart and *S*-chart using a historical mean of 45.0 and historical sigma of 3.0. It is clear from the *S*-chart that variation has been stable. On the $\overline{X}$-chart, if the only out-of-control signal recognized is the big hit, a point beyond the 3-sigma limits, then there is no signal on this chart. If other kinds of out-of-control signals are accepted (e.g., 8 points in a row on the same side of the center line), which is common practice, then the $\overline{X}$-chart would have produced out-of-control signals. If such other signals are not recognized, then the $\overline{X}$-chart simply waits for the big hit when the mean shifts.

On the EWMA chart (Figure 4.26), the full weight of each new sample mean is not used, but preceding means are allowed to contribute to the plotted value. And so, beginning with sample 19, we see that the plotted points rise steadily as the graph moves toward recentering itself at 47.1. Beginning with sample 19, the plotted points on the $\overline{X}$-chart are immediately recentered at 47.1, but the 3-sigma limits on the $\overline{X}$-chart are farther from 45.0 than the 3-sigma limits on the EWMA chart. The first out-of-control signal on the EWMA chart occurs on sample 24.

Suppose a more serious shift in the mean were to immediately yield a sample mean of 49.6 at a time when the most recently plotted value on the EWMA chart had been 45.1, say. Because the upper 3-sigma limit on the $\overline{X}$-chart is 49.50, there would be an immediate, and desirable, out-of-control signal. However, the next value plotted on the EWMA chart would be between the last value plotted and the new sample mean, namely, $.2(49.6) + .8(45.1) = 46.0$, which is not an out-of-control signal on the EWMA chart. These illustrations are consistent with a general truth. $\overline{X}$-charts tend to detect major shifts more quickly than EWMA charts and EWMA charts tend to detect smaller shifts more quickly. Of course, $\overline{X}$-charts fare better in the comparison if signals other than only the big hit are accepted. Nevertheless, if serious shifts in the mean are not a serious problem, the EWMA chart can be a helpful tool.

Obtaining EWMA Charts in MINITAB

Data should be stored in columns in either of the same two ways they would be for an $\overline{X}$-chart. Namely, either the data occupy n columns with one sample per row or all the sample values are in one column.

From the menu, choose **Stat,** then **Control Charts,** then **Time-Weighted Charts** then **EWMA. . . .** A box will appear. Indicate either that the data are in one column and identify the column or that subgroups are stored across columns and identify the columns. If the data are in a single column, either identify the subgroup size or identify the column that tells how sample values are grouped into subgroups. If the weight α is a value other than .2, enter the value in the appropriate window. If there is a target value for the process mean, click on **EWMA Options,** choose the **Parameters** tab, and enter the mean. If a historical mean is not specified, the value of $\overline{\overline{X}}$ will locate the center line and serve as the initial value A_0. If known, enter the historical sigma value. If unknown, click the tab labeled **Estimate**. This allows the user to identify which method should be used to estimate sigma. Click **OK.** The **Labels. . .** button allows the user to give the EWMA chart a title. Finally, click **OK** to obtain the chart.

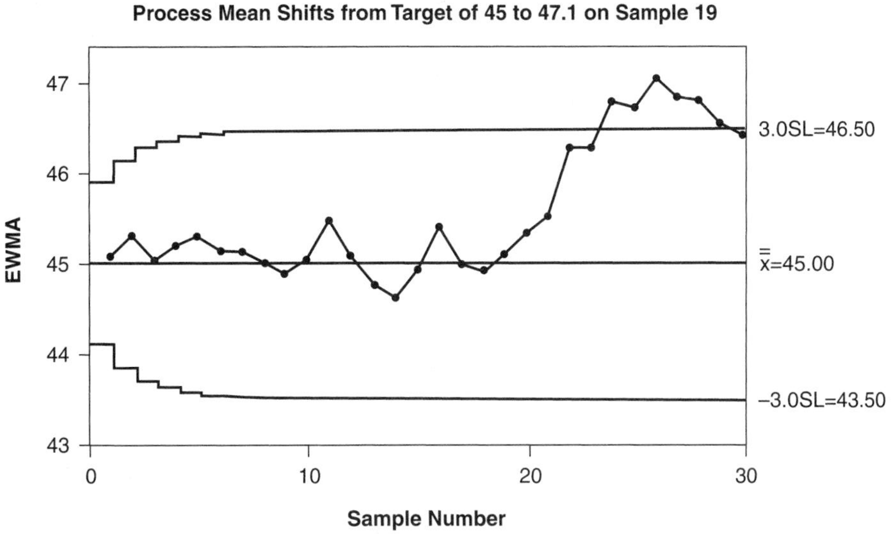

FIGURE 4.26 How an EWMA Chart Reacts when the Process Mean Shifts from Target of 45 to 47.1 on Sample 19.

PROBLEMS

1. For the example discussed in this section, the target value was 45.0 and the historical value of sigma used was 3.0. Confirm the values of the 3-sigma limits that appear in Figure 4.26.
2. Justify the statement that the 3-sigma limits on an $\overline{X}$-chart are farther from the center line than the 3-sigma limits on an EWMA chart unless $\alpha = 1$.
3. Suppose the process mean has a target value of 76, a process standard deviation of 2, and the sample size is 3. Also suppose that the last value plotted on the EWMA chart was 75, and the next few sample means are 80, 84, 80, 78, 81, and 81.
   a. Compute the next 6 values to be plotted on the EWMA chart. Use $\alpha = .2$.
   b. Compute the 3-sigma limits for an EWMA chart.
   c. Do any samples yield out-of-control signals? If so, which?
   d. What are the 3-sigma limits on an $\overline{X}$-chart?
   e. Are there points beyond the 3-sigma limits on the $\overline{X}$-chart?

4. In Problem 3, suppose the last value plotted on the EWMA chart was 75 and the next few sample means are 77.9, 77.3, 79.7, 78.9, 76.1, 75.9, 77.0, 77.5, 78.8, and 77.6 (the process mean has shifted from target to target + 1.15, a smaller shift than in Problem 3).
   a. Compute the next 10 values to be plotted on the EWMA chart. Use $\alpha = .2$.
   b. Compute the 3-sigma limits for an EWMA chart.
   c. Do any samples yield out-of-control signals? If so, which?
   d. What are the 3-sigma limits on an $\overline{X}$-chart?
   e. Are there points beyond the 3-sigma limits on the $\overline{X}$-chart?

5. Using the sample means given in Problem 4, suppose the last value plotted on the EWMA chart was 75 and more weight is placed on the sample mean by letting $\alpha = .3$.
   a. Compute the next 10 values to be plotted on the EWMA chart.
   b. Compute the 3-sigma limits for an EWMA chart.
   c. Do any samples yield out-of-control signals? If so, which?
   d. For this particular sequence of sample means, does increasing α from .2 to .3 result in an obvious speedup or slowdown (or neither) in the ability of an EWMA chart to detect the upward shift of 1.15 from the target value?

6. A chemical manufacturer checks the concentration of acid in each outgoing tanker-truck shipment of hydrochloric acid to help ensure that the process is centered at 40%. The concentrations from the last 25 outgoing trucks were:

   39.9 40.0 40.1 39.9 40.2 40.0 39.9 40.2 39.8 39.9 40.0 39.7 39.8 40.5 40.1 39.7 40.0 39.9 40.3 39.9 40.0 40.3 39.9 39.7 40.0

   Obtain an EWMA chart from these data using a moving range of length 2 to estimate σ. Does the process appear to be on target?

7. In Problem 6, suppose the process mean shifts to 39.9%, and acid concentrations for the next 50 trucks are (from left to right by rows)

39.5	40.2	40.3	40.0	39.8	39.8	39.3	40.1	39.6	39.8
39.8	39.8	39.9	39.6	40.0	39.8	39.8	39.9	39.7	40.0
39.7	39.9	39.8	39.4	39.6	39.9	39.7	39.9	39.9	39.8
40.0	39.6	39.9	39.8	39.8	39.8	40.1	39.2	39.8	39.9
39.7	39.9	40.1	39.8	39.9	40.1	39.8	40.0	39.9	40.1

   a. Does an EWMA chart indicate that the process is on target? Use moving ranges of length 2 to estimate σ.
   b. Does an individuals chart indicate that the process is on target?

8. The target diameter of a drilled hole is 1.300 cm. The historical sigma for hole diameter is .042 cm. Two holes are sampled for diameter

measurement every six hours. A worn drill bit was replaced in an effort to recenter the diameter distribution. After the replacement, the diameters from the first 10 samples are

1.33	1.31
1.33	1.26
1.24	1.32
1.34	1.31
1.31	1.32
1.21	1.30
1.29	1.27
1.33	1.28
1.31	1.22
1.33	1.29

Does the process appear to be recentered? Obtain an EWMA chart using $\alpha = .2$.

4.7 CUSUM CHARTS

CUSUM (pronounced "Q-sum") charts are particularly helpful for detecting small shifts in the process mean. When a small shift in the process mean occurs, CUSUM charts, like EWMA charts, tend to produce an out-of-control signal more quickly than an $\overline{X}$-chart will. Another similarity is that the increased sensitivity to small shifts is gained by giving up some sensitivity to larger shifts. For a manufacturing process in which the major worry is small drifts in location and keeping the process on target, a CUSUM chart can be a helpful tool. CUSUM charts do not fit the Shewhart-like *center line* $\pm$ *3 sigma* mold and will require some preparation in order to understand how they work, why they work, and how to set them up.

The term CUSUM is short for *cumulative sum*, a term that applies because the plotted value that results from each sample is tied to an accumulated sum of deviations between sample means and a value k^*, where k^* denotes the number midway between the target mean for the process and a boundary value that separates acceptable process means from process means that are too far from the target value to be acceptable. This synopsis sounds complicated, but we will identify the key quantities involved and break down the chart setup into steps.

The underlying setting is the same as the one in which an $\overline{X}$-chart is used. Samples all have the same sample size, n, which is allowed to be 1. Samples are independent. The population being sampled is normal.

We also need the notion of average run length (ARL) introduced in Section 4.5. Recall that this is the average number of samples needed for an out-of-control signal to occur.

We now identify and discuss some quantities that will be needed in order to understand a CUSUM chart:

μ_T	the target value for the process mean μ
μ_r	a value of μ that separates acceptable values of μ from unacceptable values of μ
σ	the process standard deviation
k^*	the number midway between μ_T and μ_r
k	$\|k^* - \mu_T\| \dfrac{\sqrt{n}}{\sigma}$
h	the value which, when multiplied by $\dfrac{\sigma}{\sqrt{n}}$, gives the upper control limit for the CUSUM chart

Since the target value μ_T and the process standard deviation σ mean the same things they did in our earliest work with $\overline{X}$-charts, we begin by describing μ_r. Suppose a process makes rods for which the target length is $\mu_T = 200.0$ cm. Also suppose the process mean μ is greater than 200.0 cm. If μ is only slightly greater than 200.0 cm, the manufacturer may not care. Say, hypothetically, that a statistician asks the manufacturer, "If $\mu = 200.0000001$ cm, would that be a problem?" "That's not enough difference to worry about," might be the response. This answer prompts the statistician to follow with, "If $\mu = 200.0000002$ cm, would that be a problem?" Suppose the manufacturer responds again, "That's not enough difference to worry about." So long as the manufacturer continues to give that answer, the statistician follows by asking the same question, but for a slightly incremented value of μ. A manufacturer should know what customers expect of the product and know what values of μ will cause financial or customer problems. So ultimately, the statistician will pose the question for some value and the manufacturer will say *yes*. This value is μ_r.

The hypothetical sequence of questions just discussed illustrates what μ_r represents. If a know-it-in-your-heart value of μ_r is not obvious, there is a rule-of-thumb method for arriving at its value. To detect when the process mean has shifted to an alarmingly high value, take $\mu_r = \text{USL} - 3\sigma$, illustrated in Figure 4.27. Values of μ between μ_T and μ_r produce no or few nonconforming items, while values of μ exceeding μ_r produce an appreciable percentage of nonconforming items. If $\mu = \mu_r$, management will want an out-of-control-signal within a few samples, because even a minor shift in μ from that point can result in a substantial percentage of nonconforming items.

The horizontal axis in Figure 4.27 also shows the location of k^*. In accord with the definition of k^*, this is $k^* = \dfrac{\mu_T + \mu_r}{2}$.

Figure 4.27 also illustrates that k represents how far apart μ_T and k^* are in multiples of $\sigma/\sqrt{n}$. Say $\mu_T = 200$, $\sigma = 1.2$, USL $= 204.5$, and $n = 4$.

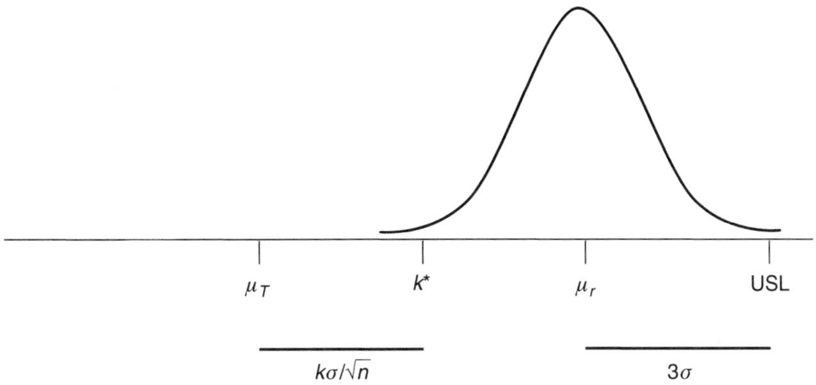

FIGURE 4.27 Graph Showing Key Quantities that are Used in Implementing a CUSUM chart.

Then $\mu_r = 204.5 - 3(1.2) = 200.9$, $k^* = \dfrac{200 + 200.9}{2} = 200.45$, and the actual distance between μ_T and k^* is .45. However, since $\dfrac{\sigma}{\sqrt{n}} = \dfrac{1.2}{\sqrt{4}} = .6$, the distance between μ_T and k^* in multiples of $\sigma/\sqrt{n}$ is $k = \dfrac{.45}{.6} = .75$. In general, $k = \dfrac{|k^* - \mu_T|}{\sigma/\sqrt{n}} = |k^* - \mu_T|\dfrac{\sqrt{n}}{\sigma}$.

This brings us to h, a value chosen by the user to determine the control limits on a CUSUM chart. For detecting when the process mean may have shifted to too high of a value, the relevant control limit is

$$\text{UCL} = h\frac{\sigma}{\sqrt{n}}$$

Soon we will see how to decide upon a value for h. For the time being, let's say we have decided to use $h = 4.5$. Then $\text{UCL} = 4.5\dfrac{1.2}{\sqrt{4}} = 2.7$.

So far, the discussion has been only in terms of the process mean shifting *upward* to a value μ_r, a value close enough to the *upper spec limit* to be of real concern. Symmetrically, the process mean is considered to be too close for comfort to the lower spec limit when $\mu = \mu_r = \text{LSL} + 3\sigma$.

We now examine how to compute the CUSUM values that are plotted on a CUSUM chart. We do so in a setting that has both upper and lower spec limits, producing what is called a two-sided CUSUM chart. Let $\overline{X}_i$ denote the mean of the ith sample. There will be two cumulative sum values that get plotted once the value of $\overline{X}_i$ is obtained. One we denote by $H(i)$, the *high-side cumulative sum,* which is used to help detect when μ is too high. The other we denote by $L(i)$, the *low-side cumulative sum,* which is used to help detect when μ is too low. If there is only one spec limit, the user would be interested in the plot of one of $H(i)$ or $L(i)$, the graph of which is called a one-sided CUSUM chart.

The value of $H(i)$ depends on $\overline{X}_i$ and the preceding high-side cumulative sum $H(i-1)$. To obtain $H(1)$, we define $H(0) = 0$. The rule for computing $H(i)$ is $H(i) = H(i-1) + (\overline{X}_i - k^*)$, unless a negative value of $H(i)$ results, in which case we set $H(i) = 0$. $H(i)$ is required to be nonnegative. All of this can be summarized into one equation:

$$H(i) = \max\{0, H(i-1) + (\overline{X}_i - k^*)\}$$

According to the formula for $H(i)$, if $\overline{X}_i > k^*$, the high-side cumulative sum increases by the distance between $\overline{X}_i$ and k^*. On the other hand, if $\overline{X}_i < k^*$, the high-side cumulative sum is reduced by the distance between $\overline{X}_i$ and k^*, subject to the limitation that $H(i)$ cannot be negative.

The larger the value of μ, the greater chance there is that a sample mean will exceed k^*, and consequently the more chance there is that the high-side cumulative sum will increase from one sample to the next. For example, if μ shifts to a value greater than k^*, more often than not, the high-side cumulative sum should increase from one sample to the next. When $H(i)$ reaches a sufficiently large value, an out-of-control signal is said to occur. Specifically, this is when $H(i) >$ UCL.

Low-side cumulative sums are computed in a similar fashion. We initialize with $L(0) = 0$ and use the iterative formula

$$L(i) = \min\{0, L(i-1) + (\overline{X}_i - k^*)\}$$

Low-side cumulative sums are required to be non-positive. An out-of-control signal occurs when $L(i) <$ LCL, where

$$\text{LCL} = -h\frac{\sigma}{\sqrt{n}}$$

After an out-of-control signal of either kind occurs, it is customary to start fresh cumulative sums by returning to $H(0) = L(0) = 0$ and treating the next sample mean as if it is the first.

We illustrate by showing the cumulative sums that result from a sequence of sample means in the setting where $\mu_T = 200$, $\sigma = 1.2$, USL = 204.5, and $n = 4$. Recall that for high-side cumulative sums, it followed that $k^* = 200.45$. Also, choosing $h = 4.5$ yielded UCL = 2.7. Suppose also there are two-sided specs: 200.0 ± 4.5. Then, for computing low-side CUSUMS, $\mu_r = \text{LSL} + 3\sigma = 195.5 + 3(1.2) = 199.1$ and $k^* = \dfrac{\mu_t + \mu_r}{2} = \dfrac{200.0 + 199.1}{2} = 199.55$ Using $h = 4.5$ gives $\text{LCL} = -h\dfrac{\sigma}{\sqrt{n}} = -2.7$.

Table 4.20 shows the CUSUM values for a sequence of sample means contrived for easy arithmetic. Remember that $H(0) = L(0) = 0$. There is an

TABLE 4.20 Computation of CUSUMs Based on Sample Means.

i	$\overline{X}_i$	$H(i) = \max\{0, H(i-1) + (\overline{X}_i - k^*)\} = \max\{0, H(i-1) + (\overline{X}_i - 200.45)\}$	$L(i) = \min\{0, L(i-1) + (\overline{X}_i - k^*)\}$ $\min\{0, L(i-1) + (\overline{X}_i - 199.55)\}$
1	199.45	$\max\{0, 0 + (-1)\} = 0$	$\min\{0, 0 + (-.10)\} = -.10$
2	200.55	$\max\{0, 0 + .10\} = .10$	$\min\{0, -.10 + 1\} = 0$
3	201.25	$\max\{0, .10 + .80\} = .90$	$\min\{0, 0 + 1.70\} = 0$
4	200.25	$\max\{0, .90 + (-.20)\} = .70$	$\min\{0, 0 + .70\} = 0$
5	201.95	$\max\{0, .70 + 1.50\} = 2.20$	$\min\{0, 0 + 2.40\} = 0$
6	201.35	$\max\{0, 2.20 + .90\} = 3.10^*$	$\min\{0, 0 + 1.80\} = 0$
7	200.65	$\max\{0, 0 + .20\} = .20$	$\min\{0, 0 + 1.10\} = 0$

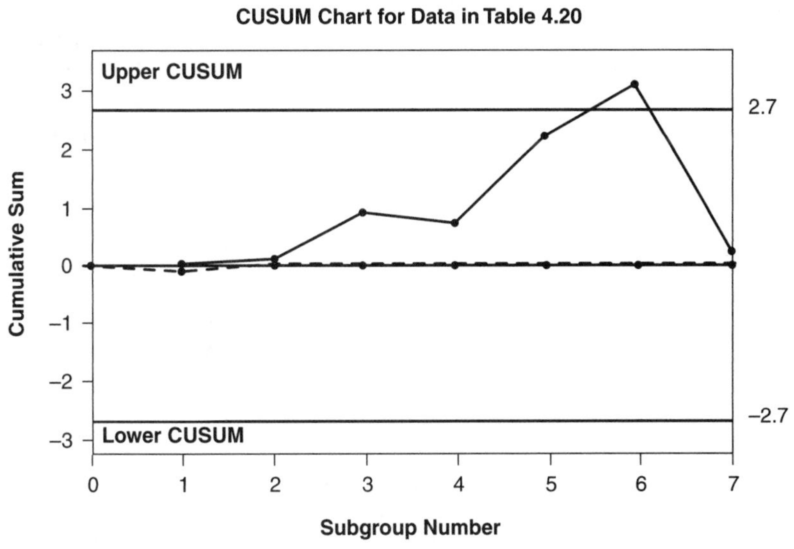

FIGURE 4.28 CUSUM Chart for Data in Table 4.20.

out-of-control signal on the sixth sample because $H(6) > 2.7$. New CUSUMS are started on the seventh sample. Figure 4.28 is the corresponding MINITAB CUSUM chart.

A CUSUM chart is designed to accomplish two things. First, when the process is on target, a large ARL is desirable. In setting up a CUSUM chart, the user can decide what this value, denoted ARL_a, should equal. Secondly, if the process mean has shifted to μ_r, management would like an out-of-control signal in relatively few samples, and a small ARL is desirable. The user also can decide what this value, denoted ARL_r, should be.

Another thing the user can choose is the sample size n. However, we shall see that ARL_a, ARL_r, and n are tied together in a way that doesn't give the user the complete freedom to choose a large value for ARL_a, a small value for ARL_r, and a small value for n, all three at the same time. The user can specify two of the three, which will determine the third.

Before learning how to set up a CUSUM chart, we need to know what kinds of values for ARL_a and ARL_r are typical. Remember that when μ shifts to μ_r, management would like to have an out-of-control signal within just a few samples. In practice, typical values of ARL_r are between 2 and 12.

Since ARL_a gives the average number of samples needed to produce a high-side (or low-side) CUSUM value beyond the control limit when the process is on target, values of ARL_a usually are chosen to be in the upper hundreds or greater. As a parallel to help us see why, consider the $\overline{X}$-chart. Suppose we view having a point fall above the upper 3-sigma limit as the indicator that the process mean may have shifted to too high a value. Recall from Section 4.5 that the concept of average run length was likened to how many tosses would be needed for a fair die to show a 5. A die is not programmed to cycle through its six faces in six tosses, and so it does not always take precisely six tosses to get the next 5. However, it can be shown that six-tosses-to-get-a-5 is the *average* behavior of the die. Note that the probability of getting a 5 on one toss is $\frac{1}{6}$, and that the average number of tosses needed to produce a 5 is $6 = \frac{1}{1/6}$. More generally it can be shown that if the probability of getting a 5 is p, then the average number of tosses needed to obtain a 5 is $1/p$. We know that on an $\overline{X}$-chart, 99.7% of the points are expected to fall between the control limits when the process is in control and on target. A more precise value is 99.73%, based on normal distributions. So the probability that a single sample mean does not fall between the 3-sigma control limits is $1 - .9973 = .0027$, and the probability it falls above the upper 3-sigma limit is .00135. Hence the average number of samples needed to produce a point above the upper 3-sigma limit is $\frac{1}{.00135} \approx 741$ when the process is in control and on target.

We now examine how to go about setting up a CUSUM chart for a process assuming that σ is known, or is estimated based on historical data. The nomogram in Figure 4.29 provides a means to relating n, ARL_a, ARL_r, h, and k.

EXAMPLE 1

Suppose 80.00 is the target value for the variable measured, spec limits are $80 \pm .53$, $\sigma = .15$, and our samples are of size $n = 9$. From the perspective of high-side CUSUMS, $\mu_r = 80.53 - 3(.15) = 80.08$, and $k^* = 80.04$. Since we already have a value for n, we can find k, the first of the four quantities the nomogram can provide:

$$k = |80.04 - 80.00|\frac{\sqrt{9}}{.15} = .80$$

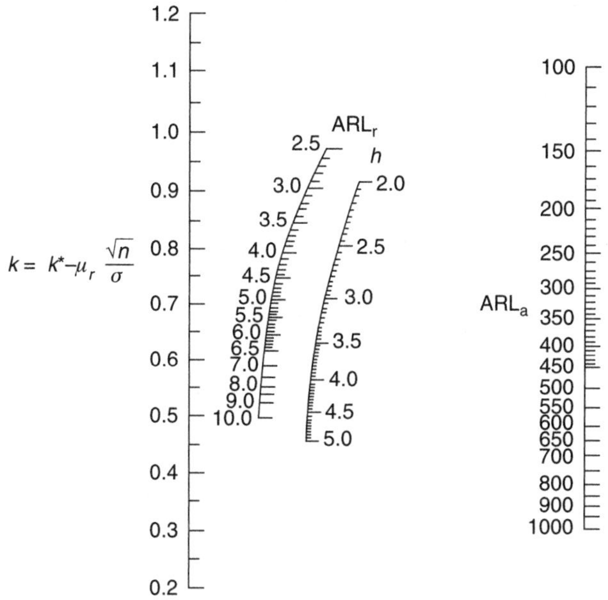

FIGURE 4.29 Nomogram for Designing CUSUM Plans, Adapted from Kemp (1962). Reprinted with Permission of Blackwell Publishing Ltd.

Because n is specified, we can choose a value for just one of ARL_a or ARL_r. Suppose we consider an acceptable value for ARL_a to be 700. By using a straight line to connect .80 on the k scale to 700 on the ARL_a scale, we see that $ARL_r = 4.7$ and $h = 3.18$, approximately. Suppose we decide that an ARL of 4.7 doesn't provide a quick enough signal when $\mu = 80.08$. We could reduce ARL_r to 3.5, say. But this will reduce h to about 2.18, bring UCL closer to zero, and make out-of-control signals easier to occur. In particular, ARL_a would equal 149, approximately. That is, a false, high-side, out-of-control signal will occur once in 149 samples, on average, when the process is on target.

Suppose management decides that the smaller value of 3.5 for ARL_r is worth the nuisance of looking for nonexistent assignable causes more frequently. Since h and k have been determined, in quality control parlance, management is said to have adopted the *CUSUM plan* $h = 2.18$, $k = .8$. In light of how the nomogram is used, specifying h and k also specifies ARL_a or ARL_r.

When the upper and lower spec limits are equally distant from the target value, as they are in this example, the values specified for h and k provide for low-side CUSUMS the same values of ARL_a or ARL_r as for high-side CUSUMS. ARLs specified by the nomogram apply to high-side and low-side CUSUMS individually. So in this

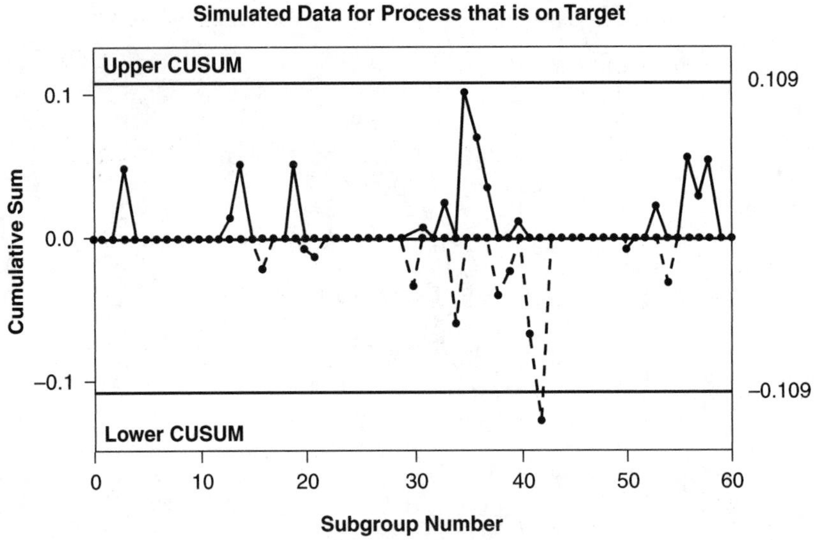

FIGURE 4.30 Simulated Data for Process that Is on Target.

example, we will average one high-side CUSUM out-of-control signal in every 149 samples, also one low-side CUSUM out-of-control signal in every 149 samples.

Figure 4.30 shows a MINITAB CUSUM chart for 60 samples of size 9 where the sample values come from a normal distribution with mean 80.00, target 80.00, process standard deviation .15, and CUSUM plan h = 2.18 and k = .8. There is one false low-side out-of-control signal. When the process is on target, CUSUM values tend to be near zero and nonzero CUSUM values tend to come back toward zero with subsequent samples.

Figure 4.31 depicts the same setting except that that the process mean has shifted to 80.08, the high-side μ_r. The ARL for high-side out-of-control signals is now 3.5.

EXAMPLE 2

Now suppose that 80.00 is the target value for the variable measured, spec limits are 80 ± .53 and σ = .15, but the sample size has not been determined. We can specify values for ARL_a and ARL_r and then determine the minimal sample size that would yield the two ARLs. Say we want ARL_a =700 and ARL_r = 3. Then h = 2.32, k = 1.09, and we can solve for n:

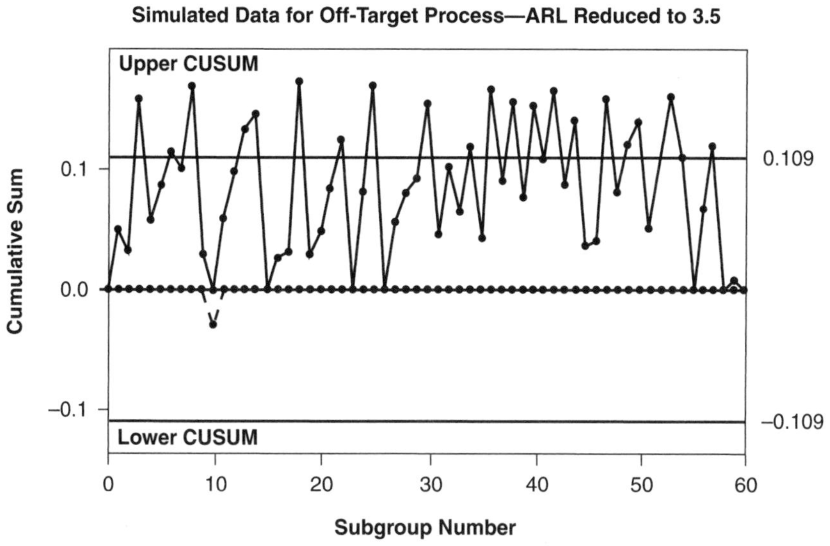

FIGURE 4.31 Simulated Data for Off-Target Process—ARL Reduced to 3.5.

$$|k^* - \mu_t|\frac{\sqrt{n}}{\sigma} = 1.09$$

$$|80.04 - 80.00|\frac{\sqrt{n}}{.15} = 1.09$$

$$\sqrt{n} = \frac{1.09(.15)}{.04} = 4.0875$$

$$n = 16.708$$

Of course, sample size must be a positive integer. But if we *could* have a fractional sample size, the nomogram is saying that we can have the values we want for ARL_a or ARL_r, but cannot do it with a sample size less than 16.708. So the smallest n we can use is 17.

When solving for *n,* if the answer is not an integer, always round *up* to the next integer.

Now, if we use $n = 17$ with the CUSUM plan $h = 2.32$ and $k = 1.09$, the values for ARL_a and ARL_r will be a slightly off-mark, because n is not 16.708. But since n is now known, we can compute the proper value of k:

$$k = |k^* - \mu_t|\frac{\sqrt{n}}{\sigma} = |80.04 - 80.00|\frac{\sqrt{17}}{.15} = 1.10$$

The nomogram shows that management can still have $ARL_a = 700$ and reduce ARL_r to 2.95. The corresponding value of h is 2.3, approximately.

Obtaining CUSUM Charts in MINITAB

Data should be stored in columns in either of the same two ways they would be for an $\overline{X}$-chart. Namely, either the data occupy n columns with one sample per row or all the sample values are in one column.

From the menu, choose **Stat,** then **Control Charts,** then **Time-Weighted Charts,** then **CUSUM. . . .** A box will appear. Indicate either that the data are in one column and identify the column or that subgroups are stored across columns and identify the columns. If the data are in a single column, identify the subgroup size. Enter the target value for the process mean. If σ is known, click the button labeled **CUSUM Options. . . ,** choose the **Parameters** tab, and enter the value. The tab labeled **Estimate** allows the user to identify which method should be used to estimate sigma. When this is completed, click the **Plan/type** tab. This allows the user to specify the cusum plan values for h and k. This also is the box in which one can put a check in the window labeled **Reset after each signal.** Finally, click **OK** to obtain the chart.

PROBLEMS

1. Find μ_r, k^*, and k if the target value is 50, there is an upper product specification of 52, the process standard deviation is .6, and the subgroup size is 4.
2. Find μ_r, k^*, and k if the target value is 100, there is a lower product specification of 96, the process standard deviation is 1, and the subgroup size is 9.
3. Find the values of the high-side cumulative sums for the sequence of sample means shown as follows if the target value is 60, the upper product specification is 60.5, the process standard deviation is $\sigma = .12$, the subgroup size is 3 and $h = 2.8$. Do any out-of-control signals occur?

Sample	1	2	3	4	5	6	7	8
Mean	60.075	60.105	60.028	60.120	60.127	60.117	60.118	60.080

4. Find the values of the low-side cumulative sums for the sequence of sample means shown as follows if the target value is 1000, the lower product specification is 962, the process standard deviation is 10, the subgroup size is 4, and $h = 3.2$. Do any out-of-control signals occur?

Sample	1	2	3	4	5	6	7	8
Mean	998	994	997	989	987	986	998	986

5. Suppose someone determines the sample size that corresponds to ARL_a = 600 and ARL_r = 4. Find values of h and k for the CUSUM plan.
6. Suppose the target value is 1,000, the lower product specification is 962, the process standard deviation is 10, and the subgroup size is 4. If it is desired to have ARL_a = 800, find k, h, and ARL_r.
7. Suppose a process has been in apparent control for 40 samples with $\bar{R}$ = .659 based on samples of size 4. The target mean is 200, spec limits are 200 ± 1.2. The sample size will be changed if need be in order to have ARL_a = 600 and ARL_r = 4.5.
   a. What is the smallest sample size that can be used?
   b. If this changes n, based on the new sample size, what is ARL_a if it is desired to keep ARL_r = 4.5?
   c. What is the CUSUM plan?
8. The current CUSUM plan for a process is h = 2.25 and k = 1.05.
   a. What does this say about ARL_r and ARL_a?
   b. The target mean is 500 and spec limits are 500 ± 14.7. The process standard deviation is 4.01. If management is content with the values of ARL_r and ARL_a from part a., what is the minimal sample size they should be using?
9. Using MINITAB or an equivalent package, obtain a CUSUM chart based on the following eight samples of spool weights if the target mean is 25, σ = .6, spec limits are 25.0 ± 2.5 and it is desired that ARL_r = 4.

Sample	x_1	x_2
1	24.50	23.89
2	25.71	24.91
3	23.71	24.59
4	24.91	24.40
5	25.73	24.30
6	23.84	23.82
7	24.15	23.93
8	23.81	23.58

4.8 SHORT-RUN CHARTS

A "given" on setting up control charts has been to have enough samples in a period of stability so that the process mean and standard deviation, or proportion of nonconforming items, or average number of nonconforming items in a sample, and so on, could be estimated. The understanding was that we would be able to remove from computations any samples that were beyond 3-sigma limits due to known assignable causes and have enough samples left to produce good estimates. Then the estimates could be used to set tentative control limits appropriate for charting future samples.

Sometimes, however, processes involve short production runs. This is common when a manufacturer routinely alters machine settings, modifies the chemical recipe, or otherwise makes adjustments to prepare the process to meet the differing product specifications that go with each new customer order. If there are just a few samples from a completed production run, estimating the process mean, proportion defective rate, and so on, is necessarily imprecise. What kinds of statistical summaries are possible when the number of samples is small? One is a *nominal* $\overline{X}$-and *S-chart* (or $\overline{X}$-and *R- chart*).

Suppose a company makes a shipment of spools of thread of a given color to fill a customer order. Four spools are sampled twice a day and tested to find the minimum tension (in centinewtons) needed to make thread pull from the spool without snagging. During the run, seven samples are tested. The target value throughout this particular order is 7.00 cn. The results appear in Table 4.21.

Instead of plotting an $\overline{X}$-chart centered at 7.00 or $\overline{X}$ (the latter is an imprecise estimate of μ because there are only seven samples), we subtract the target value from each sample value, producing the adjusted values shown in Table 4.22, and plot an $\overline{X}$-chart based on the adjusted values, with the center line located at the mean of means.

An adjusted value shows how far the sample value is from target and whether the value is above or below target. The nominal $\overline{X}$-and *S*-chart is shown in Figure 4.32.

The *S*-chart is identical to the *S*-chart one would obtain using the original data values. It raises no alarm as to variation here. The $\overline{X}$-chart shows that average tension for the seven samples was .2118 above target—that $\overline{\overline{X}} = 7.2118$ for the unadjusted data. One point is slightly below the lower 3-sigma limit. However, the issue of control has to carry less weight when there are few samples because the estimates of σ and μ have to be viewed as imprecise. From a process that is on target, one expects a mix of points plotted above and below 0 on the vertical scale. In this case, the vertical scale indicates that the point below the lower 3-sigma limit comes from the only sample to have a mean below the target value of 7. If there are spec limits,

TABLE 4.21 Tensions from Run 56, Target = 7.0.

Sample	1	2	3	4	5	6	7
Obs 1	7.17	6.98	7.31	7.18	6.73	7.34	7.31
Obs 2	7.41	7.29	7.33	7.39	6.74	7.50	7.52
Obs 3	6.96	7.29	7.13	6.83	6.99	7.36	7.76
Obs 4	6.81	7.36	7.17	7.49	7.09	7.13	7.36

TABLE 4.22 Distance from Target, Run 56, Target = 7.0.

Sample	1	2	3	4	5	6	7
Obs 1	0.17	−0.02	0.31	0.18	−0.27	0.34	0.31
Obs 2	0.41	0.29	0.33	0.39	−0.26	0.50	0.52
Obs 3	−0.04	0.29	0.13	−0.17	−0.01	0.36	0.76
Obs 4	−0.19	0.36	0.17	0.49	0.09	0.13	0.36

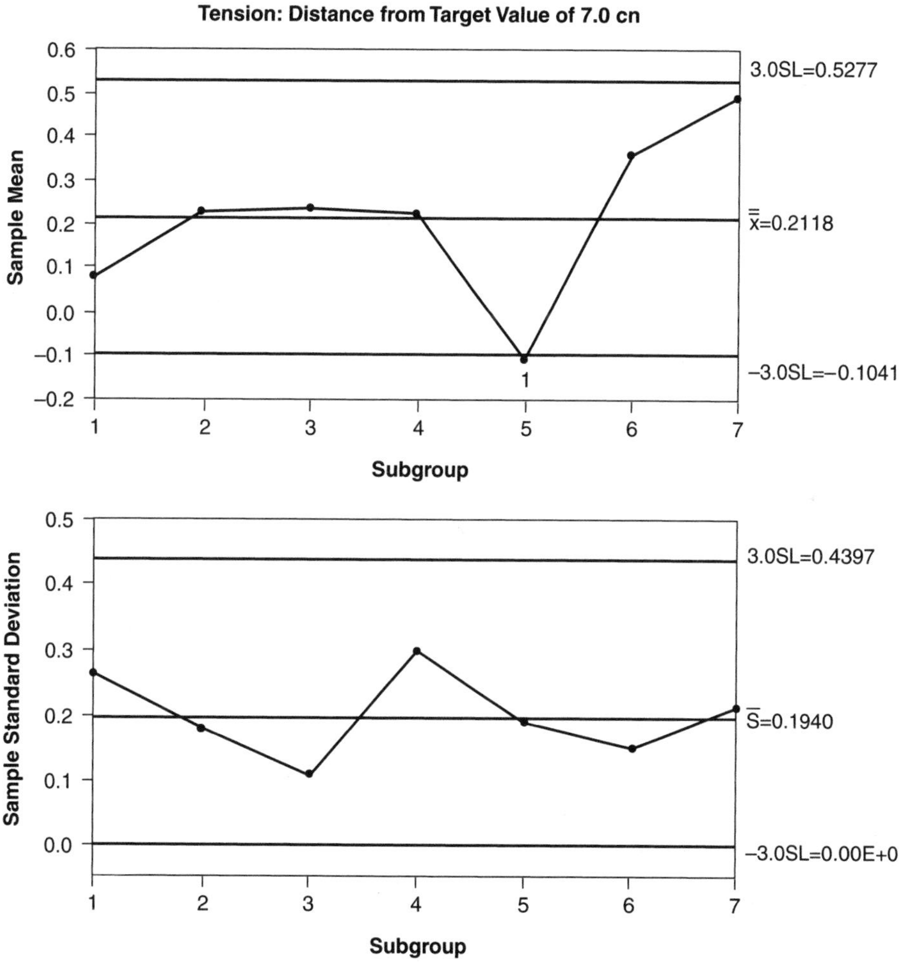

FIGURE 4.32 Short-Run Chart.

the customer or manufacturer also may want to know the value of C_{pk}, C_{pU}, or C_{pL} for this production run. Capability indices for short runs likewise must be viewed as imprecise estimates.

Using MINITAB to Show a Succession of Product Runs

The manufacturer may want to see a chart showing data from a succession of product runs, as shown in Figure 4.33. The adjusted data from all runs should be in one column, say C5. Another column, say C6, is used to identify the product run. For each value in column 5, the identifier for that

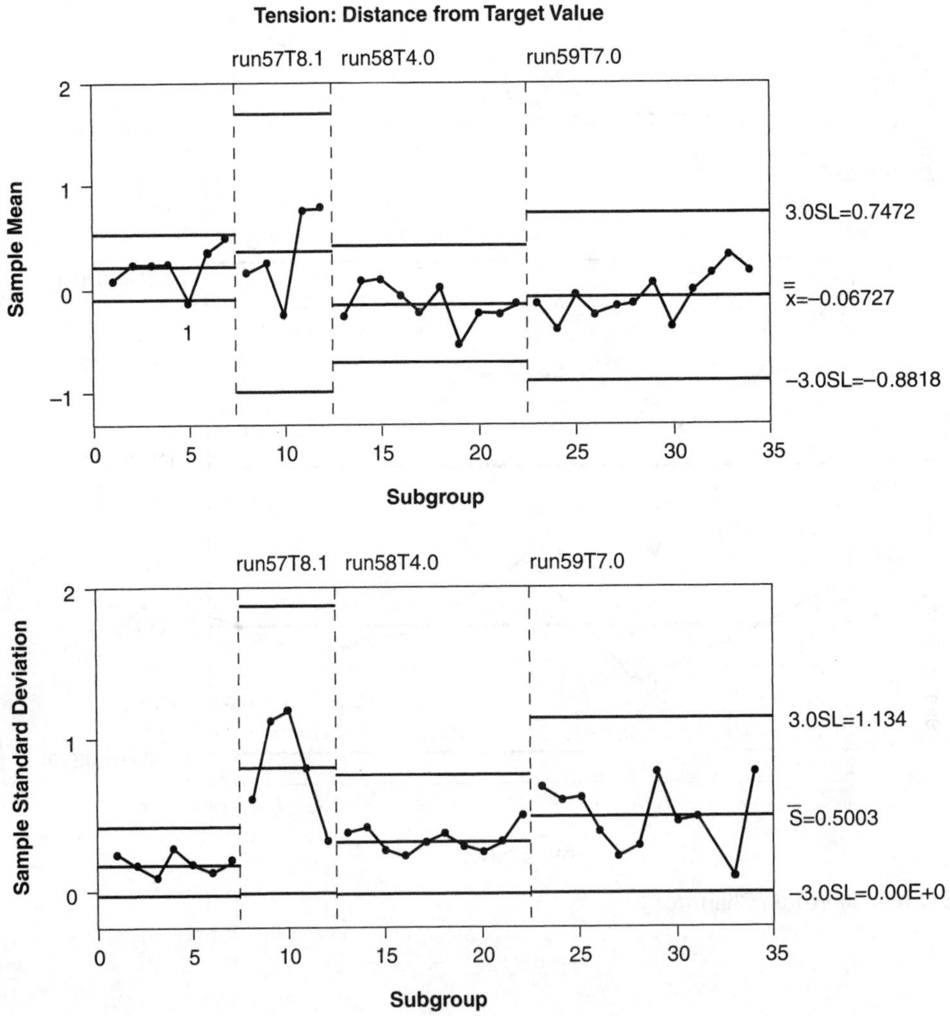

FIGURE 4.33 Short-Run Chart for Different Product Runs.

product run, for example, *run57Target8.1*, occupies the cell in the same row of column 6. A nominal $\overline{X}$-and *S*-chart or $\overline{X}$-and *R*- chart is started in the usual way. Choose **Stat,** then **Control Charts,** then **variables charts for subgroups,** then **Xbar-S. . . ,** and a box will appear. Choose the option **All observations for a chart are in one column,** and enter C5 in the window directly underneath. Then enter the subgroup size in the **Subgroup sizes:** window. Click on the **Xbar-S Options. . .** window, choose the tab labeled **Stages,** and enter C6 in the window labeled **Define Stages.** Click **OK** and proceed as usual.

3-sigma limits for a run are based on data from that run only. It would be inappropriate to pool data together from different product runs in order to obtain a common estimate of sigma for all runs because, just as targets and process means can vary between product runs, so can process standard deviations.

Using *p*-Charts with Few Points to Detect Professors' Extreme Grading Practices

Some college and university faculty are known by students to be easy graders, others to be tough. At one institution known to the author, an approach similar to the one described as follows has been used to identify faculty who are too easy or too tough. Let's say that one of a number of courses monitored is Biology 101. Many different professors teach sections of the course each semester. A variable that can be measured in each section is the percentage of students who receive an A in the course.

In an industrial environment, *p*-charts are used to track the rate at which nonconforming items are being produced. In that environment, there would be understandable interest in knowing whether different individuals or different machines produce nonconforming items at different rates. The purpose would be to know each machine's or individual's capability, with the goal of driving defective rates to zero. None of this applies to the percentage of A grades given in an academic setting because it is an accepted principle that there are some differences in how faculty grade. The institution would have no interest in knowing each faculty member's personal A *rate*, in maintaining *p*-charts on all faculty, or in wanting all faculty to have the same personal A rate.

That being said, a faculty member who bestows A grades on the masses or who literally never gives A grades (and may even take pride in this fact) is not doing right by the students. The purpose here is to detect individuals with extreme grading practices.

Consider the sections of Biology 101 that are offered every fall semester. Let us say that in the last five years, there have been no changes in course prerequisites, in the kinds of students who take the course, or in anything that would cause grade distributions to change. Further suppose that by pooling together all sections from those years, there were 614 students who got a course grade of A out of the 4,039 who took the course. So an A was received by 15.2% of the students who took the course.

We take 15.2% to be the historical mean A rate for the course, with the view that this rate is an estimate of how the average instructor would grade the average section in a fall semester. A few sections of Biology 101 are offered in spring semesters. However, they are not included in the computation above because in the spring semester, sections invariably enroll a constituent of students who failed the course in the preceding term, and a lower A rate applies. Spring sections of Biology 101 are monitored separately.

Again, no one has any interest in having all instructors grade like the average instructor. Here the interest is on determining faculty who routinely give semester grades that the average instructor would be expected to assign to only a few sections in one thousand.

The grading history of any professor who teaches a fall section in which the A rate falls outside the 2-sigma warning limits

$$.152 \pm 2\sqrt{\frac{.152(1 - .152)}{n}} = .152 \pm \frac{.718}{\sqrt{n}}$$

where n is the number of students in the section, is examined once a year. Of the 36 faculty in the department, suppose that instructors A, B, C, D, and E generate such signals in the fall of 2002. Their grading histories in the course during the last five years are given in Table 4.23.

The p-chart for A grades based on the data in Table 4.23 is shown in Figure 4.34. The historical value for p is prescribed to be .152. The chart shows 2-sigma and 3-sigma limits. Computation of the lower 3-sigma limit produces a negative number for the values of n here, and so MINITAB identifies the lower

TABLE 4.23 Grading History of Biology 101 Instructors Who Generated Warning Signals in Fall, 2002.

Instructor	A	A	A	A	A	A	B	B	B	B
Term	F 97	F 99	F 00	F 00	F 01	F 02	F 00	F01	F 02	F02
A count	13	12	10	16	12	11	7	3	5	0
n	40	40	39	38	40	40	35	36	39	32

C	C	C	C	C	C	C	C	D	D	D	E
F 99	F 99	F 00	F 00	F 01	F 01	F 02	F 02	F 98	F 01	F 02	F 02
1	0	0	2	0	1	2	0	5	7	12	11
39	39	40	42	41	35	38	38	37	40	40	39

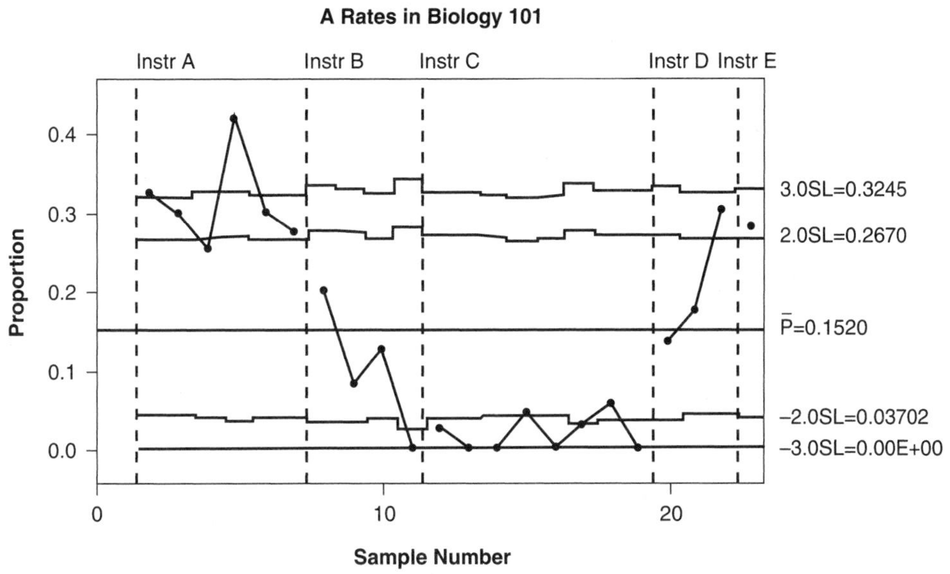

FIGURE 4.34 A Rates in Biology 101.

3-sigma limit as zero. Therefore, even if no A grades are awarded in a section, the percentage of A grades will not fall below the lower 3-sigma limit. However, for purposes of these charts, the university recognizes two signals: a point beyond the 3-sigma limits or two of three consecutive points beyond the same 2-sigma limit. Finally, in the belief that every instructor has an occasionally unusual section, only those individuals who generate two or more out-of-control signals are identified for a conversation with the department chair. These would be Instructor A, who had three signals (first section, fourth section, and sections 5 and 6), and Instructor C, who also had three (sections 1 and 2, sections 3, 4, and 5, and sections 6, 7, and 8).

PROBLEMS

1. For a company that makes vinyl sheets for seawall construction, a key quality characteristic is the thickness of a sheet. During a particular production run, the target thickness is .250 inches. Five hourly samples of size 4 are obtained, with means and standard deviations as follows:

$\bar{x}$:	.25125	.24550	.24475	.24700	.25025
s:	.00377	.00614	.00443	.00483	.00403

   a. What values are plotted on the nominal $\overline{X}$-chart?

b. Use the pooled standard deviation to estimate σ.
c. Find values for the center line and the upper and lower 3-sigma limits for the short-run $\overline{X}$-chart based on these data.

2. The target value for a hole diameter is .450 mm.

a. If the successive means for a production run plotted on a nominal $\overline{X}$-chart are .003, .000, −.001, and .005 mm, what can be said about the actual mean diameter measurements for those samples?
b. Where would the center line on the nominal $\overline{X}$-chart be located?

3. Suppose for a product run that the successive means plotted on a nominal $\overline{X}$-chart are .15, .29, .36, .31, and .41, the pooled standard deviation is $s_p = .051$, and the sample size is 4.

a. Show that there are three consecutive points, two of which fall above the upper 2-sigma limit.
b. Should this be viewed as an indication of lack of control?

5

Probability

OBJECTIVES

- Provide the basic rules of probability
- Explain the difference between independent and dependent events
- Show how to compute conditional probability
- Introduce the Fundamental Counting Principle
- Introduce and illustrate how to use hypergeometric distributions
- Introduce and illustrate how to use binomial distributions
- Introduce and illustrate how to use Poisson distributions
- Introduce and illustrate how to use normal distributions
- Define the reliability of a system

- Show how to compute the reliability of series systems, parallel systems, and other systems
- Introduce the concept of reliability function
- Introduce the concept of failure rate and failure rate function

5.1 FUNDAMENTALS OF PROBABILITY

If someone were to take a quarter and flip it in the air, you might assume, based on your experience, that the coin has the same chance of landing heads as it does of landing tails. In doing so, you are estimating *how much* of a chance there is that the particular event *heads* will occur. The amount of chance that an event will occur is measured by what is called the **probability** of the event.

Just from living life, people get experience estimating the probability that various things will happen. Many would readily say of a coin that the probability of a heads is $\frac{1}{2}$. Giving an answer of $.5 = 50\%$ really is a way of saying that the outcome *heads* constitutes 50% of all the possible outcomes of a coin toss, and all probability is divvied up between the two outcomes into identically equal shares. We should not expect that the actual probability of a head for a particular coin is precisely .5 followed by infinitely many zeros. Saying that the probability of getting a head is $\frac{1}{2}$ is an idealization, a model for how a coin should behave when given a good spinning flip. That said, we point out that the probability models and the probability rules we discuss in this chapter are known to model natural phenomena very well.

In working symbolically with probability, often it will be helpful to let capital letters stand for events. Suppose we let H represent the event: *getting a head on a coin toss.* The symbolic way of saying that *the probability of getting a head on a coin toss is* $\frac{1}{2}$ is

$$P(H) = \tfrac{1}{2}.$$

We now introduce a simple model for computing probabilities, which we call the *m-over-n rule.* Note that it leads to the result $P(H) = \frac{1}{2}$ when a coin is tossed.

Suppose one of n possible outcomes will result from some action and each of the n outcomes has the same probability of being the one that results. Let A be an event and let m denote the total number of outcomes that result in the occurrence of event A. Then

$$P(A) = \frac{m}{n}.$$

EXAMPLE 1

A company manufactures light bulbs. Each completed bulb is tested. If it doesn't work properly, it is physically examined by a lab technician who assigns one of the following as the reason for the problem: out-of-spec voltage, out-of-spec wattage, no vacuum—seal problem, no vacuum—cracked glass, contact button problem, thread problem. While a defective bulb may have more than one of these problems, the technician follows a procedure that singles out one cause. Suppose 400 bulbs were examined today with the following breakdown:

out-of-spec voltage	100
out-of-spec wattage	83
no vacuum — seal leak	68
no vacuum — cracked glass	89
contact button problem	48
thread problem	12

Suppose a light bulb is selected at random from among the 400 defective bulbs. Used in this way, the words *at random* mean that each bulb has the same chance of being the one chosen. Then $P(\text{contact button is identified as the problem}) = \frac{48}{400} = .12$. The probability for each possible problem cause is given in Table 5.1. The reader should confirm the computations.

While introducing some of the formal rules of probability, we also will introduce some terminology used for describing various kinds of relationships between events.

Two events A and B are called mutually exclusive if they cannot occur at the same time.

In tossing a coin once, the events *getting a head* and *getting a tail* are mutually exclusive. In a golf tournament, the event *Tiger Woods finishes in first*

TABLE 5.1 Probability Distribution of Defect Causes for 400 Light Bulbs.

Cause of Problem	Probability
out-of-spec voltage	.25
out-of-spec wattage	.21
no vacuum—seal leak	.17
no vacuum—cracked glass	.22
contact button problem	.12
thread problem	.03

place and the event *Tiger Woods finishes in third place* are mutually exclusive. However, the events *Tiger Woods finishes in first place* and *Tiger Woods finishes in the top three* are not mutually exclusive. Also, if Rich Beem and Tiger Woods are entered in the same tournament, the events *Rich Beem finishes in first place* and *Tiger Woods finishes in second place* are not mutually exclusive.

Return to the light bulb example in which one bulb is chosen randomly from among the 400 defective bulbs. Let the following event notation define which cause was identified:

V: out-of-spec voltage	*L*: leak in seal	*B*: contact button problem
W: out-of-spec wattage	*G*: cracked glass	*T*: thread problem

Then *V* and *W* are mutually exclusive, as are *V* and *L*, as are any different two of the six events.

Sometimes we are interested in a probability that involves two or more events. For example, what is the probability that the defective bulb selected has a vacuum-related problem? The question also could be phrased in this leading way, *What is the probability that the problem is cracked glass* **or** *a leak in the seal?* The notation for this probability is P(G or L). Following the m-over-n rule, we have

$$P(G \text{ or } L) = \frac{89 + 68}{400} = \frac{157}{400} = .39.$$

The event "*A* or *B*" is said to occur when at least one of *A* and *B* occurs.

Note that the event "A or B" occurs if just one of A and B occurs or if both occur simultaneously. Of course, the particular events G and L are mutually exclusive and cannot occur simultaneously. When two events A and B are mutually exclusive, there is a helpful rule for computing P(A or B):

P(A or B) = P(A) + P(B) when A and B are mutually exclusive events.

Earlier we computed that P(G or L) = .39 directly by using the m-over-n rule. Since G and L are mutually exclusive, we could have used Table 5.1 and applied our P(A or B) rule for mutually exclusive events: P(G or L) = P(G) + P(L) = .22 + .17 = .39. This rule should not be applied unless A and B are mutually exclusive. We will illustrate why after examining another way to relate two events.

The event "A and B" is said to occur when A and B occur simultaneously.

EXAMPLE 2

The statistics department at a university has 20 faculty members. Eleven belong to the American Statistical Association, and 8 belong to the American Society for Quality. Among these are 4 faculty who belong to both. (a) What is the probability that a faculty member picked at random from the 20 people in the department will belong to both the ASA and ASQ? (b) What is the probability that the person chosen will belong to the ASA or ASQ?

We establish some notation:

S: the person chosen belongs to the ASA
Q: the person chosen belongs to the ASQ

To answer the first question we need to find $P(S \text{ and } Q)$. From the m-over-n rule,

$$P(S \text{ and } Q) = \frac{4}{20}.$$

To answer the second question, we need to find $P(S \text{ or } Q)$. We know that $P(S) = \frac{11}{20}$ and $P(Q) = \frac{8}{20}$. But S and Q are not mutually exclusive. We will not get the correct answer by adding $P(S)$ and $P(Q)$ because the 4 people who belong to both organizations get double-counted—they contribute to the numerators of both $P(S)$ and $P(Q)$. To undo the double-count, we need to subtract $\frac{4}{20}$, the probability an individual belongs to both organizations at the same time. So

$$P(S \text{ or } Q) = \frac{11}{20} + \frac{8}{20} - \frac{4}{20} = \frac{15}{20} = .75$$

The reasoning used in this example applies generally to finding P(A or B) for any two events A and B. It is called the *Addition Rule:*

For any events A and B, $P(A \text{ or } B) = P(A) + P(B) - P(A \text{ and } B)$.

The Addition Rule applies to all settings when trying to find P(A or B). The rule that can be used when A and B are mutually exclusive, $P(A \text{ or } B) = P(A) + P(B)$, actually is just a special case of the Addition Rule. To see why, notice what happens when we use the Addition Rule to find P(G or L) in the light bulb example:

$$\begin{aligned} P(G \text{ or } L) &= P(G) + P(L) - P(G \text{ and } L) \\ &= \frac{89}{400} + \frac{68}{400} - \frac{0}{400} \\ &= \frac{157}{400} \\ &= .39 \end{aligned}$$

Whenever two events A and B are mutually exclusive, $P(A \text{ and } B) = 0$, which makes the two formulas for P(A or B) identical.

The rule for finding $P(A \text{ or } B)$ when A and B are mutually exclusive events has a natural extension to three or more events. First we need to understand that when we say *three or more given events are mutually exclusive,* we mean that every possible pair of the events is mutually exclusive, which is equivalent to saying that no two of them can occur at the same time. Suppose there are, say, four events A, B, C, and D. We would also need to understand that the event "A or B or C or D *occurs*" means that one or more of those four events occurs. The extension to four mutually exclusive events is: P(A **or** B **or** C **or** D) = P(A) + P(B) + P(C) + P(D), **whenever** A, B, C **and** D **are mutually exclusive.** The general statement would go as follows.

> If $\mathbf{A}_1, \mathbf{A}_2, \ldots, \mathbf{A}_n$ are mutually exclusive events, then $\mathbf{P}(\mathbf{A}_1 \text{ or } \mathbf{A}_2 \text{ or } \ldots \text{ or } \mathbf{A}_n) = \mathbf{P}(\mathbf{A}_1) + \mathbf{P}(\mathbf{A}_2) + \ldots + \mathbf{P}(\mathbf{A}_n)$.

So in Example 1, if a light bulb is selected at random from the 400 bulbs, the probability that the defect noted will be one of the three most common defects is $P(V \text{ or } W \text{ or } G) = .25 + .21 + .22 = .68$. Had there been two of those three events that were not mutually exclusive, the answer couldn't be determined without more information.

Table 5.1 provides the breakdown of every possible defect cause into mutually exclusive categories and the probability for each. This is called the **probability distribution** of defect causes. All probability distributions have one other characteristic.

> The sum of probabilities in a probability distribution must equal 1.

If the total is not 1, then one or more probabilities is in error or the breakdown isn't complete or two or more of the categories are not mutually exclusive.

> Let **A** be an event. The event **"A *doesn't occur,*"** denoted **A′** is called the ***complement of* A.**

We determined earlier (Table 5.1) that the probability that a light bulb would be identified as having a contact button problem is $P(B) = .12$. To find the probability that the bulb chosen won't be identified as having a contact button problem, one could apply the m-over-n rule directly:

$$
\begin{aligned}
P(B') &= \frac{\text{number of outcomes that result in other than a contact button problem}}{400} \\
&= \frac{400 - 12}{400} \\
&= \frac{400}{400} - \frac{12}{400} \\
&= 1 - P(B) \\
&= .88.
\end{aligned}
$$

It can be shown that what happened in this example—the probability of the complementary event equals one minus the probability of the event—applies to any event. Suppose that the probability that there won't be any worksite accidents next month is .3. Then the probability that there will be one or more worksite accidents next month is .7.

For any event A, $P(A') = 1 - P(A)$.

EXAMPLE 3

There will be many occasions when the probability question posed is about an event A, but it is easier to evaluate $P(A')$ and then $1 - P(A')$, rather than compute $P(A)$ directly. This is often the case when words such as *at least* appear in the question. It will be helpful to gain some experience recognizing complementary events. Here are some examples of events and their complements.

A	A′
At least one accident occurs next month	No accidents occur next month
The lot contains at least two defective items	The lot contains one or no defectives
The family has more than four children	The family has four or fewer children
Robert received at most three parking tickets	Robert received at least four tickets
Andrew is at least as tall as Susan	Andrew is shorter than Susan
The team got no wins this season	The team won at least one game

A Historical Note

In the 1930s, a Russian mathematician named A. N. Kolmogorov (1903–1987) identified three basic probability assumptions from which follow the entire structure of how we understand probability to work. Specifically, the validity of the probability rules people are taught can be demonstrated if one can make three reasonable assumptions about probability. These are called Kolmogorov's three *axioms*. In these axioms, which are stated as follows, A represents an arbitrary event and S represents the sure-to-happen event: *the outcome will turn out to be one of the possible outcomes.*

1. $P(S) = 1$
2. $P(A) \geq 0$
3. Given infinitely many mutually exclusive events $A_1, A_2, A_3, \ldots$, the probability that one of the events occurs is the sum of the individual probabilities of those events.

Since $1 = 100\%$, the first axiom says simply that there is a 100% chance that the outcome will turn out to be one of the possible outcomes. The second axiom says that a probability cannot be negative. The third axiom is similar to the rule for computing $P(A \text{ or } B)$ when A and B are mutually exclusive or to its extension when there is a finite number of mutually exclusive events, but allows for infinitely many mutually exclusive events. Again, Kolmogorov's axioms are the building blocks for probability in the sense that all the valid probability rules people use, including rules that predated Kolmogorov, can be shown to be consequences of the three axioms. A great debt is owed to him for providing a unifying theory.

Looking for Underlying Structure Can Be Helpful

It can be all too easy to use the m-over-n rule to get an impulsive and intuitive, but also incorrect, answer to a probability question. Sometimes what is needed to find the correct answer is to look for an underlying structure that shows what has to happen in order for an event to occur. We illustrate this idea with a simple game.

EXAMPLE 4

Consider a game that has two players, Player A and Player B. Each player has an ordinary six-sided die. The players toss the dice simultaneously. If the numbers shown on the resulting two die faces differ by 0, 1, or 2, Player A wins. If the faces differ by 3, 4, or 5, Player B wins. At first glance, this game may seem fair because each player has three ways to win. One might even assume that each of the differences 0, 1, 2, 3, 4, and 5 occurs with probability $\frac{1}{6}$.

But we now look more deeply to examine how these differences are produced. Player A's die is equally likely to show any of 1, 2, 3, 4, 5, or 6. The same is true for Player B's die. Refer to Table 5.2. By letting the number on Player A's die identify a row of the table and the number on Player B's die identify a column, we see that there are 36 possible outcomes to a simultaneous die toss. The table shows the difference in die faces for each of the 36 outcomes. Since it is reasonable to assume that the outcome A4,B2 has the same probability of occurring as A3,B6 or as any of the other outcomes, then each of the 36 outcomes has probability $\frac{1}{36}$. But while only two outcomes result in a difference of 5 (A6,B1 and A1, B6), there are 10 outcomes that result in a difference of 1. By using the m-over-n rule, we readily find the probability distribution for the difference in die faces (Table 5.3). It now becomes apparent that the probability Player A wins is

$$P(\text{difference} = 0 \text{ or } 1 \text{ or } 2) = \frac{6}{36} + \frac{10}{36} + \frac{8}{36} = \frac{24}{36} = \frac{2}{3}$$

TABLE 5.2 Difference Between Faces When Two Dice Are Tossed.

		Player B					
		1	2	3	4	5	6
	1	0	1	2	3	4	5
	2	1	0	1	2	3	4
	3	2	1	0	1	2	3
Player A	4	3	2	1	0	1	2
	5	4	3	2	1	0	1
	6	5	4	3	2	1	0

TABLE 5.3 Probability Distribution for the Difference Between Two Die Faces.

Difference	0	1	2	3	4	5
Probability	$\frac{6}{36}$	$\frac{10}{36}$	$\frac{8}{36}$	$\frac{6}{36}$	$\frac{4}{36}$	$\frac{2}{36}$

A distinction between the probability distributions in Table 5.1 and Table 5.3 must be noted. The distribution in Table 5.1 is based on empirical data obtained from 400 light bulbs. There are no empirical data leading to the probability distribution in Table 5.3. The distribution is a *model*, giving probabilities for the difference between die faces for idealized, perfectly balanced dice.

Rules for Dependent and Independent Events

In Example 4, consider the two events

A: Player A wins

S: Player B rolls a 6

We know that $P(A) = \frac{2}{3}$ and $P(S) = \frac{1}{6}$. The latter may be observed in either of two ways. One is that Player A's die has 6 equally likely faces, one of which is "6." The other is that when the players toss their dice simultaneously, there are 36 equally likely outcomes and on six of them, Player B rolls a 6 (refer to Table 5.2).

We now consider a question that asks us to find what is called a ***conditional probability.*** *When Player A wins, what is the probability that Player B rolls a 6?* This is a different question than "*What is the probability that Player*

TABLE 5.4 The 24 Outcomes Possible When A Occurs.

		Player B						
		1	2	3	4	5	6	
	1	0	1	2				
	2	1	0	1	2			
	3	2	1	0	1	2		
Player A	4		2	1	0	1	2	← $S \mid A$
	5			2	1	0	1	
	6				2	1	0	

B rolls a 6?" We already know that the answer to the latter question is $P(S) = \frac{1}{6}$. The former question asks us to find the probability of S subject to a condition—namely, that the event A is given to have occurred. The notation for this conditional probability is

$$P(S \mid A)$$

Before giving a general formula for how to compute a conditional probability such as $P(S \mid A)$, we introduce a line of reasoning for evaluating $P(S \mid A)$ that motivates the general formula. Table 5.4 shows those 24 outcomes that result in Player A winning. These are the only outcomes possible, once we impose the condition *Player A wins.* Since the original 36 outcomes are equally likely, then the 24 should be equally likely to be the outcome when Player A wins. Three of them result in Player B rolling a 6. By the m-over-n rule, $P(S|A) = \dfrac{3}{24} = \dfrac{1}{8}$. In other words, in the games that Player A wins, Player B has a 1 in 8 chance of rolling a 6.

Before introducing the general rule for computing conditional probabilities, we observe that $P(S \mid A)$ may be viewed as a comparative ratio of the *chance* that S occurs *with* A to the *chance* that the given event A occurs:

$$P(S \mid A) = \frac{3}{24} = \frac{\frac{3}{36}}{\frac{24}{36}} = \frac{P(S \text{ and } A)}{P(A)}$$

By having the general rule for computing conditional probabilities be defined in terms of a ratio of probabilities, we will be able to compute conditional probabilities in settings where outcomes are not as easy to enumerate as in this example.

Let E and F be two events for which $P(F) > 0$. The ***conditional probability*** of E given F, denoted $P(E \mid F)$, is defined by

$$P(E \mid F) = \frac{P(E \text{ and } F)}{P(F)}.$$

EXAMPLE 5

A college has used the high percentage of females in its student body in various advertising efforts. Recently, 60% of the students attending the college have been female. The 60% breaks down this way: 8% have been foreign females, 10% have been out-of-state—U.S. females, and 42% have been in-state females. The overall breakdown is: 70% of all students have been in-state, 20% have been out-of-state—U.S., and 10% have been foreign. What is the probability that the next foreign student admitted will be female? Compare this to the probability that the next applicant will be female.

First we establish some notation:

F applicant is female
G applicant is foreign

The answer to the first question is

$$P(F|G) = \frac{P(F \text{ and } G)}{P(G)} = \frac{.08}{.10} = .8,$$

which really is the statement that we project for the future what happened in the past—80% of the foreign students have been female. We are to compare this value to $P(F) = .6$. Notice that once the condition G is imposed, the probability that F will occur increases. When the occurrence of one event affects the probability that another event will occur, we say that the two events are *dependent*:

Two events **A** and **B** are said to be ***dependent*** when $\mathbf{P(A\,|\,B) \neq P(A)}$ or $\mathbf{P(B\,|\,A) \neq P(B)}$. **A** and **B** are said to be ***independent*** when they are not dependent.

In trying to determine whether two events are dependent by checking whether either of these inequalities hold, we point out that if one of the inequalities holds, so will the other. So only one need be checked. To illustrate, suppose that we know that $P(A\,|\,B) \neq P(A)$. Then $P(B|A) = \frac{P(B \text{ and } A)}{P(A)} = \frac{P(A|B)P(B)}{P(A)} \neq P(B)$ because $P(A\,|\,B)$ and $P(A)$ do not cancel.

In Example 5, are the events F: *applicant is female* and I: *applicant is an in-state student* dependent? Compare $P(F\,|\,I)$ to $P(F) = .6$. $P(F|I) = \frac{P(F \text{ and } I)}{P(I)} = \frac{.42}{.70} = .60 = P(F)$. So F and I are independent. Note: It also must be that $P(I\,|\,F) = P(I) = .70$. Why can it not be that $P(I\,|\,F) \neq P(I)$? When two events are independent, the occurrence of one does not affect the chance that the other will occur.

Another way of determining whether two events A and B are independent is to compare $P(A \text{ and } B)$ to $P(A)P(B)$. Again recall that

$P(A \text{ and } B) = P(A|B)P(B)$. If A and B are independent, then the equation becomes $P(A \text{ and } B) = P(A)P(B)$.

Two events A and B are independent if P(A and B) = P(A)P(B).
A and B are dependent if P(A and B) ≠ P(A)P(B).

Had we used this alternative method to determine whether I and F (Example 5) are independent or dependent, we would have computed $P(I)P(F) = .7(.6) = .42$ and compared to $P(I \text{ and } F) = .42$. Because the results are equal, I and F are independent.

EXAMPLE 6

We now consider a classic problem in sales and inventory: items sold (or items in stock) are made by different machines, or purchased from different suppliers, or fabricated at different plants, and so on, that have different product defect rates. What is the overall defect rate for the products stocked or sold? Consider that a high volume discount store sells every television set it can stock. To meet demand, it orders its sets from three manufacturing plants—Plant A, Plant B, and Plant C. The discount store is aware that 10% of the sets made at Plant A need warranty repairs, as do 7% from Plant B and 3% from Plant C. So it orders as many as possible from Plant C, then as many as possible from Plant B, and lastly from Plant A. Specifically, 60% of its sets come from Plant C, 25% from Plant B, and 15% from Plant A. What percentage of the sets it sells will need warranty repairs?

This can be viewed as a probability problem. From its stock, pick a television set at random. Ask instead, *What is the probability that the set will need warranty repairs?* We establish some notation:

W: the television set chosen will need warranty repairs
A: the set chosen was made at Plant A
B: the set chosen was made at Plant B
C: the set chosen was made at Plant C

Let's first write what we know:

Product mix	$P(A) = .15$	$P(B) = .25$	$P(C) = .60$
Repair rates for sources	$P(W\|A) = .10$	$P(W\|B) = .07$	$P(W\|C) = .03$

Refer to Figure 5.1. When a television set needs warranty repairs, exactly one of three mutually exclusive things has happened: it needs repairs and was made at Plant A, *or* it needs repairs and was made at Plant B, *or* it needs repairs and was made at Plant C. In other words, W may be decomposed into three mutually exclusive events:

$$W = (W \text{ and } A) \text{ or } (W \text{ and } B) \text{ or } (W \text{ and } C).$$

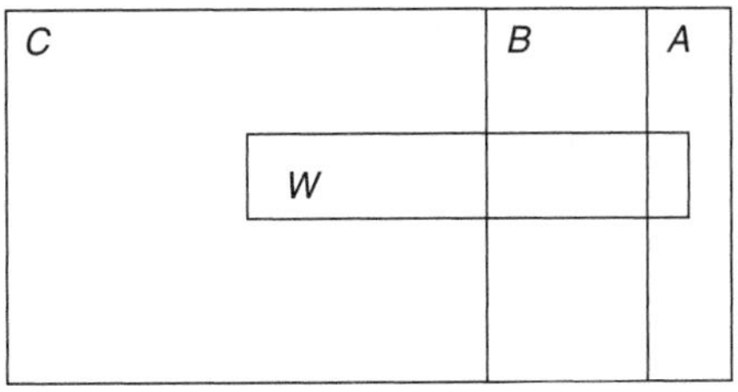

FIGURE 5.1 *W* Breaks Down into Three Mutually Exclusive Sets.

Because the three events W and A, W and B, and W and C are mutually exclusive, then

$$P(W) = P(W \text{ and } A) + P(W \text{ and } B) + P(W \text{ and } C)$$
$$= P(W|A)P(A) + P(W|B)P(B) + P(W|C)P(C).$$

What is important here is that being able to decompose W into mutually exclusive events allows us to write the overall defect rate in terms of percentages we know—the mix of sources and the source defect rates. Specific to the known source mix and warranty repair rates for the three plants, we have

$$P(W) = .10(.15) + .07(.25) + .03(.60) = .0150 + .0175 + .0180 = .0505$$

The previous equation, written for the three product suppliers A, B and C, readily extends to accommodate any number of suppliers. Suppose there are *n* suppliers. Let A_i denote the event: *product comes from Supplier i*. The extension of the equation to *n* suppliers, known in more general settings as the *Law of Total Probability*, is:

$$P(W) = P(W | A_1)P(A_1) + P(W | A_2)\ P(A_2) + \ldots + P(W | A_n)P(A_n)$$

We now turn to a second classic probability result that relates to this one.

EXAMPLE 7

In Example 6, the discount store tried to reduce the number of warranty repairs by limiting the percentage of sets it receives from Plant A. But because of the high warranty repair rate associated with Plant A's process, TV sets from Plant A still might account for the lion's share of the discount store's warranty repairs. What fraction of the television sets that are returned for warranty repairs do we expect to have been made at Plant A? To answer, we find $P(A | W)$.

The tie to Example 6 is that we will need the Law of Total Probability, $P(W) = P(W|A)P(A) + P(W|B)P(B) + P(W|C)P(C)$. Now, $P(A|W) = \frac{P(A \text{ and } W)}{P(W)}$, which becomes

$$P(A|W) = \frac{P(W|A)P(A)}{P(W|A)P(A) + P(W|B)P(B) + P(W|C)P(C)}$$

$$= \frac{.10(.15)}{.10(.15) + .07(.25) + .03(.60)}$$

$$= .297$$

So while TV sets from Plant A make up only 15% of the store's stock, they are expected to account for 30% of the store's warranty repairs.

If we let A_i denote the event: *product comes from Supplier i*, the formula for the probability that a defective item comes from Supplier i just mentioned extends to n suppliers.

Bayes' Rule:

$$P(A_i|W) = \frac{P(W|A_i)P(A_i)}{P(W|A_1)P(A_1) + P(W|A_2)P(A_2) + \cdots + P(W|A_n)P(A_n)}$$

Counting Concepts Used in Probability

Suppose an employee who is scheduling a job skill audit finds that the audit has to be on Monday, Tuesday, or Wednesday and must start at one of: 8 A.M., 9 A.M., 1 P.M., or 2 P.M. Additionally, the audit will be conducted either by Auditor A or Auditor B. In how many ways can the audit be scheduled? In getting the answer, note that there are three decisions to be made. Let's say that first we will choose a day, then pick a time slot, and finally we, or someone, will pick an auditor. The diagram in Figure 5.2 is called a *tree*. Each branch segment in the tree represents one way of performing a given task. The three highest segments of the figure represent the three ways to make the first decision, choosing a day. The next highest segments represent time slot choices. Since for each of the 3 ways to choose a day there are 4 ways to choose a time slot, then there are 12 ways to select a day and time slot together, each represented by one of the 12 segments at the second-highest level. The last choice—and lowest level of the tree—represents auditor choices. For each of the 12 choices of day and time slot, there are 2 choices of auditor. So the lowest level has 24 segments, representing the 24 ways of making all three choices together, that is, the 24 different ways to schedule an audit.

There would have been 24 segments at the lowest level even if the highest level had represented choices for auditor, the middle level choices for time of day, and the lowest level choices for day. In that case, there would have been 2 segments at the highest level, 8 at the middle level (8 = 2

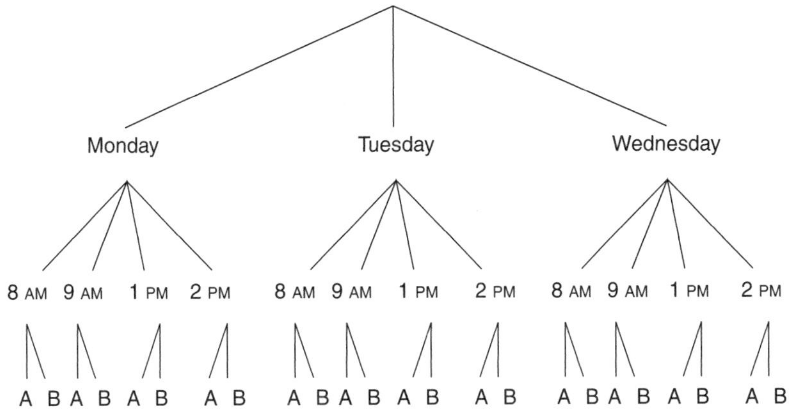

FIGURE 5.2 The Branches of This Tree Show That There are 24 ways to Schedule a Meeting with One of the Two Auditors.

times 4), and 24 at the lowest level (24 = 8 times 3). The important thing here is that $24 = 3 \times 4 \times 2$, regardless of the order in which the factors 3, 4, and 2 are multiplied. The reasoning used to find the answer 24 extends to cover general situations:

> Fundamental Counting Principle—Given k tasks to perform, if the first task can be performed in n_1 ways, and for each way of performing the first task, the second task can be performed in n_2 ways, and for each way of performing the first two tasks, the third task can be performed in n_3 ways, . . . , and for each way of performing the first $k - 1$ tasks, the kth task can be performed in n_k ways, then all k tasks can be performed in $n_1 n_2 n_3 \ldots n_k$ ways.

EXAMPLE 8

In a horse race with four horses, how many orders of finish are possible if there are no ties?

Name the horses A, B, C, and D. A brute-force approach would be to list all possible orders of finish—ABCD, ABDC, ACBD, and so on—and then count them. It is easier to use the Fundamental Counting Principle. The first place finisher could be any of 4 horses. Once first place is decided, the second place finisher could be any of the remaining 3 horses. Once the first two places are decided, the third place finisher could be either of the remaining 2 horses. And once the first three places are decided, there is only one possible horse that can come in last. So there are $4 \times 3 \times 2 \times 1 = 4! = 24$ orders of finish.

A permutation of n distinct objects is an ordered arrangement of those objects.

In Example 8, we reasoned that there must be 4! = 24 permutations of the four letters A, B, C, and D. An important thing about permutations is that order matters. The horse race outcomes ABDC and BACD use the same letters, but represent two different permutations out of the 24 possible permutations of those letters (or orders of finish of the horses). The reasoning that produced the answer 4! may be applied to find the number of permutations of any number of objects.

The number of permutations of n distinct objects is $n!$.

The application of permutations to probability can be direct. Suppose the four horses in Example 8 are of equal ability. Then all of the 4! possible orders of finish are equally likely. So the probability that the order of finish is, say, DCBA would be $\frac{1}{24}$.

EXAMPLE 9

In Section 3.2, we discussed a number of different kinds of out-of-control signals that can be used with control charts. One pattern that is no longer accepted as an out-of-control signal is the so-called *monotone* pattern for seven points in a row—either

$$\overline{X}_k < \overline{X}_{k+1} < \overline{X}_{k+2} < \overline{X}_{k+3} < \overline{X}_{k+4} < \overline{X}_{k+5} < \overline{X}_{k+6}$$

or

$$\overline{X}_k > \overline{X}_{k+1} > \overline{X}_{k+2} > \overline{X}_{k+3} > \overline{X}_{k+4} > \overline{X}_{k+5} > \overline{X}_{k+6}$$

Such a pattern appears on the $\overline{X}$- < chart in Figure 5.3. The chart indicates that the process was centered and well within the 3-sigma control limits during the period the 25 samples were taken. Yet the monotone pattern that ended on sample 20 is noticeable when taken in isolation, which is why it used to give pause. In Section 3.2 it was pointed out that in order for a pattern of successive points to qualify as a good out-of-control signal for detecting a particular change in the process, it should have only a small chance of occurring when the process is in control, and be far more likely to be observed if the change occurs.

Consider a basic out-of-control signal—a point above the upper 3-sigma limit. According to the Empirical Rule, we expect to average roughly 1.5 such points per 1,000 when a process is on target, based on normal distributions. Using the m-over-n rule, this is equivalent to saying that the probability a single point will fall above the upper 3-sigma limit is $\frac{1.5}{1000} = .0015$. When we learn a bit more about normal distributions, we will see that a more precise figure is .00135. Suppose that, unknown to us, the process mean shifts to 6.85, the upper

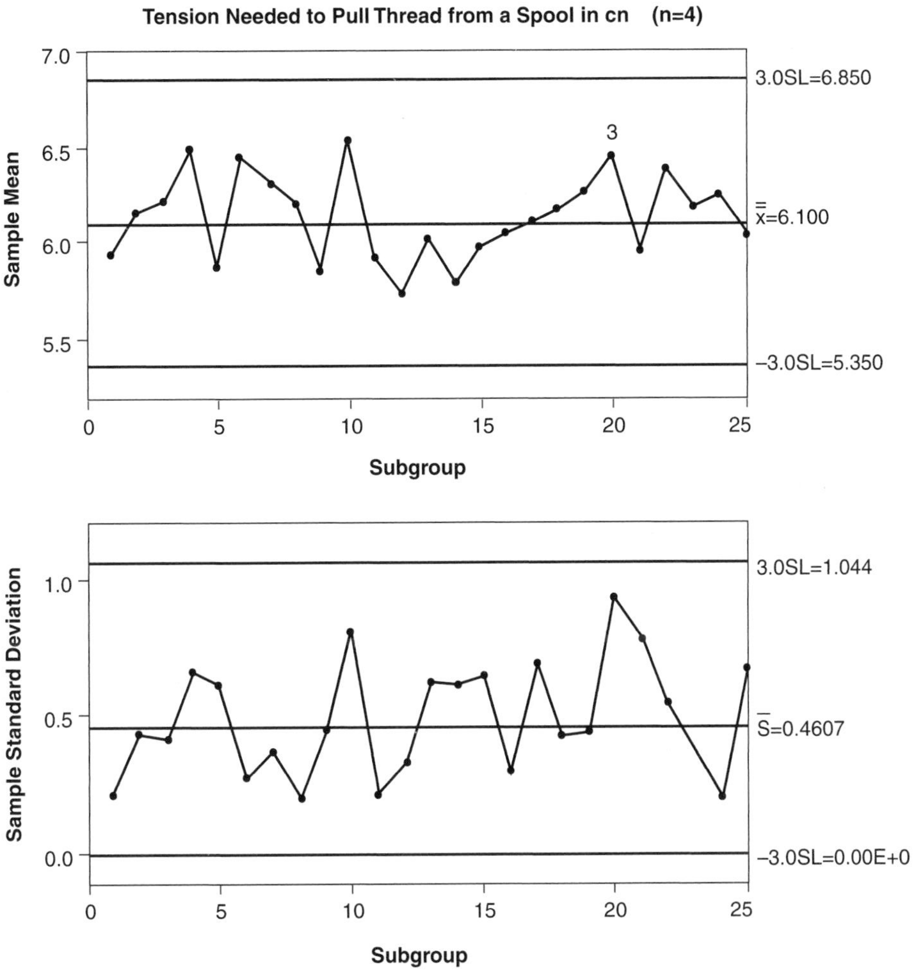

FIGURE 5.3 Monotone Increasing Pattern for Seven Points in a Row.

3-sigma limit. After the shift, the probability that a single point will fall above the upper 3-sigma limit of 6.85 is $\frac{1}{2}$.

Will a monotone increasing pattern do a similarly good job of detecting an upward shift of such magnitude? Assume there are no ties. Then the next seven sample means constitute a random sample of seven distinct values from the population of means. First we find the probability that the next seven means have the monotone increasing pattern $\overline{X}_k < \overline{X}_{k+1} < \overline{X}_{k+2} < \overline{X}_{k+3} < \overline{X}_{k+4} < \overline{X}_{k+5} < \overline{X}_{k+6}$ when the process is in control. Whatever the particular values the next seven sample means might happen to be, every possible permutation of those seven values is equally likely to be the sequence in which those means are observed. There are $7! = 5040$ permutations of the seven values, so $P(\overline{X}_k < \overline{X}_{k+1} < \overline{X}_{k+2} < \overline{X}_{k+3} < \overline{X}_{k+4} < \overline{X}_{k+5} < \overline{X}_{k+6}) =$

$\frac{1}{5{,}040} = .000198$. Now suppose the process mean shifts to 6.85.

After the shift, the probability that the next seven sample means have a monotone increasing pattern is still $\frac{1}{7!} = .000198$. Of course, one could reason that the monotone pattern would more likely begin with the last sample taken before the shift, because it is quite likely that the first sample mean after the shift will be greater than the preceding sample mean. Even so, for the complete signal to register, the means of the first six samples taken after the shift occurs would have to have a monotone increasing pattern, and the probability of that is $\frac{1}{6!} = .00139$. The reality is that a monotone pattern for seven points is simply a rare occurrence, regardless of where the process is centered. Taking such a pattern to be an out-of-control signal will lead to a lot of noise chasing.

Sometimes the focus of a problem is on the number of ordered arrangements of some, but not all, of n distinct objects. In a 10-horse race, how many ways are there for horses to finish 1-2-3? According to the Fundamental Counting Principle, the answer is $10 \times 9 \times 8 = 720$, referred to as *the number of permutations of 10 objects taken 3 at a time.* The notation for this value is ${}_{10}P_3$. That is, ${}_{10}P_3 = 10 \times 9 \times 8 = 720$. The 3 and 10 in ${}_{10}P_3$ identify, respectively, the number of factors in the product and the largest of those factors.

The number of permutations on n objects taken r at a time is

$${}_nP_r = n \times (n-1) \times (n-2) \times \cdots \times (n-r+1) = \frac{n!}{(n-r)!}$$

In some problems, order doesn't matter. Suppose we just received a box of 50 computer chips and will pick four of them at random for testing. How many different combinations (i.e., groups) of four chips are there? The order in which the four are chosen is irrelevant here. In counting parlance, we are trying to find *the number of combinations of 50 objects taken 4 at a time,* denoted ${}_{50}C_4$. The value ${}_{50}P_4 = \frac{50!}{46!} = 50 \times 49 \times 48 \times 47 = 5{,}527{,}200$ is larger than the answer because it counts the number of ways there are to choose four different computer chips *in order.* Suppose the 50 computer chips are labeled A, B, C, . . . , Z, a, b, c, . . . , x. The four chips C, W, V, and w constitute one combination, and W, V, w, and C constitute the same combination because order does not matter. If order were taken into account, each combination would generate $4! = 24$ of the 5,527,200 possible permutations of 50 objects taken 4 at a time. In the spirit of the Fundamental Counting Principle, we can generate all permutations by performing two tasks: first pick a combination of four chips, then arrange those four chips in some order. The first task can be performed in ${}_{50}C_4$ ways, and for each way of performing the first task, the second task can be performed in 24 ways. So the number of ways to perform both tasks is ${}_{50}P_4 = {}_{50}C_4 \times 4!$, so that

$$ {}_{50}C_4 = \frac{{}_{50}P_4}{4!} = \frac{5{,}527{,}200}{24} = 230{,}300. $$

The number of combinations of n objects taken r at a time is

$$ {}_nC_r = \frac{{}_nP_r}{r!} = \frac{n!}{r!(n-r)!} $$

Using the Texas Instruments TI-83 Calculator to Obtain ${}_nP_r$ and ${}_nC_r$

To evaluate ${}_{50}P_4$, turn on the calculator and key in the number 50 and hit the **MATH** key. Press the right-arrow key repeatedly so that the cursor is on **PRB.** Then hit the down-arrow key repeatedly so that the cursor in on **nPr.** Hit the **ENTER** key, then key in the number 4 and hit the **ENTER** key. The value 5527200 will show on the display. The instructions for evaluating ${}_{50}C_4$ are similar, except that the down-arrow key should be hit repeatedly to move the cursor to **nCr.**

Using Excel to Obtain ${}_nP_r$ and ${}_nC_r$

To evaluate ${}_{50}P_4$, open an Excel file and click on the button $\boldsymbol{f_x}$. Choose **Statistical** from the Function category window, scroll down the Function name window to **PERMUT,** then hit **OK.** A box with two windows will appear. In the box labeled **Number,** key in 50. In the box labeled **Number chosen,** key in 4. The value 5527200 will appear in the gray area of the box. To evaluate ${}_{50}C_4$, similarly click on the button $\boldsymbol{f_x}$, but choose **Math & Trig** from the Function category window, scroll down the Function name window to **COMBIN,** and hit **OK.** In the box that appears, key in 50 in the window labeled **Number,** and key in 4 in the window labeled **Number chosen.** The value 230300 will appear in the gray area of the box.

The Mean and Standard Deviation of a Probability Distribution

In our early work with sample data, in preparation for setting up and understanding $\overline{X}$-charts, we met two kinds of means and two kinds of standard deviations. The symbol $\overline{X}$ denoted the mean of a sample, and μ denoted the mean of the population from which the sample was taken. Similarly, S denoted the standard deviation of the sample and σ the population standard deviation. We observed that, sometimes, a population would have to be viewed as conceptual. Suppose a company makes mending plates and one of the quality variables being measured is the diameter of the screw holes drilled in the plates. The population of diameter measurements of all of the holes that *could* be drilled in plates by the existing equipment in its current state has to be conceptual. Nevertheless, we could evaluate μ and σ, at least in theory, if we only had access to that population of measurements. In this section, we see how we could evaluate μ and σ if we knew the probability distribution of the variable. In the next section we discuss probability distributions that are used to model some of the populations frequently encountered in statistical process control. Although a probability model is an idealization of the population, in practice the models used do a good job of approximating populations. Because of this, values of μ and σ from probability distribution models are the basis for locating center lines and control limits for many of the control charts discussed earlier and for solving a number of problems frequently encountered in statistical process control.

Suppose the quality variable of interest is discrete (see Section 1.2). While it is not a quality variable, suppose for sake of a simple example that the variable of interest is the difference between die faces when two fair dice are tossed. Any time the dice are tossed, the value the variable will equal must

TABLE 5.5 Probability Distribution for the Difference of Die Faces.

Difference	0	1	2	3	4	5
Probability	$\frac{6}{36}$	$\frac{10}{36}$	$\frac{8}{36}$	$\frac{6}{36}$	$\frac{4}{36}$	$\frac{2}{36}$

be one of the integers 0, 1, 2, 3, 4, or 5. In Example 4, we obtained the following probability distribution model for the difference in die faces:

Of course, real dice are not perfectly balanced like our idealized dice are, but the probabilities given by the model in Table 5.5 should be very close to the actual probabilities.

We now introduce some notation that will be helpful when working with the probability distribution. Suppose x represents a value the variable can equal. The probability the value x will occur is denoted $p(x)$. For example, when two perfectly balanced dice are tossed, the probability that the difference in faces will equal 2 is $p(2) = \frac{8}{36} = .222$.

Now imagine a population described by the distribution in Table 5.5. The population consists of zeros, ones, twos, threes, fours, and fives, where the proportion of each number in the population is equal to the corresponding probability specified by the distribution. As a tangible illustration, consider the population that consists of the 36 numbers in this list:

0 0 0 0 0 0 1 1 1 1 1 1 1 1 1 1 2 2 2 2 2 2 2 2 3 3 3 3 3 3 4 4 4 4 5 5

Instead of actually tossing the dice to obtain a difference between the two faces, it is equivalent to pick at random one of the 36 numbers in the list. The population mean for the difference in die faces would then equal

$$\begin{aligned}
\mu &= \frac{0+0+0+0+0+0+1+1+\cdots+4+5+5}{36} \\
&= \frac{6(0)+10(1)+8(2)+6(3)+4(4)+2(5)}{36} \\
&= 0\left(\frac{6}{36}\right)+1\left(\frac{10}{36}\right)+2\left(\frac{8}{36}\right)+3\left(\frac{6}{36}\right)+4\left(\frac{4}{36}\right)+5\left(\frac{2}{36}\right) \\
&= \frac{70}{36} \\
&= 1.944.
\end{aligned}$$

The important thing to notice is what is going on in the third-to-last expression. It can be shown that for any discrete variable, the population mean μ is a weighted average of the values that the variable can equal, where the weight multiplying each value is the probability it is accorded by the probability distribution. That is, the population mean value of the variable is given by $\mu = \sum xp(x)$, where the sum is taken over all possible values of the variable. The population mean also is referred to as the *mean of the distribution* or the *expected value of the variable.* And so when we actually toss two ordinary (not idealized) dice and observe the difference in faces, the number we get comes from a population that has a mean of 1.944, approximately.

The mean of the probability distribution of a discrete variable is $\mu = \sum xp(x)$, where the sum is taken over all possible values of the variable.

Along similar lines, when we know the probability distribution of a discrete quality variable, we can obtain the population standard deviation, also called the standard deviation of the probability distribution.

The standard deviation of the probability distribution of a discrete variable is $\sigma = \sqrt{\sum (x - \mu)^2 p(x)}$, where the sum is taken over all possible values of the variable.

The population standard deviation of the difference in die faces is

$$\sigma = \sqrt{\left(0 - \frac{70}{36}\right)^2 \frac{6}{36} + \left(1 - \frac{70}{36}\right)^2 \frac{10}{36} + \left(2 - \frac{70}{36}\right)^2 \frac{8}{36} + \left(3 - \frac{70}{36}\right)^2 \frac{6}{36} + \left(4 - \frac{70}{36}\right)^2 \frac{4}{36} + \left(5 - \frac{70}{36}\right)^2 \frac{2}{36}}$$

$$= \sqrt{.630 + .248 + .001 + .186 + .469 + .519}$$

$$= \sqrt{2.053}$$

$$= 1.433.$$

PROBLEMS

1. The distribution of first letters of the first names of people listed on a page in a telephone directory is: A-10, B-16, C-16, D-25, E-6, F-4, G-12, H-9, I-2, J-44, K-18, L-22, M-31, N-10, O-0, P-18, Q-0, R-45, S-29, T-17, U-0, V-4, W-14, X-0, Y-0, and Z-0. If one of these people is selected at random, what is the probability that their first name begins with
   a. a vowel (A, E, I, O, or U)?
   b. a consonant?
   c. one of the letters D, J, L, M, R, or S?
2. The age distribution of the workforce in a company is

Age	**Frequency**
25–29	8
30–34	6
35–39	11
40–44	7
45–49	0
50–54	6
55–59	7
60–64	1

An employee is selected at random. Consider the events F: *Employee is at least* 35, S: *Employee is under* 60, and T: *Employee is in his or her twenties*. Find

   a. P(F)
   b. P(F′)
   c. P(S)
   d. P(F and S)
   e. P(F or S)
   f. P(F or T)

3. Consider the events

   A there will be no accidents at the plant next month, and
   B there will be precisely one accident at the plant next month.

   a. Is it possible to have $P(A) = .6$ and $P(B) = .8$?
   b. Is it possible to have $P(A) = .6$ and $P(B) = .1$?
   c. Suppose $P(A) = .6$ and $P(B) = .08$. What is the probability that there will be at least one accident next month?

4. Suppose the probability that a company makes at least 5 defective items in a day is .03. Identify which of the following are necessarily true:

   a. P (at least 6 defectives in a day) $\leq .03$
   b. P (at most 5 defectives in a day) $= .97$
   c. P (at most 4 defectives in a day) $= .97$
   d. P (at least 6 defectives in a day) $< .03$

5. In a process that produces mildew wash, each jug has water and bleach dispensed into it. The machinery is programmed to dispense .65 gallons of water and .35 gallons of bleach. Suppose the actual amount of water dispensed is equally likely to be .63, .64, .65, .66, or .67 gallons, and the actual amount of bleach dispensed is equally likely to be .33, .34, .35, .36, or .37 gallons. If water and bleach are dispensed independently:

   a. Find the probability distribution for the number of gallons of liquid in a jug after it has been filled.
   b. What is the probability that a jug contains more than 1 gallon of liquid?

6. Two fair dice are tossed.

   a. Find the probability distribution for the sum of die faces. Find the probability that the two faces sum to
   b. 12
   c. 11 or more
   d. 7
   e. at most 5
   f. an even number

7. In Problem 5, consider the events

   E: Jug contains at least 1.02 gallons of liquid
   F: Jug contains exactly .36 gallons of bleach
   G: Jug contains exactly 1.00 gallons of liquid
   H: Jug contains exactly .96 gallons of liquid

   Find:

   a. $P(F \mid E)$
   b. $P(E \mid F)$

TABLE 5.6 Homicides by Race and Sex—Charleston, South Carolina January 1, 1996 through June 30, 2002.

Perpetrator	Victim					
	B/M	B/F	W/M	W/F	O/M	O/F
B/M	30	4	2	1		
B/F	1					
W/M	1			3		
W/F						
O/M						
O/F						
UNKNOWN	10	1			1	

(*Source:* City of Charleston Police Department, 2002.)

c. $P\,(E \mid E)$
d. $P\,(F \mid G)$
e. $P\,(G \mid F)$
f. $P\,(H \mid F)$

8. Table 5.6 breaks down homicides in Charleston, South Carolina between January 1, 1996 and June 30, 2002 by race and sex of the victim and race and sex of the perpetrator. The letters B, W, O, M, and F stand for black, white, other, male, and female, respectively. There were 54 homicides during this period.

   Find the requested probability or conditional probability

   a. *P* (victim is black | perpetrator (perp) is unknown)
   b. *P* (victim is white | perp is unknown)
   c. *P* (perp is unknown)
   d. *P* (perp is known)
   e. *P* (perp is female | perp is known)
   f. *P* (victim is female)
   g. *P* (perp is black | perp is known)
   h. *P* (victim is black)
   i. *P* (victim is black | perp is known to be black)
   j. *P* (victim is white | perp is known to be white)
   k. *P* (victim is white female | perp is known to be white male)
   l. *P* (victim is black male | perp is known to be black male)
   m. *P* (victim and perp come from the same racial group | perp is known)

9. A bowl contains 7 red, 1 white, and 2 blue marbles. Three marbles are drawn in succession without replacement. What is the probability that the third marble is red given that the first two marbles are red?

10. At a particular university, 50% of the students participate in intramural sports, 40% of all students are male, and 35% of all students are males who participate in intramural sports. If a student is selected at random from the student body, what is the probability that the student is male given that the student participates in intramural sports?
11. In Problem 7, are the events F and G independent? The events F and E?
12. In Problem 10, are the events *being male* and *participating in intramural sports* independent?
13. Suppose two fair dice are tossed. For computing probabilities, it will be helpful to pretend that one is red and one is white. Consider the events:

    T sum of faces is 10 or more
    S red die shows a 6
    E die faces are equal

Are the following pairs of events independent or dependent?

   a. T and S
   b. T and E
   c. S and E

14. A company makes five models of lawnmowers—A, B, C, D, and E. The Table 5.7 shows both the percentage of the total number of sales attributable to each model and the percentage of purchases of the model that result in customer claims.
    a. Pooling all sales together, what proportion of sales results in customer claims?
    b. If a customer claim is chosen randomly from the list of all customer claims, what is the probability that the mower purchased was model E?
15. A fair die is tossed as many times as necessary until a 4 occurs. What is the probability that at least two tosses result?
16. In Problem 8, consider only the 42 homicides for which the perpetrator is known. Would you say that the events V_f: *victim is female* and P_w: *perpetrator is white* are independent?
17. Ken, Jo Ann, and Mike are the three agents at a real estate office. Ken makes 25% of all the sales, Jo Ann 40%, and Mike 35%. After each

TABLE 5.7 Sale and Complaint Data.

Model	A	B	C	D	E
Proportion of all sales	40%	25%	21%	10%	4%
Customer complaints for the model	.5%	1%	2%	4%	9%

sale, the client fills out a satisfaction survey. Suppose 40% of Ken's customers say they are "very pleased" with their treatment by the sales staff, 30% of Jo Ann's customers do likewise, and 25% of Mike's so indicate. What proportion of customers who indicated that they were "very pleased" had Ken as their agent?

18. How many binary code words of length 4 are there? That is, how many four-character words can be made, where each character is either 0 or 1? First get the answer by constructing a tree with four levels of branch segments. Then compare to the answer obtained by the Fundamental Counting Principle.

19. Suppose the upper 3-sigma limit of an $\overline{X}$-chart is 6.85. What is the probability that the out-of-control signal *point above* 6.85 will occur at least once in the next seven samples taken after the process mean shifts to 6.85?

20. Find the mean and standard deviation for the result of one toss of a fair die. Use the probability distribution for $p(x) = \frac{1}{6}$ for $x = 1,2,3,4,5,6$.

5.2 SOME IMPORTANT PROBABILITY DISTRIBUTIONS

In Section 1.2, we introduced the concepts of discrete and continuous variables. Later we learned how to use various kinds of control charts, mostly of the Shewhart type, to track or monitor the typical kinds of variables used to measure quality characteristics. We also examined various capability indices. In order to understand how to apply such charts, what is accomplished by 3-sigma control limits, and what kinds of capability index values are *good*, we did not need to know much about probability. We did need one key statistics result. The Empirical Rule, sometimes called the 68/95/99.7 percent rule, provided enough of a foundation to allow us to understand and begin to use control charts and capability indices.

In this section we examine some of the important probability distribution models associated with quality variables. Familiarity with these distributions will provide a deeper understanding of how control-chart sensitivity to process change depends upon the magnitude of the change. The distributions also are helpful for answering sample size and other questions frequently encountered in the quality arena.

Several of these probability distributions have formulas that involve notation for combinations. In this setting, we use the more popular symbol $\binom{n}{r}$ to denote the number of combinations of n things taken r at a time, rather than the symbol ${}_nC_r$, which is more frequently seen in counting problems.

Hypergeometric Distributions

One distribution frequently encountered in sampı *hypergeometric distribution.* Suppose there are N items n lots is the lot, that D of them have a particular characteristic (e.g., kind in a fective), and that the other $N - D$ items do not have that em is de- We will select a random sample n of the N items without reteristic. By this we mean that every combination of n different items has ment. probability of being the n items we select. The variable of interest me the number of items with the characteristic in the sample. This varis has what is called a hypergeometric distribution. Familiarity with hype geometric distributions will allow us to answer a question such as, *What is the probability that in the sample there will be exactly k items with the characteristic?*

For completeness, we observe that N must be a positive integer, that D must be one of the integers $0, 1, 2, \ldots, N$ and that the sample size n is one of the positive integers $1, 2, 3, \ldots, N$. Let X denote the number of items with the characteristic in the sample. At first glance, one might assume that the range of allowable values for X must be integers between 0 (no items in the sample have the characteristic) and n (every item in the sample has the characteristic). However, if the sample size n exceeds $N - D$, then the sample would have to contain at least $n - (N - D)$ items with the characteristic. And if $D < n$, the maximum possible value of X would have to be D. Summarizing the setting we have:

N	the number of items in the lot
D	the number of items with the characteristic in the lot
n	the sample size
X	the number of items with the characteristic in the sample

Range of values for X: integers from $\max\{0, n - (N - D)\}$ to $\min\{n, D\}$

Let k be a value in the range of values of X. To find $P(X = k)$, the probability that the sample contains exactly k items with the characteristic, we use the m-over-n rule. Since all samples are equally likely, the denominator will be the total possible number of samples, $\binom{N}{n}$. The numerator will be the number of possible samples that contain exactly k items with the characteristic. Since there are $\binom{D}{k}$ ways to choose k items with the characteristic, and for each such way, the other $n - k$ sample items that don't have the characteristic may be chosen in $\binom{N - D}{n - k}$ ways, the number of possible samples containing x items is $\binom{D}{k}\binom{N-D}{n-k}$. So the probability that there will

be exactl ns with the characteristic in the sample is

$$= k) = \frac{\binom{D}{k}\binom{N-D}{n-k}}{\binom{N}{n}}, \textit{ when k is in the range of values for X.}$$

The mean and standard deviation of the hypergeometric distribution are $\mu = n\frac{D}{N}$ and $\sigma = \sqrt{n\frac{D}{N}\left(1 - \frac{D}{N}\right)\frac{N-n}{N-1}}$. Sometimes μ and σ are referred to as the expected value of X and the standard deviation of X, respectively.

EXAMPLE 1

From a shipment of 20 auto batteries, 8 are selected at random for testing. If at most one battery in the sample is defective, the shipment will be accepted. Otherwise the shipment will not be accepted. (a) If 4 batteries in the shipment are defective, what is the probability that the shipment will be accepted? (b) What is the expected number of defective batteries in the sample?

The variable being observed is X: *the number of defective batteries in the sample.* X has a hypergeometric distribution with N = 20, D = 4, and n = 8.

a. There are two mutually exclusive ways the shipment will be accepted—either there are no defectives or there is exactly one defective. So

$$\begin{aligned} P(\text{shipment is accepted}) &= P(X = 0) + P(X = 1) \\ &= \frac{\binom{4}{0}\binom{16}{8}}{\binom{20}{8}} + \frac{\binom{4}{1}\binom{16}{7}}{\binom{20}{8}} \\ &= \frac{1 \times 12{,}870}{125{,}970} + \frac{4 \times 11{,}440}{125{,}970} \\ &= .102 + .363 \\ &= .465 \end{aligned}$$

b. The expected number of batteries in the sample is

$$\mu = 8 \times \frac{4}{20} = 1.6.$$

Binomial Distributions

Another distribution that has frequent applications in sampling is the *binomial distribution.* The binomial distribution model also is the basis for the center line and 3-sigma limits on *np*-charts (Section 4.2). As with hypergeometric distributions, each item in the sample either has a particular characteristic (e.g., it does not conform to specifications) or does not have the characteristic. As before, we let X denote the total number of items with the characteristic in the sample and n denote the sample size. What is different is that every item in the sample is viewed as having the same probability p of having the characteristic, *independent* of whether other items in the sample have the characteristic. A typical example would be the number of defective items in a sample of n items produced by an assembly line for which the process defect rate is p.

Interestingly, even though it is natural in many applications of binomial distributions to let p denote the probability that a single item is defective, sometimes p is referred to as the *success probability*. In manufacturing there is a focus on minimizing defects. In many applications outside this arena, it is natural to view an item with the characteristic as a success, and an item not having the characteristic as a failure. Suppose someone playing roulette makes n bets in a row on red. Viewing a bet as an *item*, the number of wins in n bets can be viewed as the total number of items possessing the *win* characteristic in a sample of n items. If n subjects who contracted hepatitis B are given an experimental drug, the total number of subjects cured would be another example.

The key distinction between the hypergeometric and binomial models is the independence requirement of the binomial model. Consider a lot of 20 items, four of which are defective. The defect rate for the lot is $\frac{4}{20} = 20\%$, and the first item taken from the lot has probability $\frac{4}{20} = .2$ of being defective. The second item taken for the sample also has probability $\frac{4}{20}$ of being defective, because before any sample items are drawn, each of the original 20 items has the same chance to be the second item chosen. Similarly, each sample item has probability $\frac{4}{20}$ of being defective. What we cannot say is that every sample item has probability $\frac{4}{20}$ of being defective *independent of whether other items are defective*. Suppose we know the first item chosen is defective. Because the sampling is without replacement, the conditional probability that the second item is defective becomes $\frac{3}{19} = .158$. If the first item chosen is not defective, the conditional probability that the second item is defective becomes $\frac{4}{19} = .211$. In the binomial model, every sample item has the same probability p of being defective, regardless of whether any other sample items are known to be defective, which makes it an appropriate model for the number of defectives in a sample of n items from a production stream.

Summarizing this setting we have:

n	the sample size
p	the probability each single item has the characteristic

Whether an item has the characteristic is independent of whether other sample items do.

X the number of items with the characteristic in the sample

Range of values of X: $0, 1, 2, \ldots, n$

Under the independence condition just described, X is said to have a *binomial distribution*.

EXAMPLE 2

In which of the following settings does the variable have a binomial distribution and in which does it have a hypergeometric distribution?

a. X: The number of boys in the next 10 births at Roper Hospital

b. Y: The total number of 5's rolled in 8 tosses of a fair die

c. W: The number of defective computer chips in a sample of 50 chips from a production stream having a .2% rate for defectives

d. V: The number of hearts in a five-card hand from an ordinary 52-card deck

a. A baby's gender should be independent of other babies' genders. Taking the probability that a baby is a boy to be $\frac{1}{2}$, then X has a binomial distribution with $n = 10$ and $p = .5$.

b. The outcomes on different tosses are independent. So Y has a binomial distribution with $n = 8$ and $p = \frac{1}{6}$.

c. It is reasonable to assume that whether a chip is defective is independent of whether other chips are. So W has a binomial distribution with $n = 50$ and $p = .002$.

d. Since sampling is without replacement from a deck having 13 hearts, V has a hypergeometric distribution with $N = 52$, $D = 13$, and $n = 5$.

As noted, the number of 5's in 8 tosses of a fair die has a binomial distribution with $n = 8$ and $p = \frac{1}{6}$. We now find $P(X = 2)$, the probability of getting exactly two 5's in 8 tosses. Doing this will make it easier to understand the general formula for $P(X = k)$ when X has a binomial distribution with given values for n and p.

Take the view that getting a 5 on a toss is a success and any other number is a failure. Let S_i denote a success on the ith toss and F_i denote a failure on the ith toss. One way to have two 5's is to have the 5's be on the first two tosses, followed by six non-fives—that is, to have the event

$$S_1 \text{ and } S_2 \text{ and } F_3 \text{ and } F_4 \text{ and } F_5 \text{ and } F_6 \text{ and } F_7 \text{ and } F_8$$

occur. Because outcomes on different tosses are independent, the probability of getting the two successes and six failures in this particular order is $(\frac{1}{6})^2(\frac{5}{6})^6$. Another way of getting the two 5's is for the event

$$S_1 \text{ and } F_2 \text{ and } F_3 \text{ and } F_4 \text{ and } S_5 \text{ and } F_6 \text{ and } F_7 \text{ and } F_8$$

to occur. The probability of this event is $\frac{1}{6} \times \frac{5}{6} \times \frac{5}{6} \times \frac{5}{6} \times \frac{1}{6} \times \frac{5}{6} \times \frac{5}{6} \times \frac{5}{6} = \left(\frac{1}{6}\right)^2 \left(\frac{5}{6}\right)^6$, as before. Likewise, any given sequence of 2 successes mixed with 6 failures has probability $\left(\frac{1}{6}\right)^2 \left(\frac{5}{6}\right)^6$. Since there are as many sequences of 2 successes and 6 failures as there are ways to choose which two of the eight tosses will be the successes, the probability of getting two 5's in eight tosses is $P(X = 2) = \binom{8}{2}\left(\frac{1}{6}\right)^2 \left(\frac{5}{6}\right)^6$. This reasoning may be used to obtain the general formula for binomial distributions:

> Let a variable X have a binomial distribution with n and p as previously described. Then
>
> $$P(X = k) = \binom{n}{k}(p)^k(1 - p)^{n-k}, \textit{ for } k = 0, 1, 2, \ldots n.$$

It can be shown that the mean and standard deviation of a binomial distribution are $\mu = np$ and $\sigma = \sqrt{np(1 - p)}$, respectively.

EXAMPLE 3

In a study done at a large hospital, 15% of the patients experienced errors in having drugs administered to them. Typical errors included incorrect dosages, drugs being given at the wrong time, and even the wrong drug being administered. Out of a sample of 25 patients at the hospital, what is the probability that two or fewer experience drug administration errors? What is the expected number of patients who will experience such errors? What is the standard deviation of the number of such patients?

Assuming that patients experience errors independently of one another, then the number of patients experiencing such errors would have a binomial distribution with $n = 25$ and $p = .15$. The probability that two or fewer patients experience drug administration errors is

$$P(X \leq 2) = P(X = 0) + P(X = 1) + P(X = 2)$$

$$= \binom{25}{0}.15^0.85^{25} + \binom{25}{1}.15^1.85^{24} + \binom{25}{2}.15^2.85^{23}$$

$$= .0172 + .0759 + .1607$$

$$= .2538.$$

The expected value of the number of patients who experience drug administration errors is $\mu = np = 25 \times .15 = 3.75$. The standard deviation is $\sigma = \sqrt{np(1-p)} = \sqrt{25 \times .15 \times .85} = 1.79$.

Using Tables or MINITAB to Find Cumulative Binomial Probabilities

In Example 3, instead of computing each of $P(X = 0)$, $P(X = 1)$, and $P(X = 2)$ directly from the formula and adding to get the cumulative probability, we could have used Table V in the Appendix. Entering Table V using $n = 25$ and $p = .15$, we see that the probability of having two or fewer patients experience drug administration errors is .254. Had we wanted the probability that exactly 2 of 25 patients would experience such problems instead of the probability of 2 or fewer, we still could use Table V. Observe that $P(X = 2) = P(X \leq 2) - P(X \leq 1) = .254 - .093 = .161$.

If the given value of n or p is not included in the Table V, the reader can use a statistics package such as MINITAB to find binomial probabilities. Although this was discussed briefly at the end of Section 4.2, we illustrate by finding the probability there will be a total of 4 or fewer 5's when a fair die is tossed 41 times. The number of 5's in 41 tosses of a fair die has a binomial distribution with $n = 41$ and $p = \frac{1}{6}$. Enter a 4 in the top cell of C1. From the menu, choose **Calc,** then **Probability Distributions,** then **Binomial . . . .** A window will appear. Click the window labeled **Cumulative probability.** In the window labeled **Number of trials:,** enter the value 41. In the window labeled **Probability of success:,** enter .166667. In the window labeled **Input column:,** enter C1. In the window labeled **Optional storage:,** enter c2, say, as the column into which the cumulative probability will be placed. Click **OK.** The answer .164033 will appear at the top of C2.

EXAMPLE 4

A customer has just received a shipment of 4,000 widgets from a manufacturer that maintains a p-chart on the proportion of widgets that don't meet specifications. This particular shipment is one of many that were produced during a period when the manufacturer's p-charts show that the process was in control with a widget nonconforming rate of $\bar{p} = .01$. The customer finds the 1% nonconforming rate to be acceptable. Nevertheless, the customer maintains an inspection plan whereby 100 widgets are selected from a shipment and the shipment is rejected if three or more items fail to meet specs. The customer's reasoning is that the sample should be representative of the shipment, and that three or more nonconforming items in the sample is an indication that the rest of the shipment has a high proportion of nonconforming items.

Suppose the sample contains three widgets that don't meet specs. If a new sample of 100 widgets is drawn from the remaining 3,900 widgets, what is the probability that the new sample contains exactly k widgets that don't meet specs, $k = 0, 1, 2, \ldots, 100$? What does the answer say about using a sampling plan for making shipment acceptance decisions?

The process that made the items was in a state of control with a nonconforming rate of 1%, and whether one item meets specs is independent of whether other items meet specs. So the probability that the second sample will contain exactly k items that don't meet specs is $\binom{100}{k}.01^k.99^{100-k}$, the same as the probability that there will be k nonconforming widgets in any sample of 100 widgets taken from a process that is in control with a defect rate of 1%. Because the vendor's process is in control, sampling to determine shipment acceptability is pointless. The number of defective widgets in the rest of the shipment is independent of how many defective widgets are in the sample.

EXAMPLE 5

It was pointed out in Section 3.2 that a pattern of points often used with an $\overline{X}$-chart to detect a possible upward shift in the process mean is 8 points in a row above the center line. Another signal sometimes used is 10 or more points above the center line out of 11 consecutive points. Find the probability of each of these signals when the process is in control.

There are several natural ways to approach the first question. Based on normal distributions, the probability that one sample mean falls above the center line when the process is in control is .5. Because samples are independent, whether one sample mean falls above the center line is independent of whether other sample means do. If we let A_i denote the event that the ith sample mean falls above the center line, then the probability 8 points in a row fall above the center line is

$$\begin{aligned}
&P(A_1 \text{ and } A_2 \text{ and } A_3 \text{ and } A_4 \text{ and } A_5 \text{ and } A_6 \text{ and } A_7 \text{ and } A_8) \\
&\quad = P(A_1)P(A_2)P(A_3)P(A_4)P(A_5)P(A_6)P(A_7)P(A_8) \\
&\quad = .5^8 \\
&\quad = .0039.
\end{aligned}$$

As another approach, let X be the number of points above the center line out of 8 consecutive points. Then X has a binomial distribution with $n = 8$ and $p = .5$. The probability that all eight points will fall above the center line is

$$P(X = 8) = \binom{8}{8} .5^8 .5^0 = 1 \times .5^8 \times 1 = .0039.$$

Also, one could use Table V (in the Appendix) with $n = 8$ and $p = .5$: $P(X = 8) = 1.000 - .996 = .004$.

To answer the second question, let Y be the number of points that fall above the center line out of 11 consecutive points. When the process is in control, Y has a binomial distribution with $n = 11$ and $p = .5$. In order to use Table V, we consider complementary events. The probability that at least 10 points fall above the center line equals 1 − P(9 or fewer points fall above the center line):

$$\begin{aligned} P(Y \geq 10) &= 1 - P(Y \leq 9) \\ &= 1 - .994 \\ &= .006. \end{aligned}$$

Binomial Distributions Approximate Hypergeometric Distributions When *n* Is Substantially Less Than *N*

Recall that in Example 1, a lot of 20 auto batteries had four defectives. The defect rate for the lot was $\frac{4}{20} = 20\%$, and each of the 8 batteries in the sample had a 20% chance of being defective. But because we could not say that every battery in the sample had probability $\frac{4}{20}$ of being defective *independent of whether other batteries were defective*, we observed that X, the number of defective batteries in the sample, had a hypergeometric distribution with $N = 20$, $D = 4$, and $n = 8$, not a binomial distribution with $n = 8$ and $p = .2$.

As observed earlier, the conditional probability that the next battery chosen to be in the sample will be defective depends on which batteries have already been sampled. Indeed, computed probabilities using both distributions can produce glaringly different answers. The probability of having exactly two defective batteries in the sample is

$P(X = 2) = \dfrac{\binom{4}{2}\binom{16}{6}}{\binom{20}{8}} = \dfrac{6 \times 8{,}008}{125{,}970} = .381$. The incorrect binomial probability is much smaller: $\binom{8}{2}.2^2.8^6 = 28 \times .04 \times .2621 = .294$. But now suppose we are sampling $n = 8$ batteries from a lot of $N = 200$ batteries, with the same per-lot defective rate of 20%. There are 40 defective batteries in the lot, and because sampling is without replacement, if the first battery chosen is defective, the conditional probability that the second item is defective becomes $\frac{39}{199} = .196$, very close to the 20% per-lot defective rate. The correct answer, based on a hypergeometric distribution with $N = 200$, $D = 40$, and $n = 8$, is $P(X = 2) = \frac{\binom{40}{2}\binom{160}{6}}{\binom{200}{8}} = \frac{780 \times 2.119 \times 10^{10}}{5.510 \times 10^{13}} = .300$, very close to the binomial distribution approximation just obtained, .294.

Because the sample size n is much smaller than the population size N, the unsampled part of the lot has nearly the same defect rate as the lot itself, regardless of which batteries may have been sampled. **When n is substantially less than N, binomial probabilities may be used to approximate hypergeometric distribution probabilities.** For a given proportion of defectives in the sample, the larger the value of N, the better the binomial approximation gets. For example, if $N = 2{,}000$, $D = 400$, and $n = 8$, the correct answer is

$$P(X = 2) = \frac{\binom{400}{2}\binom{1{,}600}{6}}{\binom{2{,}000}{8}} = \frac{79{,}800 \times 2.308 \times 10^{16}}{6.261 \times 10^{21}} = .294,$$ which equals the binomial probability approximation to three decimal places.

Poisson Distributions

The Poisson distribution model is the basis for the center line and 3-sigma limits on c-charts (Section 4.3). As with binomial distributions, the variable of interest is the number of times a certain characteristic is observed in a sample—for example, the number of defects in the sample. A Poisson distribution model works well when the sample provides lots of opportunities for the characteristic to occur, but the probability that any single opportunity produces the characteristic is small. To help illustrate when Poisson distributions may be applied, we first consider several examples that seem quite different. We then make an observation that shows why the examples really are alike.

Suppose 1,100 spot welds are inspected daily and X, the variable of interest, is the number of nonconforming welds observed per day. Lately the mean rate for nonconforming welds has been 6.7 per day. There are lots of opportunities to have nonconformities, but the probability that any one opportunity will yield a nonconformity is small.

As a second example, suppose a sample consists of 5,000 meters of insulated copper wire. The variable of interest is X, the number of insulation holes in a sample. Recently, the mean rate has been 1.5 holes per 5,000 meters. If the wire in a sample were viewed as consisting of 500,000 one-centimeter segments, there are lots of segments in which holes can occur, but the probability than any one segment will have a hole is small.

The number of murders in Montana in a year is another example. Suppose that Montana has averaged 23 murders/yr in recent years. Consider a year as being composed of a succession of one-minute time intervals. Each such time interval can yield one or more murders, but the probability that a murder will occur in a single one-minute interval is small.

The variable of interest could also be for example, the number of chocolate chips in a cookie, where lately, the cookies made by a company have averaged 3.2 chocolate chips per cookie. Take the view that each cookie is formed by dropping into a pile a given number of droplets of batter, each droplet having the volume of one chocolate chip. Each droplet either is a chocolate chip or it is not. Thus the making of a cookie presents lots of opportunities for chocolate chips to occur, but the probability that a chocolate chip occurs on any single opportunity is small.

In order to understand when Poisson distributions may be applied, it will be helpful to consider a frame of reference in which the variable of interest is the total number of times an event of interest occurs during a time interval of a given length. The number of murders in Montana in a year fits this setting. So do the other examples if modified slightly. Suppose a machine inspects all 1,100 spot welds at a rate of one weld per second. Or a machine detects insulation holes in wire by examining a 5,000-meter sample of wire linearly at a rate of 57 cm/sec. Or a cookie is made by dropping the chocolate-chip-sized droplets of batter into a pile at a rate of 1 droplet per .1 seconds. Then each variable can be recast as the total number of times an event is observed during a time interval of a given length. It is in this setting that we discuss the assumptions needed in order to apply a Poisson distribution.

Let X denote the number of times an event occurs during a time interval of a given length. Suppose properties 1, 2, and 3 all hold:

1. During each very-short-time subinterval, the probability that exactly one event occurs is proportional, approximately, to the length of the subinterval.
2. During each very-short-time subinterval, the probability that two or more events occur is negligible.
3. The number of events that occur in nonoverlapping time intervals are independent.

Before deciding to use a Poisson distribution for any variable, the reader should make a judgment that the properties are reasonable for that variable. Suppose X is the number of murders in Montana in a year. We discuss the reasonableness of points 1, 2, and 3 under the supposition that Montana averages roughly 23 murders/year.

Consider a time subinterval of the year having short length, say one second. The probability that exactly one murder will occur during a one-second time subinterval has to be small. Suppose we double the length of the subinterval. The probability that there will be exactly one murder during any two-second period still must be small, but ought to be approximately twice the probability that there will be exactly one murder in any one-second time interval, because we have *doubled the opportunity* for a murder to occur. Similarly, the probability of exactly one murder occurring during a time interval of length 2.3 seconds should be approximately 2.3 times the probability that exactly one murder occurs during a one-second time interval. Assumption 1 is reasonable here. That is, it is reasonable to assume that there is some proportionality constant $\mu > 0$ for which

$$P(\text{exactly one murder occurs during a short time subinterval of length } t) \approx \mu t$$

Now consider assumption 2. However small may be the probability that exactly one murder will occur during a time subinterval of length one second, the probability that two or more murders will occur during that short a period of time would have to pale in comparison—that is, be negligible.

According to assumption 3, the number of murders that occur between, say, 3:00 P.M. on June 17 and 5:32 P.M. on June 29 will neither influence nor be influenced by the number of murders that occur during any nonoverlapping time period, for example, between 1:15 A.M., October 9 and 1:16 A.M., December 12. As long as the time intervals don't overlap, independence is a reasonable assumption. Therefore, so all three assumptions are reasonable to make when the annual murder rate is only 23/yr. Accordingly, a Poisson distribution will be a good model for the number of murders that occur during a given time interval.

In order to compute probabilities for a Poisson distribution, one need know only two things—the length of the time interval and the value of the proportionality constant μ. Fortunately in many applications, μ is known or known approximately. It can be shown that μ is the average number of events that occur in one unit of time. We now give the formula for computing Poisson probabilities and illustrate with some examples.

Let X denote the number of times a particular kind of event occurs during a time interval of length L. Let $\mu > 0$ denote the expected number of events to occur in a time interval of length one. When conditions 1, 2, and 3 all hold, it can be shown that $P(X = k) = \dfrac{(\mu L)^k e^{-\mu L}}{k!}$ for $k = 0,1,2,\ldots$.

The mean of the probability distribution is $\mu_C = \mu L$ and the standard deviation is $\sqrt{\mu_C}$.

Note that the formula $\mu_C = \mu L$ is consistent with the fact that the expected number of events during a time interval of length $L=1$ is $\mu_C = \mu L = \mu \times 1$.

EXAMPLE 6

A sample consists of 5,000 meters of insulated copper wire. The variable of interest is X, the number of insulation holes in a sample. Recently, the mean rate has been 1.5 holes per 5,000 meters. Assuming the process stays in control, what is the probability that a single sample will have three holes or less? What is the standard deviation of the number of holes in a sample? What is the probability that a short sample of 2,000 meters of wire will have three holes or less? What is the standard deviation of the number of holes in such a sample?

To answer the first question, define one unit length of wire to be 5,000 meters. Then $\mu = 1.5$ holes/unit. The answer to the first question is

$$\begin{aligned} P(X \leq 3) &= P(X = 0) + P(X = 1) + P(X = 2) + P(X = 3) \\ &= \frac{1.5^0 e^{-1.5}}{0!} + \frac{1.5^1 e^{-1.5}}{1!} + \frac{1.5^2 e^{-1.5}}{2!} + \frac{1.5^3 e^{-1.5}}{3!} \\ &= .2231 + .3347 + .2510 + .1255 \\ &= .9343 \end{aligned}$$

Alternatively, cumulative Poisson probabilities may be found in Table VI in the Appendix. The table shows that when $\mu_c = 1.5$ holes/unit, $P(X \leq 3) = .934$.

The standard deviation of the number of holes in a sample is $\sqrt{1.5} = 1.225$ holes/unit (i.e., 1.225 holes/5,000m).

Let Y denote the number of holes in a 2,000-meter sample of wire. To answer probability questions about Y, first we need to find the expected number of holes in a 2,000-meter sample. A two thousand meter sample is 40% of full sample length (40% = 2,000/5,000). That is, $L = .4$. According to the Poisson probability formula just given, the expected number of holes in such a sample is $\mu_C = \mu L = 1.5 \times .4 = .60$. The probability of three or fewer holes when .6 are expected may be found from Table VI in the Appendix—the answer is .997. Alternatively, one may use direct computation:

$$\begin{aligned} P(Y \leq 3) &= P(Y = 0) + P(Y = 1) + P(Y = 2) + P(Y = 3) \\ &= \frac{.6^0 e^{-.6}}{0!} + \frac{.6^1 e^{-.6}}{1!} + \frac{.6^2 e^{-.6}}{2!} + \frac{.6^3 e^{-.6}}{3!} \\ &= .5488 + .3293 + .0988 + .0198 \\ &= .9967 \end{aligned}$$

The standard deviation of Y is $\sqrt{.6} = .775$ holes/2,000 m.

EXAMPLE 7

Suppose 1,100 spot welds are inspected daily and values of X, the number of nonconforming welds observed per day, are plotted on a c-chart (Section 4.3). The mean rate for nonconforming welds has lately been 6.7 per day. The probability distribution of X should be approximately Poisson with $\mu_C = 6.7$. The company sets the center line at $\mu_C = 6.7$, but rather than use 3-sigma limits of $\mu_C \pm 3\sqrt{\mu_C}$, the company decides to use *exact probability limits* instead. As described in Section 4.2, lower and upper probability limits, LPL and UPL, are of the form *integer* + .5, chosen so that the probability a point from an in-control process will fall outside a given probability limit is as close to .005 as possible, but not greater than .005. From Table VI in the Appendix, when $\mu_C = 6.7$, we see that .005 is between $.001 = P(X \leq 0)$ and $.009 = P(X \leq 1)$. Thus, LPL = 0.5. Also, .995 is between $.991 = P(X \leq 13)$ and $.996 = P(X \leq 14)$. So $P(X \geq 14) = 1 - .991 = .009$, while $P(X \geq 15) = 1 - .996 = .004$. Hence, UPL = 14.5.

When Poisson Probabilities May Be Used to Approximate Binomial and Hypergeometric Probabilities

Suppose a variable X has a binomial distribution where n is large and p is small. A Poisson distribution having mean $\mu_C = np$ may be used to approximate $P(X = k)$.

EXAMPLE 8

A worker can check one out a needed tool from the company's tool crib. Tools are supposed to be returned at the end of the worker's shift. Because missing tools have been a problem, the company decides to establish a control chart for the number of missing tools. Each tool has an identifying bar code. Since bar codes are numerically sequential, a random sample of 25 tools is determined by having a statistics package generate 25 random bar codes. This is done each Wednesday. Records for the 25 tools identified by the bar codes are checked to determine if each tool is accounted for. Any tool not in the crib and not checked out during the current shift is designated as missing. The variable being charted is X, the number of missing tools in a sample. After 28 samples, a total of 14 tools are identified as missing, and a time series plot shows that the process is stable. What kinds of probability distribution models are appropriate for X? What kinds of control charts are appropriate?

Certainly a binomial model is appropriate, where $n = 25$ and p has to be estimated. Since the sample size is always the same, one chart to consider is the *np*-chart (see Section 4.2). The estimate of p, the population proportion of tools missing, is $\bar{p} = \frac{14}{25 \times 28} = .02$. The value for the center line of the *np*-chart is $n\bar{p} = 25 \times .02 = .5$, which estimates the expected number of missing tools per sample.

As pointed out in Section 4.2, it is reasonable to use 3-sigma control limits when $n\bar{p} \geq 5$ and $n(1 - \bar{p}) \geq 5$. Since the first condition is not met, exact probability limits should be used. Relevant cumulative binomial probabilities from Table V in the Appendix are shown in Table 5.8. The table indicates that there is no lower probability limit and that UPL = 3.5.

Because each sample provides 25 opportunities to have a missing tool but each opportunity has only a small chance of doing so, a Poisson model also provides a reasonable approximation, which makes a *c*-chart an appropriate tool (see Section 4.3). All that is needed is an estimate of μ_C, the expected number of missing tools in one sample. This is $\mu_C \approx \bar{C} = \frac{14}{28} = .5$, which locates the center line on the *c*-chart. The values for the center lines on the two charts have to be the equal because

$$\bar{C} = \frac{\text{number of nonconformities (missing tools) in all samples combined}}{\text{number of samples}}$$

$$= n \times \frac{\text{number of nonconformities in all samples combined}}{n \times \text{number of samples}}$$

$$= n\bar{p}$$

For comparative purposes, relevant cumulative Poisson probabilities from Table VI in the Appendix are also shown in Table 5.8. Here, too, there is no lower probability limit and UPL = 3.5.

TABLE 5.8 Poisson Probabilities Approximate Binomial Probabilities when n is Large and p is Small.

Cumulative Binomial Probability			Cumulative Poisson Probability
n	k	$p = .02$	$\mu_C = .5$
25	0	0.603	0.607
	1	0.911	0.910
	2	0.987	0.986
	3	0.999	0.998
	4	1.000	1.000

The point is that when n is large and p is small, binomial probabilities approximately equal probabilities based on a Poisson distribution with $\mu_C = np$. A comparison of the two cumulative distributions given in Table 5.8 illustrates.

Generally, a table of binomial probabilities will not be available for n large and p small. Suppose $n = 3{,}600$ and $p = .0025$. Some calculators and some statistics packages might not be able to compute the needed binomial probabilities. A Poisson distribution with $\mu_C = 9.0$ can be used instead.

We already have observed that when n is substantially less than N, binomial probabilities may be used to approximate hypergeometric probabilities. If for the binomial distribution it is the case that n is large and p is small, it follows that Poisson probabilities can also approximate hypergeometric probabilities.

EXAMPLE 9

In a lot of 2,000 computer chips, suppose 70 are defective. Fifty different chips are sampled for testing. What is the probability there will be exactly one defective chip in the sample? Compare the answer to the approximation using a Poisson distribution.

Let X denote the number of defective chips in the sample. Based on a hypergeometric distribution with $N = 2{,}000$, $D = 70$, and $n = 50$,

$$P(X = 1) = \frac{\binom{70}{1}\binom{1{,}930}{49}}{\binom{2{,}000}{50}} = .306,$$

a computation that is beyond the range of many calculators. The average number of defective chips in a sample is

$\mu = n \times \dfrac{D}{N} = 50 \times \dfrac{70}{2{,}000} = 1.75$. Using a Poisson approximation with $\mu_C = 1.75$ gives

$$P(X = 1) \approx \frac{1.75^1 e^{-1.75}}{1!} = .304.$$

Normal Distributions

Having become familiar with $\overline{X}$-charts in Chapter 1, we already know some facts about normal distributions. We know that normal distributions do a very good job of modeling many variables in nature, particularly variables that are a sum of random inputs of other variables. We know that when a random sample is drawn from a population that has a normal distribution with mean μ and standard deviation $\sigma > 0$, the population of means has a normal distribution with mean μ and standard deviation $\sigma/\sqrt{n}$. We know that when a random sample is drawn from an arbitrary population with mean μ and standard deviation $\sigma > 0$, the population of means has a distribution with mean μ and standard deviation $\sigma/\sqrt{n}$, *and that if* n *is large*, has an approximate normal distribution. We know that the Empirical Rule is the basis for 3-sigma limits on many different kinds of control charts. We even have been introduced (Section 3.3) to procedures commonly used for determining whether a collection of data values could have come from population with a normal distribution. Just knowing the Empirical Rule helped us get the basics of control charts. But when it came to finding probabilities based on normal distributions, problems had to be limited to ones in which the values 68%, 95%, or 99.7% could be applied. We now remove this limitation by examining normal distributions in more detail.

Finding probabilities for variables with normal distributions requires a different approach than was used with hypergeometric, binomial, and Poisson distributions. The latter three distributions are models for discrete variables. Recall that a discrete variable has gaps between the values the variable can equal. If the variable being observed is X, the number of defective items in a sample, X can equal 2 or 3, but not 2.4. Whenever X is discrete, there is a rule or formula for computing $P(X = k)$ that produces a positive probability when k is in the range of values of X. Further, when k is not in the range of X, it is understood that $P(X = k) = 0$, as in $P(X = 2.4 \text{ defectives}) = 0$.

Normal distributions are models for a large number of continuous variables, variables that do not have gaps. Theoretically, if the resistance of an electrical part can be 2 = 2.000. . . ohms or 3 = 3.000. . . ohms, the resistance also could equal 2.4 = 2.4000. . . ohms, 2.4333. . . ohms, or any real number between 2 and 3 ohms. Probabilities for continuous variables are determined in a different way—by finding areas under curves. We illustrate how to do this for normal variables immediately after Example 10. Example 10 is *optional reading*, intended for those who may be curious about why the method of finding probabilities for continuous variables has to be different.

EXAMPLE 10 (*OPTIONAL*)

Suppose a process makes electrical parts that have a resistance of at least 2.0 ohms but less than 3.0 ohms. Assume that each possible number in the continuum of real numbers between 2 ohms and 3 ohms is equally likely to be the resistance of a part. A reasonable model for generating W, the resistance of a random part, is to flick the spinner in Figure 5.4 and observe the number to which the spinner points when it stops.

Because the spinner is *fair*, it will be easy to find the probability that W will be in a given interval. For example, it must be that

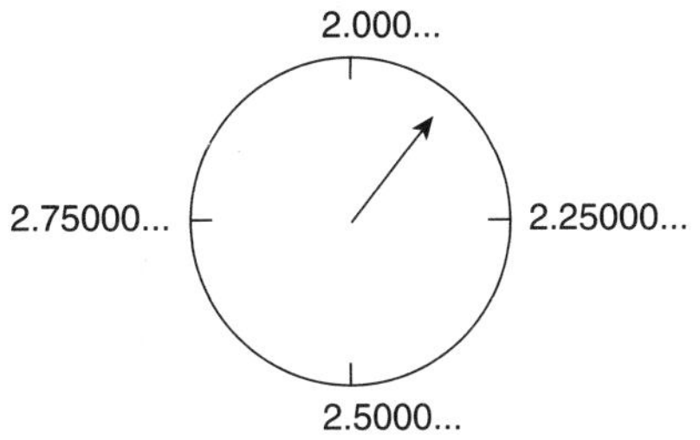

FIGURE 5.4 Model for the Resistance of an Electrical Part When All Values Between Two and Three Ohms Are Equally Likely.

$P(2 \leq W < 2.5) = .5$, since that interval makes up half the circumference. Similarly, $P(2 \leq W < 2.05) = .05$, $P(2 \leq W < 2.005) = .005$, $P(2 \leq W < 2.0005) = .0005$, and so on. From our probability axioms, we know that $P(W = 2.000\ldots)$, the probability the spinner will come to rest pointing precisely at the number 2.000. . . , cannot be negative. Further, $P(W = 2.000\ldots)$ has to be less than each of $P(2 \leq W < 2.5)$, $P(2 \leq W < 2.05)$, $P(2 \leq W < 2.005)$, and so on. The only way out is to have that $P(W = 2.000\ldots) = 0$. And since all possible resistances are equally likely, $P(W = k) = 0$ for all real numbers k between 2 and 3.

Whenever X, the variable being observed, is continuous, it can be shown that $P(X = k) = 0$ for all values of k (k rendered to *all* decimal places). So we cannot get one by *adding* together, in the usual way, values of $P(X = k)$ for all k in the continuum of values that X can equal. Mathematical theory tells us that answering probability questions about X involves finding areas under an appropriate curve.

The Standard Normal Distribution

Suppose Z, the variable being observed, has a normal distribution with mean 0 and standard deviation 1, a so-called *standard normal distribution*. Being able to answer probability questions for Z will allow us to answer probability questions about any normal distribution. Computing probabilities about Z involves finding areas under the *standard normal curve* (Figure 5.5), which is the graph of the equation $y = \dfrac{e^{-z^2/2}}{\sqrt{2\pi}}$.

Let k denote a real number. The cumulative probability $P(Z \leq k)$ is equal to the area that is under the curve and to the left of k. Table VII in the Appendix may be used to find this area. For example, $P(Z \leq 2.10)$ is equal to the area of the shaded region shown in Figure 5.6. From Table VII, we see that this area equals .9821. That is, $P(Z \leq 2.10) = .9821$.

To find $P(a \leq Z \leq b)$ for two numbers a and b, where $a < b$, we subtract the area that is under the curve and to the left of a from the area that is

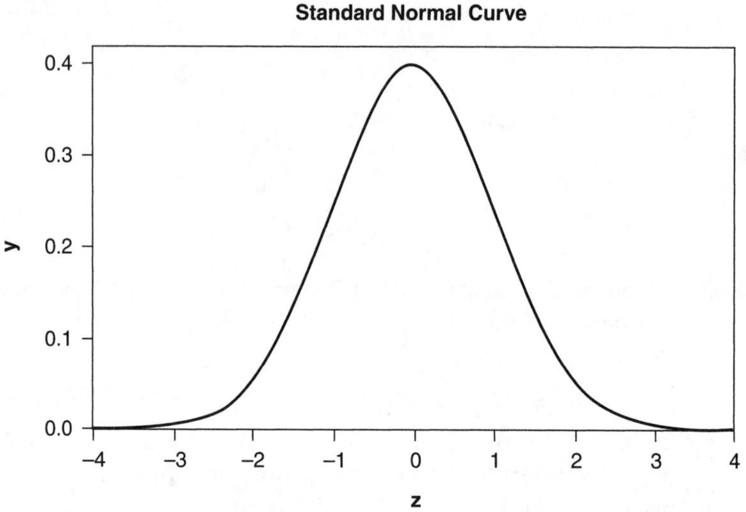

FIGURE 5.5 Standard Normal Curve.

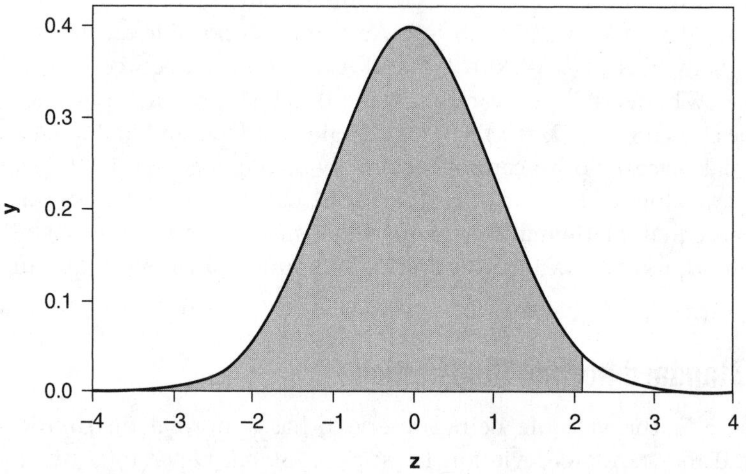

FIGURE 5.6 $P(Z \leq 2.10)$.

under the curve and to the left of *b*. This difference is the area between *a* and *b*. For example, $P(-1.03 \leq = Z \leq 2.10)$ equals the area of the shaded region shown in Figure 5.7. According to Table VII in the Appendix the area to the left of $-$ 1.03 is .1515 and the area to the left of 2.10 is .9821. So the area between $-$ 1.03 and 2.10, illustrated in Figure 5.7, equals .9821 $-$ .1515 = .8306.

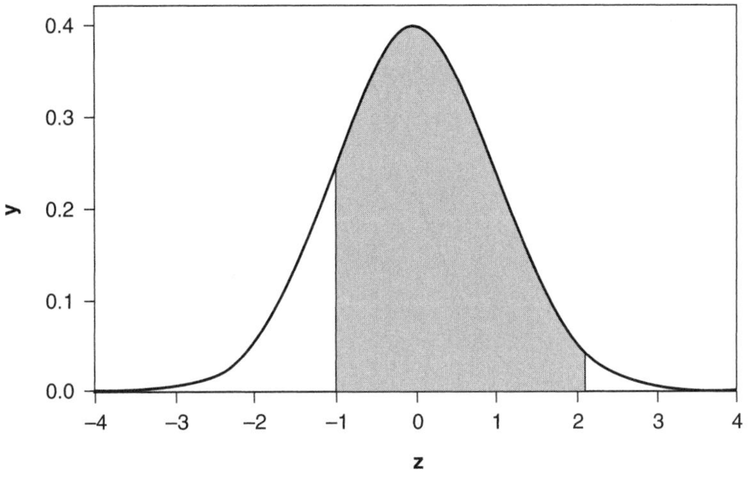

FIGURE 5.7 $P(-1.03 \le Z \le 2.10)$.

Because the standard normal variable Z is continuous, $P(Z = k) = 0$ for each value k. Therefore all of the following probabilities are equal: $P(a \le Z \le b)$, $P(a < Z \le b)$, $P(a \le Z < b)$, and $P(a < Z < b)$. That is, when computing a probability for a continuous variable, whether a given vertical line segment is included as part of the shaded region does not change the area of the region. Similarly, Table VII in the Appendix provides values of both $P(Z \le k)$ and $P(Z < k)$ since $P(Z \le k) = P(Z < k)$. Likewise, $P(Z \ge k) = P(Z > k)$.

We now show how Table VII in the Appendix can be used to find probabilities of the form $P(Z \ge k)$ or $P(Z > k)$. We illustrate with $P(Z > 1.26)$. The region that is under the standard normal curve and determined by values of Z greater than 1.26 is shaded in Figure 5.8. Table VII is designed to provide the area to the left of 1.26, which gives $P(Z \le 1.26)$ or $P(Z < 1.26)$ directly. But finding $P(Z \ge 1.26)$ or $P(Z > 1.26)$ means finding the area that is under the standard normal curve and to the right of the value k.

One way of using Table VII in the Appendix to find a probability such as $P(Z > 1.26)$ is to first find the probability of the complementary event $P(Z \le 1.26)$. From Table VII, $P(Z \le 1.26) = .8962$. Then

$$P(Z > 1.26) = 1 - P(Z \le 1.26) = 1 - .8962 = .1038.$$

A second approach, one that avoids having to subtract, takes advantage of the symmetry of the standard normal curve. The area under the curve and to the right of 1.26 equals the area under the curve and to the left of -1.26. See Figure 5.9. Then $P(Z \ge 1.26) = P(Z \le -1.26) = .1038$, obtained directly from Table VII using the value -1.26 for k.

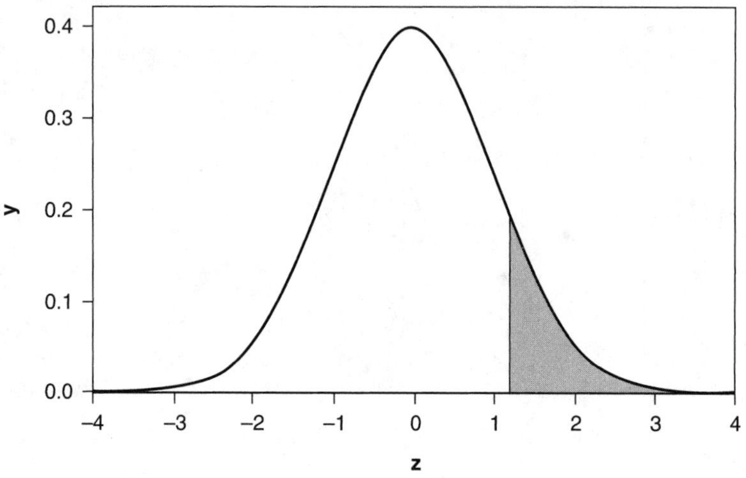

FIGURE 5.8 $P(Z \geq 1.26)$.

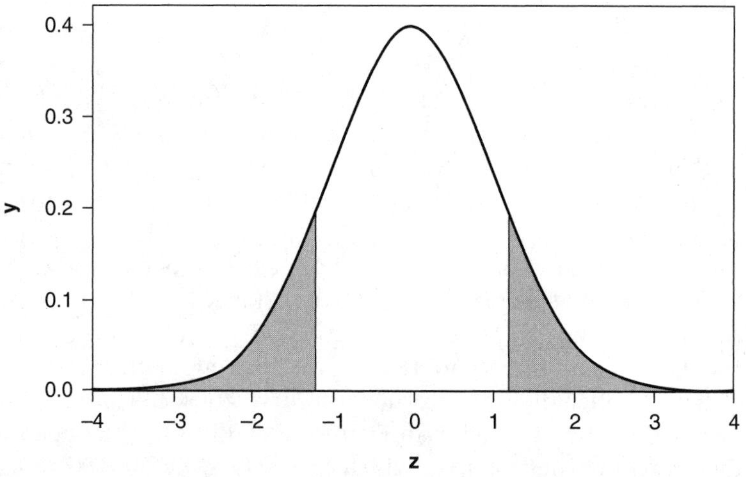

FIGURE 5.9 $P(Z \geq 1.26) = P(Z \leq -1.26)$.

Finding Probabilities for Other Normal Distributions

Suppose X, the variable being measured, has a normal distribution with mean μ and standard deviation σ. For completeness, we point out that the equation of the normal curve for this distribution is

$$y = \frac{e^{\frac{(x-\mu)^2}{2\sigma^2}}}{\sqrt{2\pi}}.$$

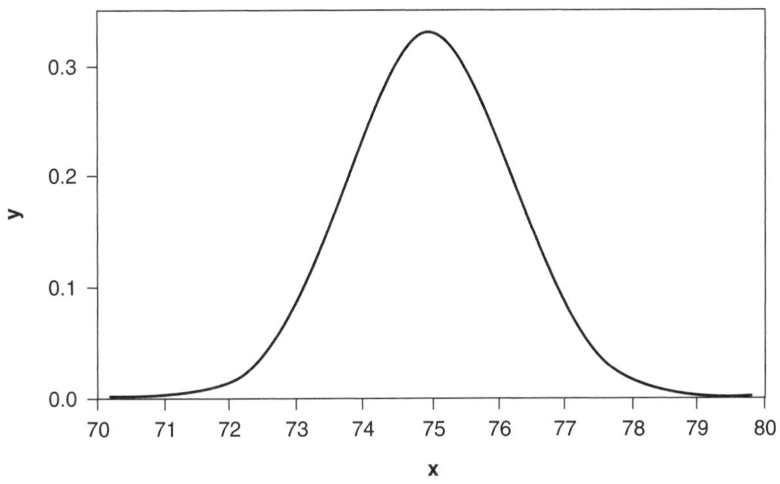

FIGURE 5.10 Normal Curve for Which $\mu = 75$ and $\sigma = 1.2$.

Figure 5.10 shows a normal curve for which $\mu = 75$ and $\sigma = 1.2$. It would not be possible to provide tables of cumulative probabilities for all possible values of μ and σ. However, one can evaluate probabilities such as $P(a \le X \le b)$ by restating the probability in terms of z-scores. To do this, all we need is to accept one result from probability theory. **If X has a normal distribution, the z-score of X has a standard normal distribution.** We illustrate how to use this fact in several examples.

EXAMPLE 11

Suppose the resistance in ohms of the electrical components made by a company has an approximate normal distribution with mean 75.0 and standard deviation 1.2. Roughly what percentage of components produced have a resistance between 75.3 ohms and 77.0 ohms? Let X denote the resistance of a randomly selected component. In Figure 5.11, the area of the shaded region under the curve and between 75.3 and 77.0 equals the probability we seek, $P(75.3 \le X \le 77.0)$. We convert the question into one about z-scores so that we can use Table VII in the Appendix. The z-scores for 75.3 and 77.0 are

$$\frac{75.3 - \mu}{\sigma} = \frac{75.3 - 75.0}{1.2} = .25 \text{ and } \frac{77.0 - \mu}{\sigma} = \frac{77.0 - 75.0}{1.2} = 1.67$$

respectively. Let Z denote the z-score of a random component. Having a resistance measurement X fall between 75.3 and 77.0 is the

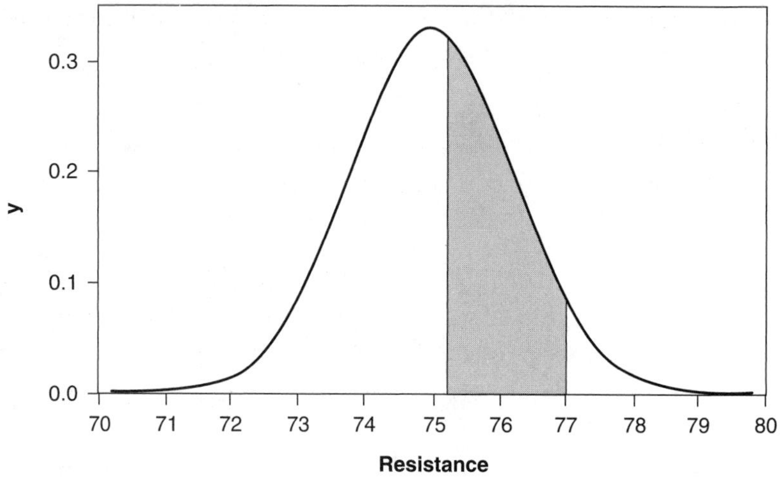

FIGURE 5.11 $P(75.3 \le X \le 77.0)$.

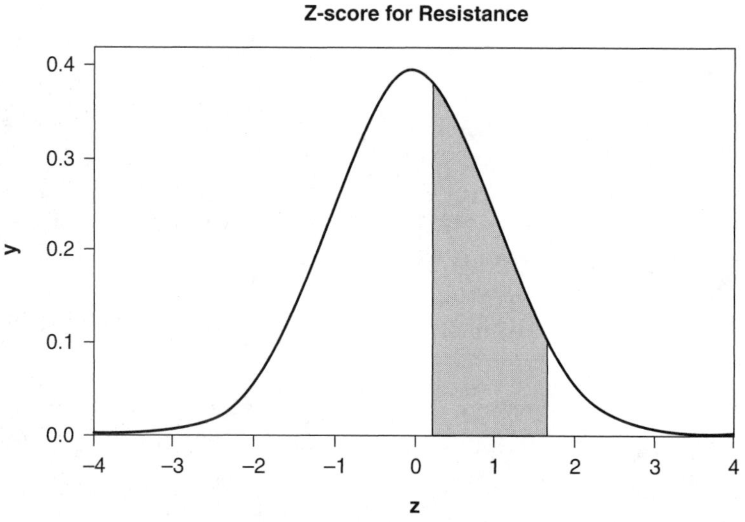

FIGURE 5.12 $P(.25 \le Z \le 1.67)$.

same as having its z-score Z fall between .25 and 1.67. That is, $P(75.3 \le X \le 77.0) = P(.25 \le Z \le 1.67)$. The latter area is depicted in Figure 5.12. From Table VII,

$$\begin{aligned} P(.25 \le Z \le 1.67) &= \text{area to left of } 1.67 - \text{area to left of } .25 \\ &= .9525 - .5987 \\ &= .3538. \end{aligned}$$

So approximately 35.38% of components produced will have a resistance between 75.3 ohms and 77.0 ohms.

EXAMPLE 12

If the quality characteristic being measured on an item has a normal distribution, the Empirical Rule says that approximately 99.7% of the item measurements will fall within three standard deviations of μ. This is equivalent to saying that approximately 99.7% of the measurements will have *z*-scores between -3 and 3. Table VII in the Appendix allows us to give a more precise percentage:

$$\begin{aligned} P(-3 < Z < 3) &= \text{area to the left of 3 minus the area to the right of } -3 \\ &= .99865 - .00135 \\ &= .9973. \end{aligned}$$

EXAMPLE 13

In Example 11, the resistance of electrical components had a normal distribution with process mean $\mu = 75.0$ and standard deviation $\sigma = 1.2$. Suppose the sample size is $n = 9$. Assuming that the process stays in control, what is the probability that the mean of one sample will fall between 74.5 and 75.5 ohms?

We know that the population of means has mean $\mu_{\bar{X}} = \mu = 75.0$ ohms and standard deviation $\sigma_{\bar{X}} = \frac{\sigma}{\sqrt{n}} = \frac{1.2}{\sqrt{9}} = .4$ ohms. Further, even though the sample size is small, since individual resistance measurements have a normal distribution, sample means also have a normal distribution. In short, sample means have a normal distribution with mean 75.0 and standard deviation .4. To find $P(74.5 \leq \overline{X} \leq 75.5)$, we first translate 74.5 and 75.5 into *z*-scores relevant to sample means:

$$\frac{74.5 - \mu_{\overline{X}}}{\sigma_{\overline{X}}} = \frac{74.5 - 75.0}{.4} = -1.25$$

$$\textit{and} \quad \frac{75.5 - \mu_{\overline{X}}}{\sigma_{\overline{X}}} = \frac{75.5 - 75.0}{.4} = 1.25.$$

Then

$$\begin{aligned} P(74.5 \leq \overline{X} \leq 75.5) &= P(-1.25 \leq Z \leq 1.25) \\ &= .8944 - .1057 \\ &= .7887. \end{aligned}$$

EXAMPLE 14

Three metal plates are sampled every half hour and a key thickness measurement taken from each plate. An $\overline{X}$-chart is kept, centered on the target value 86.50 mm. Sample means have an approximate normal distribution and the process standard deviation is $\sigma = .29$ mm. The process has been in control. Suppose the process mean experiences an unwanted upward shift of .30 mm. (a) What is the probability that the first sample taken after the shift will generate an out-of-control signal? (b) What is the probability there will be at least one such signal in the first five samples? (c) What is the average run length after the shift?

a. The 3-sigma limits for the $\overline{X}$-chart are

$$\mu_{\text{target}} \pm 3\frac{\sigma}{\sqrt{n}} = 86.50 \pm 3 \times \frac{.29}{\sqrt{3}} = 86.50 \pm .50, \text{ so that}$$

$UCL_{\overline{X}} = 87.00$ and $LCL_{\overline{X}} = 86.00$. Sample means still come from an approximate normal distribution for which

$$\sigma_{\overline{X}} = \frac{\sigma}{\sqrt{n}} = \frac{.29}{\sqrt{3}} = .1674. \text{ After the upward shift in the}$$

process mean, $\mu_{\overline{X}}$ will equal 86.80. The probability the mean of one sample will produce an out-of-control signal is $P(\overline{X} \geq 87.00) + P(\overline{X} \leq 86.00)$. Z-scores for the 3-sigma limits are

$$\frac{87.00 - \mu_{\overline{X}}}{\sigma_{\overline{X}}} = \frac{87.00 - 86.80}{.1674} = 1.19 \text{ and}$$

$$\frac{86.00 - \mu_{\overline{X}}}{\sigma_{\overline{X}}} = \frac{86.00 - 86.80}{.1674} = -4.78. \text{ Then}$$

$P(\overline{X} \geq 87.00) = P(Z \geq 1.19) = P(Z \leq -1.19) = .1170.$

Also, $P(\overline{X} \leq 86.00) = P(Z \leq -4.78) \approx 0$. So the probability of an out-of-control signal on a single sample is approximately $.1170 + 0 = .1170$.

b. The number of out-of-control signals in five samples has a binomial distribution with $n = 5$ and $p = .1170$. Then

P(at least one out-of-control signal) =

1 − P (no out-of-control signals)

$$= 1 - \binom{5}{0}(.1170)^0(.8830)^5$$

$$= 1 - .5368$$

$$= .4632.$$

c. The average run length (see Section 4.5) is

$$\text{ARL} = \frac{1}{p} = \frac{1}{.1170} = 8.55.$$

PROBLEMS

1. In Example 1, what is the range of values of X? We already computed that $p(0) = .102$ and $p(1) = .363$. Using the formula for hypergeometric probabilities, find the probability of each of the other values in the range of X and show that they sum to 1, except possibly for roundoff error. Use the probabilities you've found to show that $\mu = 1.6$, except possibly for roundoff error.
2. A company having financial problems needed to downsize. Nine employees were terminated, the youngest of whom was 60 years old (see Norton, 1990). The company claimed that the selection process had no statistical connection to age. In the decision process, 144 employees were considered for termination. Their age breakdown is given in Table 5.9. To model a termination selection process that is age-neutral, suppose the nine who are chosen are determined by blind draw. (a) What is the probability that all nine would be in the 60–64 age bracket? (b) Among the nine selected, how many would be expected to be in the 60–64 age bracket?
3. Four of the cards in an ordinary 52-card deck are *jacks*. Suppose a 5-card hand is dealt from the deck. Let X denote the number of jacks in the hand. What is the range of values of X? Find $P(X = k)$ for every possible value of k.
4. A committee of three is to be chosen by drawing straws from a department of 15 people in order to establish a committee to hear a grievance.
   a. What is the probability that the committee will consist of Fred, Rose, and Tom?

TABLE 5.9 Age Breakdown of a Company that Downsized.

	Under 60	60 to 64	
Not Terminated	129	6	135
Terminated	0	9	9
	129	15	144

b. What is the probability that Fred will be on the committee?
c. If five people in the department are women, what is the expected number of women on the committee?

5. Suppose there are 7 defective resistors in a box of 50. If any resistors in a sample taken from the box are defective, the box will not be accepted. What should the sample size be in order to have at least a 75% chance of rejecting such a box?

6. Suppose the defect rate for a process that makes resistors is 2%.
   a. What is the probability that in a sample of 20 resistors taken from production, at most 2 will be defective? Use Table V in the Appendix.
   b. Use the formula for $P(X = k)$.
   c. Find the probability that exactly 2 of 20 will be defective using Table V.

7. A pattern of points sometimes used with an $\overline{X}$-chart to detect a possible downward shift in the process mean is 16 or more points above the center line out of 20 consecutive points. Use Table V in the Appendix to find the probability of such a signal when the process is in control.

8. Suppose there is one chance in 200 that a 35-year old woman will give birth to a baby with Down's syndrome. Suppose records in a particular hospital show that of the last 320 births by 35-year old women, 6 babies had the syndrome. Is the event *6 or more have the syndrome* unusual? (Compare to .05.)

9. Suppose that, historically, 35% of all Mondays at a production facility incur downtime. To shake up the system, process changes are instituted for 15 weeks (i.e., 15 Mondays). It is decided to make the changes permanent if at most 2 of the 15 Mondays experience downtime.
   a. Suppose the changes do not really affect the Monday-downtime rate. What is the probability that the changes will be made permanent?
   b. Suppose the changes actually reduce the rate of Mondays with downtime to 6%. What is the probability that the changes are not adopted?

10. Suppose 24% of all customers who return their product purchase/registration form also enclose the customer satisfaction card with the box *very pleased* checked. To enhance customer satisfaction, sales personnel undergo a required training course. If 12 or more out of the next 25 product purchase/registrations received are accompanied by satisfaction cards with the *very pleased* box checked, the training will be deemed successful. What is the probability the training will be viewed as successful if the training has no actual impact?

11. The standard regimen for treating a given disease is successful for 30% of all subjects treated. Suppose a new treatment regimen will be viewed as having demonstrated that it is superior to the standard regimen if it is successful on 15 or more of the next 25 subjects treated. If there is no difference in success rates for the two treatment regimens:
    a. Find the mean and standard deviation of the number of subjects successfully treated with the new regimen.
    b. What is the probability that the new regimen is incorrectly declared to be superior?
12. A manufacturer's inspection procedure is designed so that if the proportion of second-quality widgets reaches 20%, a second-quality widget will show up in a sample at least 95% of the time.
    a. What does Table V in the Appendix say about the smallest sample size n that may be used?
    b. Use a formula or statistics software package to improve on the answer from part a.
13. A dangerous intersection averages 1.8 accidents a day during the workweek. Use Table VI in the Appendix to find:
    a. The probability there will be at most 3 accidents on a workday.
    b. The probability there will be more that 5 accidents in a workday.
    c. The probability there are at least 16 accidents during a workweek.
14. a. If a city has been experiencing 5.5 burglaries a week, should a week in which there is at most one burglary be considered unusual? Answer by comparing the probability to .05.
    b. A week with 11 or more burglaries?
    c. A month (four weeks) having 33 or more burglaries?
15. A company is known for claiming about its customers that only one in 1,000 complains. Last week the company had 3,302 sales, and you know that customer service logged 12 customer complaints. Assuming that the company's claim is correct, and based on 3,302 sales:
    a. What, approximately, is the probability that at least 12 complaints would occur?
    b. What is the standard deviation of the number of customer complaints?
    c. How many standard deviations from the mean is 12?
16. If a company that makes spandex fiber averages 8.8 knots per million meters of fiber, what is the probability that in a warp that uses 750,000 meter sample of fiber, there will be at most two knots?
17. The thickness of mending plates made by a process is normally distributed, currently with a process mean of 2.31 mm and standard deviation .0092 mm. Specification limits are 2.30 ± .03 mm.

a. What percentage of plates made by the process exceed the upper specification limit?
b. What percentage of plates fall below the lower specification limit?
c. What percentage of plates do not meet specifications?
d. What percentage of plates would not meet specifications if the process mean could be kept at the target value of 2.30 mm?

18. The current process mean lifetime of a filter used in a manufacturing process is 457 hours with a standard deviation of 29.5 hours. If filter lifetime has a lower specification limit of 415 hours and filter lifetime is normally distributed:
a. What percentage of filters do not meet specifications?
b. What should be the minimal process mean in order to have almost all filters meet specifications?

19. An $\overline{X}$-chart indicates that the process has been in control and the historical mean is 12.7. The subgroup size is 4 and the process standard deviation is .16. Suppose the process mean experiences a sudden downward shift of .10.
a. What is the probability that the first sample taken after the shift will generate a point outside the 3-sigma limits?
b. What is the probability that there will be at least one such point in the first 10 samples drawn after the shift occurs?
c. After the shift, what proportion of product units will not meet specifications of $12.8 \pm .6$?

5.3 RELIABILITY

Everyone understands what reliability in a product means in the nonstatistical sense. A product is reliable if it can be counted on to perform its intended function when needed, even under conditions that might cause similar products to not work well or to fail. Someone who has owned several cars is likely to have fond memories of a particularly dependable automobile—it would always start no matter how cold it got, or it never required much more than routine maintenance, or it lasted 30 years, or it lasted over 300,000 miles, and so on. Through experience, most people develop a sense for what reliability means for many different kinds of products, not just big-ticket items such as cars. Someone might have a favorite ink pen because refills last a long time, or because it writes well on slick surfaces, or writes on surfaces that are nearly vertical.

Can reliability be measured? If so, just what, exactly, would one measure? Understanding how reliability is measured requires a basic knowledge of probability. This section introduces several commonly used measures of reliability.

The **reliability** of a product during a given time period is the probability that it will not fail during the time period.

Imagine a device that consists of various components that function as a system. The reliability of the device could be defined to be the probability that the device will not fail during the four-year warranty period. A single use of the product also could be considered as the time period. For example, the reliability of a torpedo is the probability that the torpedo will not fail when it is used.

We first consider the reliability of a system of components connected *in series*. This means that for the device to work, every component must work. We assume that whether each individual component fails is independent of whether other components do so. We consider two examples—torpedoes and Christmas tree lights.

What is the probability that the torpedo will arm itself, home-in on the target, and explode? Suppose the working definition of what it means for a torpedo to not fail is that each of these three component mechanisms must not fail. Figure 5.13 illustrates this series system. One cannot travel the path from beginning to end if any of the components fail. Suppose we use the following notation for events:

A: the arming device is successful

H: the homing device is successful

D: detonation is successful

Also, let R_A denote the reliability of the arming device, R_H the reliability of the homing device, R_D the reliability of detonation, and R the reliability of the system. That is,

$$\begin{aligned} R_A &= P(A) \\ R_H &= P(H) \\ R_D &= P(D) \\ R &= P(\text{torpedo is successful}) \end{aligned}$$

The reliability of the system is

$$\begin{aligned} R &= P(A \text{ and } H \text{ and } D) \\ &= P(A)P(H)P(C) \\ &= R_A R_H R_D. \end{aligned}$$

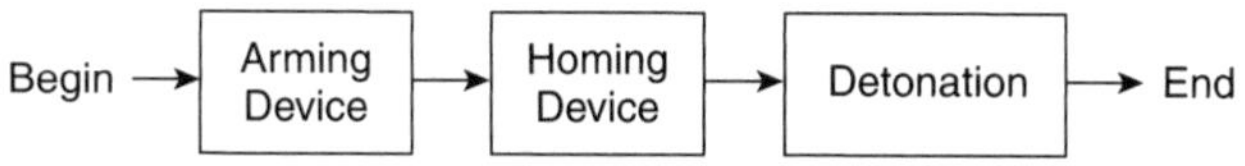

FIGURE 5.13 Series System for a Torpedo.

So if arming devices work successfully 99.9% of the time, homing devices work successfully 99% of the time, and detonation upon impact occurs 98% of the time, the reliability of the system is $R = .999 \times .99 \times .98 = .969$.

If a system consists of k independent components in series and the reliability of component i is R_i, then the reliability of the system is $R = R_1 R_2 \ldots R_k$.

EXAMPLE 1

The first strands of Christmas tree lights ever made were connected in series. If one bulb burned out, the whole strand went out. Suppose such a strand consists of 24 lights and the probability that an individual bulb burns for one month is .97. Assume bulb life spans are independent. If the working definition of reliability of a strand is that no bulbs need to be replaced during a one-month period, the reliability of a strand is $R = .97^{24} = .48$.

A system of components is said to be connected *in parallel* if for the system to work, all that needs to happen is that at least one component works. We again assume that whether each individual component fails is independent of whether other components do so. A system of four components that work in parallel is shown in Figure 5.14.

EXAMPLE 2

Mick owns three cars, an 11-year-old Honda Accord, a 31-year-old Volkswagen Superbeetle (his first car), and a 30-year-old Ford LTD (once owned by his wife's father). All three cars are available for Mick to drive to work. The probability that the Honda will start is .999, that the Volkswagen will start is .80, and that the Ford will start is .90. What is the probability that Mick can leave for work in the morning? This system has three components in parallel. If at least one vehicle will start, the system does not fail.

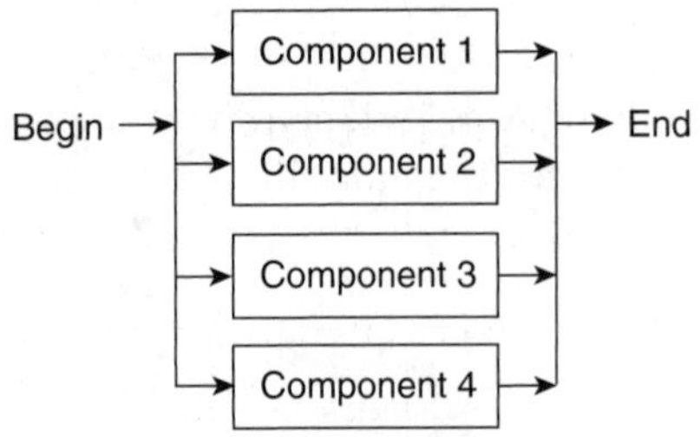

FIGURE 5.14 Parallel System of Four Components.

Let H, V, and F denote respectively that the Honda, Volkswagen, and Ford starts. Let R_H, R_V, and R_F denote the reliability of the Honda, Volkswagen, and Ford, respectively, and R the reliability of the system. Then

$$\begin{aligned} R &= P(\text{at least one vehicle starts}) \\ &= 1 - P(\text{all vehicles fail to start}) \\ &= 1 - P(H' \text{ and } V' \text{ and } F') \\ &= 1 - P(H')P(V')P(F') \\ &= 1 - (1 - R_H)(1 - R_V)(1 - R_F) \\ &= 1 - (1 - .999)(1 - .8)(1 - .9) \\ &= 1 - .00002 \\ &= .99998. \end{aligned}$$

If a system consists of k independent components in parallel and the reliability of component i is R_i, then the reliability of the system is $R = 1 - (1 - R_1)(1 - R_2)\ldots(1 - R_k)$.

EXAMPLE 3

A patient has an infection that is difficult to treat with antibiotics. The doctor knows that Antibiotic A will kill the infection 60% of the time, that Antibiotic B, which is more powerful than Antibiotic A, will work 65% of the time, and that Antibiotic C, the most powerful of the three, will work 99% of the time. The doctor treats such infections beginning with the mildest antibiotic, then escalates upward in sequence if necessary. What is the reliability of this treatment protocol?

This is a parallel system of three components. To not fail, all that has to happen is that at least one drug will work. However, components are not independent. For example, when Antibiotic B fails on a patient, this means that Antibiotic A also fails, and when Antibiotic A is successful on a patient, Antibiotic B also will be successful. Because the component independence requirement is not met, the formula just given for the reliability of a parallel system does not apply. For this particular treatment protocol, the reliability is

$$\begin{aligned} R &= P\,(\text{at least one antibiotic works}) \\ &= P\,(\text{Antibiotic C works}) \\ &= .99 \end{aligned}$$

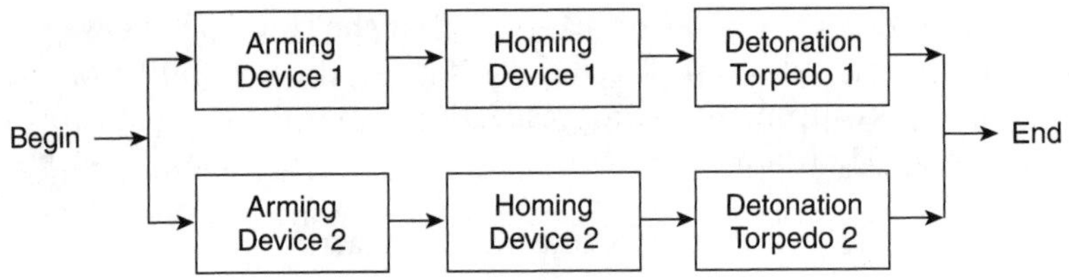

FIGURE 5.15 Two Series Systems in Parallel.

EXAMPLE 4

Some systems are a mix of components in series and parallel. Regarding the torpedoes discussed earlier, suppose the protocol in attacking a target is to fire two torpedoes. The definition of success is that at least one torpedo is successful. As before, let R_A denote the reliability of an arming device, R_H the reliability of a homing device, R_D the reliability of a torpedo's detonation upon impact, and R the reliability of the two-torpedo system. Also let R_T denote a torpedo's reliability. Figure 5.15 illustrates this system. Assume that all six components function independently.

The reliability of the two-torpedo procedure is

$$\begin{aligned} R &= P\,(\text{at least one torpedo is successful}) \\ &= 1 - (1 - R_T)(1 - R_T) \\ &= 1 - (1 - R_A R_H R_D)^2 \end{aligned}$$

We know from before that $R_T = R_A R_H R_D = .969$, so that $R = 1 - (1 - .969)^2 = .999$.

Some Commonly Used Measures in Reliability

One frequently cited measure in reliability is the failure rate of a product. The units for this measure are failures/unit-of-time or failures/cycle. In order to understand this measure, let's suppose that each of n new units of the product is used until it fails or until it has been used a total of T hours (or days, months, years, cycles, etc.), whichever occurs first. Let T_i denote the total amount of time item i is used. Thus T_i equals the time until failure of item i if item i fails, or equals T if item i does not fail. **The failure rate of the n product units during the first T hours of use is**

$$\lambda_{\text{estimate}} = \frac{\text{total number of items that fail}}{\sum_{i=1}^{n} T_i}$$

The word *estimate* in $\lambda_{\text{estimate}}$ emphasizes that the failure rate is (a) obtained from a sample of items, and (b) estimates λ, the failure rate for the first T hours of use that would be computed if one knew the value of T_i for every product unit in the population of all product units that could be made by the manufacturing process with the capability it possesses at the time the units in the sample are produced.

EXAMPLE 5

Sixty solar-powered patio lights are monitored for two years. Six power cells failed, having failure times of .099 years (i.e., five days), .019 years, 1.337 years, .088 years, 1.989 years, and 1.877 years. The first-two-year failure rate for these lights is

$$\lambda_{\text{estimate}} = \frac{6}{.099 + .019 + 1.337 + .088 + 1.989 + 1.877 + 54(2)}$$
$$= \frac{6}{113.409} = .0450 \text{ failures/yr.}$$

The concept of failure rate during a given time period may be extended to failure rate as a function of time. Actually, it is natural to introduce two related functions at this point—the reliability function and the failure rate function (hazard function).

Let T be the life span of a system or component. The *reliability function* of the system or component is the function $R(t) = P(T > t)$, the probability that the lifetime of the system or component will exceed time t.

EXAMPLE 6

Suppose the complete life spans of 20 solar-powered patio lights are monitored. The life spans in years, sorted from shortest to longest, are:

0.008 0.027 0.041 0.052 0.551 0.989 0.995 1.014 1.164 1.200 1.485 1.551
1.759 1.858 1.888 1.910 1.948 2.014 2.107 2.279

One can estimate the reliability function $R(t)$ at $t=0$ and at each of the values just listed by using the proportion of lights in the sample whose life spans exceed t. Thus we estimate that $R(0) \approx \frac{20}{20} = 1$, since 100% of the lights in the sample survived beyond time $t = 0$. Similarly,
$R(.008) \approx \frac{19}{20} = .95$, $R(.027) \approx \frac{18}{20} = .90 \ldots R(2.279) \approx 0$ (see Figure 5.16). In Figure 5.16, consecutive points are connected by line segments, forming a polygonal curve that approximates a

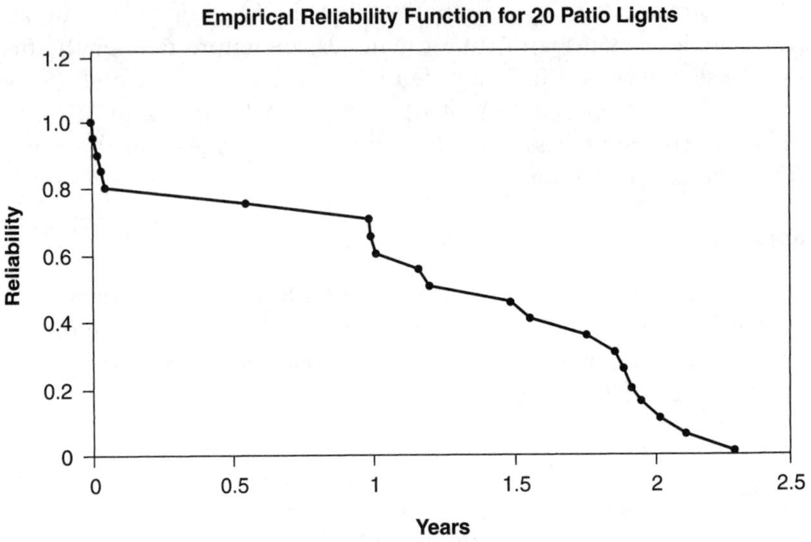

FIGURE 5.16 Empirical Reliability Function for 20 Patio Lights.

smooth curve that is the graph of the unknown, true reliability function $R(t)$.

Another function of interest is the failure rate function, or hazard function, which we define momentarily. To understand what this function represents, it will be helpful to first examine its empirical approximation. Suppose $t_1, t_2, t_3, \ldots, t_k$ represent potential failure time values, where $0 < t_1 < t_2 < t_3 < \ldots < t_k$, and that all observed failure times are less than t_k. Failure rates are computed for each of the intervals from 0 to t_1, t_1 to t_2, t_2 to t_3, . . . , and t_{k-1} to t_k. How the failure rates change across these time intervals approximates how failure rate changes as a continuous function of time.

The *failure rate function* or *hazard function* of a system or component is $h(t)$, the instantaneous failure rate of a system or component that has survived to time t.

Just as the empirical reliability function had some serious kinks in it because plotted points were based on data, the graph of the empirical failure rate can have some serious and misleading spikes in it when a failure rate is computed for a time interval of short duration, which can happen if the time interval begins at one observed life span value and ends at the next larger life span value. To help avoid spikes when graphing the empirical failure rate, it will be helpful to use time intervals that are of equal or approximately equal length (or an equal number of cycles) and which are

TABLE 5.10 Failure Data for Patio Lights.

Month	1/00	2/00	3/00	4/00	5/00	6/00	7/00	8/00	9/00	10/00
Failures	20	5	0	1	0	1	1	2	1	0
Month	11/00	12/00	1/01	2/01	3/01	4/01	5/01	6/01	7/01	8/01
Failures	1	2	5	3	4	0	2	3	3	2
Month	9/01	10/01	11/01	12/01	1/02	2/02	3/02	4/02	5/02	
Failures	2	8	10	8	6	4	2	3	1	

not too narrow (i.e., $t_i - t_{i-1}$ is not too close to zero). Some trial and error may be appropriate.

EXAMPLE 7

Suppose 100 solar-powered patio lights are put into service and, at the end of each month, the number of lights that failed during the month is recorded. The data appear in Table 5.10. Estimate the failure rate function $h(t)$.

We will measure failure rate in terms of failures/yr and consider each month to be of duration $\frac{1}{12} = .08333$ yr. Treat a patio light that fails during a given month as if it fails when half the month has elapsed. The failure rate computations for the first four months are shown in Table 5.11, and all failure rates are summarized in Table 5.12.

Figure 5.17 is the graph of the empirical failure rates shown in Table 5.12. A smooth curve that *fits* these failure rates has been superimposed on the graph. The smooth curve is an approximation of what the graph of $h(t)$ should look like.

The trends indicated by the plotted points or by the curve are typical of failure rate functions of many products, and even of human mortality (think of the mortality rate for humans as the failure rate of a product). Early in a product's life, there is a phase-in period with a high but rapidly decreasing failure rate. Failures during this period often are caused by parts or subassemblies of poor quality, poor assembly of the product, substandard installation of the product, or misuse of the new product. It is common for a product that survives the phase-in period to next undergo a long period during which the failure rate is relatively low and constant. Failures during this period usually are thought of as attributable to chance—accidents, poor maintenance by the owner, parts or subassemblies of marginal quality, and so on. The third typical phase is a wear-out period attributable to the advancing age/use of the product. During this period, the failure rate is increasing, gradually at first, but then more rapidly with time.

TABLE 5.11 Empirical Failure Rate Function Computations.

t	Number of failures in period ending at time t	Number of units on test at beginning of period	Estimate of failure rate $h(t)$ units = failures/yr
.083	20	100	$\frac{20}{20(.5)(.08333) + 80(.08333)} = 2.67$
.167	5	80	$\frac{5}{5(.5)(.08333) + 75(.08333)} = .77$
.250	0	75	$\frac{0}{0(.5)(.08333) + 75(.08333)} = 0$
.333	1	75	$\frac{1}{1(.5)(.08333) + 74(.08333)} = .16$

TABLE 5.12 Failure Rate Data for Patio Lights.

t	.083	.167	.250	.333	.417	.500	.583	.667	.750	.833
$h(t)\approx$	2.67	.77	0	.16	0	.16	.17	.34	.17	0
t	.917	1.000	1.083	1.167	1.250	1.333	1.417	1.500	1.583	1.667
$h(t)\approx$	.18	.36	.94	.61	.86	0	.45	.71	.76	.53
t	1.750	1.833	1.917	2.000	2.083	2.167	2.250	2.333	2.417	
$h(t)\approx$	.56	2.53	4.14	4.80	5.54	6.00	4.80	14.40	24.00	

Another measure frequently encountered in the reliability arena is the *mean time to failure* (MTTF) of a product or system. Suppose the quality characteristic measured on an item is its life span. *The mean time to failure of an item is μ, the population mean of all the life spans of all the items the system could produce.* This, of course, is a value that is unknown, but which can be estimated using the sample mean of all the life spans in a random sample of items. The mean time to failure also is called the *average life(span)* or *expected life(span).*

Suppose we place k items on test and let $T_1, T_2, \ldots, T_k$ be the life spans of those k items. Notice that the mean time to failure for the sample is

$$\text{MTTF}_{estimate} = \overline{T} = \frac{\sum_{i=1}^{k} T_i}{k} = \frac{1}{k/\sum_{i=1}^{k} T_i} = \frac{1}{\lambda_{estimate}}.$$

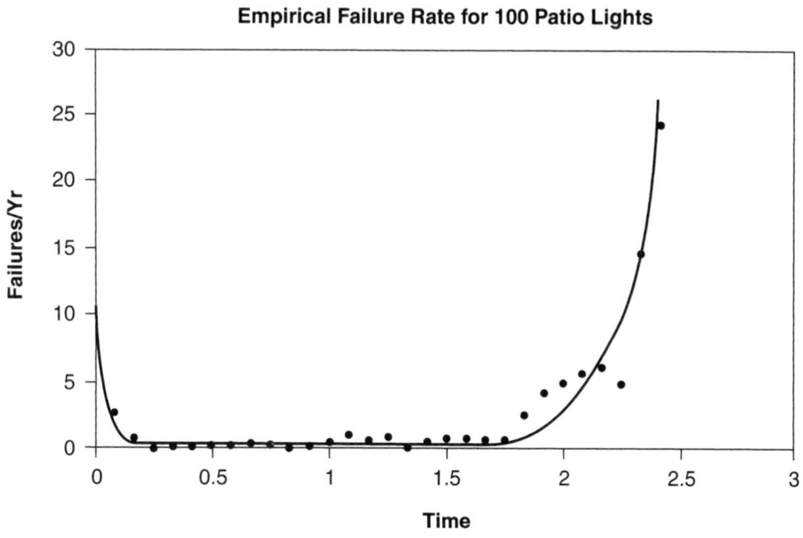

FIGURE 5.17 Empirical Failure Rate for 100 Patio Lights.

The analogous relationship $\mu = \frac{1}{\lambda}$ holds for the unknowns μ and λ. So if the failure rate of an item is 2.5 failures/year, then the MTTF for the product is .4 years. To aid in understanding why this relationship is natural, it may be helpful to take the view that when an item fails, it is immediately replaced by a new item. If the person who oversees the replacements averages 2.5 replacements in a year, then it is reasonable that the average life span of a product unit should be .4 years because 2.5 back-to-back time intervals of this length comprise one year.

PROBLEMS

1. Refer to Example 4. Suppose the protocol in attacking a target is to fire two torpedoes. The definition of success is that both torpedoes are successful. As before, let R_A denote the reliability of an arming device on a torpedo, R_H the reliability of a homing device, and R_D the reliability of one detonation. Sketch a diagram of and find the reliability R of this two-torpedo system.
2. The four welds that hold two components together are critical for the integrity of the product unit. All four must pass a visual inspection.
   a. If the probability each weld will pass an inspection is .98, what is the probability that a unit will pass inspection?

b. If 95% of units are to pass inspection, the probability that an individual weld will pass inspection must be improved to what value?

3. Two inspectors visually inspect one component of a fuel injector. If a flaw is present, one inspector will notice it with probability .95, the other with probability .9. A *no go* by either inspector causes the part to be scrapped.
   a. What is the reliability of this system for identifying a flaw?
   b. Suppose training can increase the reliability of each inspector at detecting flaws. What should be the reliability of each of two equally adept inspectors if the system of inspection is to have a reliability of .9999?
4. The probability of a power outage on any given day is .007. In the event of an outage, a generator that works with probability .98 will be used. What is the reliability of this system?
5. What is the reliability of the four-component system diagrammed in Figure 5.18 if the reliability of each component is .96?
6. What is the reliability of the six-component system diagrammed in Figure 5.19 if the reliability of each component is .96?
7. In Problem 6, what should the reliability of each component be in order to make the reliability of the system be .99999?
8. Twenty batteries were tested for three hours. Five batteries died, with life spans of 2.92, 2.93, 2.75, 2.95, and 2.93 hours. Estimate the failure rate for this kind of battery during the first three hours of use.
9. Refer to Table 5.12. Confirm the estimated values of the failure rate $h(t)$ for $t = .417$, $t = .500$, and $t = .583$.
10. Using Figure 5.17, estimate $h(2.25)$, the instantaneous failure rate of a patio light that has not failed after two years and three months of use. Then estimate the mean time to failure for this light.

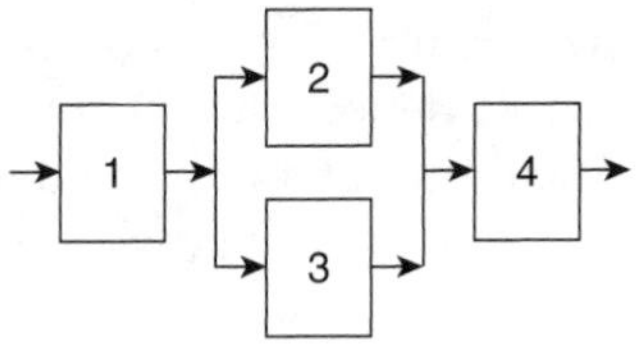

FIGURE 5.18 Diagram of Four-Component System.

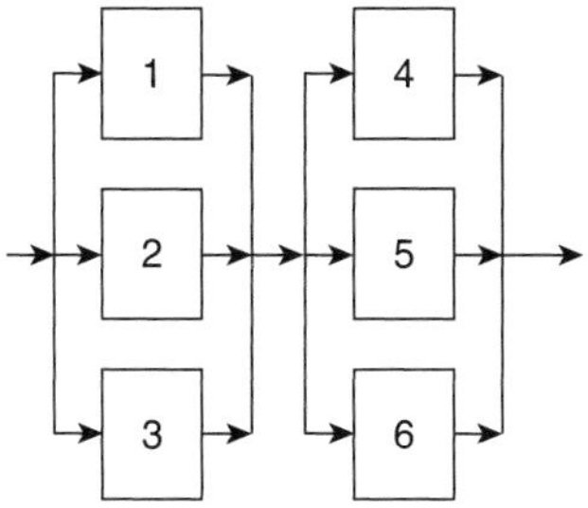

FIGURE 5.19 Diagram of a Six-Component System.

11. Using Figure 5.16, estimate the probability that an individual patio light will last at least (a) 6 months; (b) 1 year; (c) 2 years.

6

Topics in Quality

OBJECTIVES

- Introduce Deming's 14 points
- Sketch the backgrounds and contributions of quality arena visionaries
- Introduce the use of Pareto charts, cause-and-effect diagrams, scatter diagrams, flowcharts, and other problem solving tools
- Introduce the concept of total quality management
- Introduce the Six Sigma concept

6.1 THINKING QUALITY

Up to this point the focus has been on statistics. Knowing how to measure location and variation in a quality variable, how to estimate reliability and mean time to failure, how to identify when process changes may have occurred, how to measure process capability, and so on, are essential elements in the knowledge base of statistical process control. However, those who work in the quality arena understand that while statistical tools are important, there is more to the subject of quality than its statistical side. This chapter is devoted to examining quality as a point view. We begin this section by examining some of the principles espoused by W. Edwards Deming (1900–1993), viewed by many as the greatest quality pioneer of the twentieth century. We then close by identifying other widely recognized leaders and thinkers in the quality arena.

Deming had a Ph.D. in physics, a master's degree in mathematics and physics, and studied electrical engineering as an undergraduate. In 1927, while working at the U. S. Department of Agriculture, he became aware of the statistical tools developed by Walter Shewhart, and later became a collaborator of Shewhart's. In 1936, he went to London to study statistics under Ronald A. Fisher, a visionary later knighted for his contributions to that discipline. In 1939, Deming went to work for the National Bureau of the Census, which was involved in preparing for the 1940 census. Beginning in 1942, Deming contributed to the quality of war material used in World War II by training industries in how to use statistics principles in manufacturing.

At war's end, American industry was robust, and the United States was the world's major producer of manufactured goods. Japanese industry, on the other hand, had been decimated by war's end. The manufactured goods they produced for overseas consumption after the war were viewed as they had been prior to the war—as cheap and of extremely low quality. Unlike the United States, Japan was not blessed with natural resources. The export of goods had to be an integral part of its economy. For those goods to compete, Japan felt it needed to establish a reputation for manufacturing high-quality products. The Japanese were aware of Deming's contributions to the Allies during the war, and in 1950 he was invited to Japan by the Union of Japanese Scientists and Engineers. Industry began following Deming's principles, and Japanese products earned that high-quality reputation within a few short years. To this day, Japan awards the Deming Prize, that country's highest award, to organizations in the area of quality.

In the United States, awareness of Deming's principles surged after the showing of a June, 1980 NBC documentary, "If Japan Can, Why Can't We?" American manufacturers of automobiles, radios, televisions, and other products had taken a hit from Japanese imports and, in an effort to compete and regain its market loss, began implementing statistical quality control tools and rethinking some of its quality-related management philosophies.

Deming's management philosophy can be summarized by his famous 14 points.

Deming's 14 Points

1. Create constancy of purpose toward improvement of products and services, with a plan to become competitive, stay in business, and provide jobs.
2. Adopt the new philosophy. We are in a new economic age. Western management must awaken to the challenge, must learn their responsibilities, and take on leadership for change.
3. Cease dependence on inspection to achieve quality. Eliminate the need for inspection on mass basis by building quality into the product in the first place.
4. End the practice of awarding business on the basis of price tag. Instead, minimize total cost. Move toward a single supplier for any one item, on a long-term relationship of loyalty and trust.
5. Improve constantly and forever the system of production and service, to improve quality and productivity, and thus constantly decrease costs.
6. Institute training on the job.
7. Institute leadership. The aim of supervision should be to help people and machines and gadgets to do a better job. Supervision of management is in need of overhaul, as well as supervision of production workers.
8. Drive out fear, so that everyone may work effectively for the company.
9. Break down barriers between departments. People in research, design, sales, and production must work as a team, to foresee problems of production and in use that may be encountered with the product or service.
10. Eliminate slogans, exhortations, and targets for the work force asking for zero defects and new levels of productivity. Such exhortations only create adversarial relationships, as the bulk of the causes of low quality and low productivity belong to the system and thus lie beyond the power of the work force.
11. a. Eliminate work standards (quotas) on the factory floor. Substitute leadership.
    b. Eliminate management by objective. Eliminate management by numbers, numerical goals. Substitute leadership.
12. a. Remove barriers that rob the hourly worker of his right to pride of workmanship. The responsibility of supervisors must be changed from sheer numbers to quality.
    b. Remove barriers that rob people in management and in engineering of their right to pride of workmanship. This means, *inter alia*, abolishment of the annual or merit rating and of management by objective.

13. Institute a vigorous program of education and self-improvement.
14. Put everybody in the company to work to accomplish the transformation. The transformation is everybody's job. (Deming, 1986)

The 14 points, taken from Deming's *Out of the Crisis* (1986), and the following brief elaboration of each point, adapted from that book, are used with the permission of The MIT Press. Additionally, Neave (1990) has insightful elaborations on each point. We now discuss each of the points briefly.

Point 1—Constancy of Purpose. It is easy for a company to become too caught up in contemporary issues at the expense of improvement of product and service. Resources must be made available for long-term planning aimed at making a better product, for finding new markets, for reducing production costs, for improving production methods, for training and educating workers and managers, for improving all aspects of marketing and customer service, for making a product that the customer is pleased to use, for engaging in research, and so on. These activities should be never-ending, to help ensure that a company achieves what should be its main goals: staying in business, providing jobs, and providing marketable goods and services that help people live better. A company should have a statement of its constancy of purpose that is made known to customers, suppliers, and employees. There should be a policy that an employee will not lose his or her job for making a contribution to quality.

Point 2—Adopt the New Philosophy. That which needlessly lowers the quality of life should not be tolerated. Some examples are: a workplace that is not clean, commonly accepted levels of mistakes and defects, tools and materials that are not appropriate for the job, workers who are not properly trained, workers who cannot be relied upon to do what they say or to arrive when they said they would, job training that is not effective, inadequate supervision, and employees who are not empowered to make judgments when a situation "outside the box" occurs. Deming believed that in order to help ensure that managers have a sense of ownership and the knowledge to make good decisions, companies should promote from within when filling management positions.

Point 3—Cease Dependence on Mass Inspection. A willingness to do 100% inspection (and even pay for it) is the wrong approach to quality. While there can be occasions when such inspection is necessary to ferret out items that should not be allowed to reach the customer, the best approach is to keep the process in control and reduce variation to the point where inspection is not needed.

Point 4—End the Practice of Awarding Business Based on Price Tag Alone. It is best to have a long-term relationship between purchaser and supplier. Having more than one supplier is not a good idea because it adds variation between suppliers to variation between lots produced by one supplier. Of course, practical issues may rule out the possibility of having a sole supplier. What is important is to establish and maintain a good relationship with a supplier, one in which there is a willingness to work together. A supplier who is continually afraid of losing business to a lower bidder will be less willing to invest in long-term improvements.

Point 5—Improve Constantly and Forever the System of Production and Service. Rather than have the focus be on meeting product specs, the constant focus should be on increasing product uniformity by reducing variation about the target value. There must be a constant effort to improve test methods and to understand how the customer uses and abuses the product. A company should learn from its mistakes. Design flaws reported by customers should be eliminated in future versions of the product.

Point 6—Institute Training on the Job. One role of management is to ensure that employees learn and understand their jobs. Managers should have had enough experience with the company to understand the problems of production, procurement, accounting, sales, distribution, and so on. Training must include conveying an understanding of what constitutes acceptable work and what does not. Further, the training itself should not be second-rate, because bad habits, once learned, are hard to unlearn.

Point 7—Institute Leadership. The job of management is leadership, not supervision. Management to meet specs, objectives, or quotas, and management by numbers must be abolished. Understanding variation, understanding that workers should not be rewarded or punished for common cause variation, and understanding flawed statistical thinking—for example, that everyone can be above average—are crucial to good management. However, a leader should *use* data in order to know how workers are doing and know which workers may need special help. A leader is more coach than fault finder. A leader understands what his or her work unit needs to do to help the company achieve its goals, and creates an atmosphere in which workers enjoy working.

Point 8—Drive out Fear. The idea of empowering workers threatens some managers because they think they will lose power. Employees should not be afraid to make suggestions or to help another employee, even though that other employee might chance into a better evaluation/raise than them as a result. Actually, Deming believed that annual evaluations and quotas were both bad ideas. Workers may choose to take inappropriate actions in order to meet a quota (e.g., accept product units that barely fall outside spec limits, defer needed maintenance, etc.). Having annual evaluations induces workers to focus on short-term projects at the expense of taking appropriate long-term action on a process, and even to stilt data. Instead of fear there should be trust. Conducting annual evaluations is one of what Deming refers to as the *Seven Deadly Diseases* of Western management. These diseases are explained in *Out of the Crisis* (1986).

Point 9—Break Down Barriers Between Staff Areas. Departments should think of other departments as their customers. Just as teamwork is essential within a department, so is teamwork across departments. Decisions that are best for one's own department should not be made at the expense of the greater good, constancy of purpose, the product, and so on. Quality circles, that is, groups that meet regularly and consist of people from different units, departments, or interacting groups, including customers, are a good way to facilitate cross-unit exchange of information.

Point 10—Eliminate Slogans, Exhortations, and Targets for the Workforce. Slogans mouthed by management or placed on posters in the workplace do not change the system, improve operating procedures, or introduce training. The adoption of a slogan such as *Do it right the first time* or *zero defects* assumes that employees will do their jobs better because they see the slogan posted everywhere, as if they weren't trying as hard to do a good job before the slogan was posted. A slogan might even be used as an admonition when a worker makes a mistake. Slogans become acceptable only as part of a package that trains or helps workers.

Point 11a—Eliminate Numerical Quotas for the Workforce. Quotas do not bring process improvements, improve the product, improve production, or help workers do their jobs better.

Point 11b—Eliminate Management by Objective. Numerical goals and quotas must be abolished. An established goal—for example, reduce defectives next month by 5%—without a plan for how to accomplish the goal is a waste of time. The only numbers managers are allowed to bring up should be tied to survival. For example, management can inform inspectors of new criteria that must be implemented immediately because, if the percentage of defects is not reduced by 10% compared to the last shipment, the company will lose its biggest customer.

Point 12—Remove Barriers That Rob People of Pride of Workmanship. All workers want to feel a sense of accomplishment in what they do and that the product or system is better off for their contributions. An unfair merit rating system, rush jobs, antiquated tools, and management decisions that lessen quality, that make it more difficult for employees to do their job, or that keep them from doing good work, and so on, have a demoralizing impact on workers.

Point 13—Encourage Education and Self-Improvement for Everyone. An organization needs good people who are constantly improving. It is in the organization's interest to facilitate employee improvement, and in an employee's interest as well.

Point 14—Take Action to Accomplish the Transformation. Top management should see to it that enough people understand the first 13 points so that those management principles are inculcated.

While Deming's 14 Points fully justify lengthy consideration, there are other leaders who had a great impact on the field of quality. We now acknowledge some of them.

Walter Shewhart (1891–1967) originated the concepts of (and coined the terms) *assignable cause* and *chance cause* to describe these two sources of variation, and in 1924 proposed the control chart to his superiors as a tool for maintaining statistical control and for determining when assignable cause variation might be present.

The motivating problem for this was that engineers at Western Electric wanted to reduce the failures and repairs of their underground transmission systems. Shewhart received a doctorate in physics from Berkeley in 1917, was an engineer at Western Electric from 1918 to 1924, and from 1925 to 1956 worked at Bell Telephone Laboratories. He also lectured on quality control and applied statistics, acted as a consultant to the War Department, to the United Nations, and to the government of India. Shewhart believed that statistical theory should serve the needs of industry, also that quality had a subjective side—value in use, cost, esteem, and exchange. He believed in the constant evaluation of management practices and is well known for his corresponding PDSA cycle (Plan-Do-Study-Act). Shewhart also was well known for showing interest in and for inspiring others. He had a mentor impact on later luminaries in the field of quality, including W. Edwards Deming.

Harold F. Dodge (1893–1976) had a bachelor's degree in electrical engineering and a master's degree in physics and mathematics. From 1917 to 1958 he worked at Bell Laboratories where he associated with Walter Shewhart and other early pioneers in quality. One of his best known early accomplishments was the creation of acceptance sampling plans that controlled for sampling risks. In an age prior to calculators and computers, Dodge promoted the use of a sample size of 5 as better than sample size 4 because computing a mean using long division was easier. He similarly encouraged the use of the sample range for measuring variation because it was easier to compute than the standard deviation. He worked with Harry G. Romig in the 1930s and in 1940 they published the famous Dodge-Romig Sampling Inspection Tables. During World War II, Dodge developed sampling tables for and was a quality control instructor for Army Ordinance. He also was a consultant to the U.S. Secretary of War. From 1958 until 1970, he was a professor of applied mathematical statistics at Rutgers University. Dodge also was one of the founding members of the American Society for Quality.

Kaoru Ishikawa (1915–1989) majored in applied chemistry, received a doctorate in engineering, and was a professor of engineering at Tokyo University. Among the things he is famous for is the **cause-and-effect diagram** (also known as the *fishbone diagram* or as the *Ishikawa diagram*), a tool for finding root causes of problems (for example, see Figure 6.3). Ishikawa believed that quality in the workplace required the support, understanding, and leadership of top management. He promoted the idea that quality should be the focus during the life of a product, not just during its production, and that customer satisfaction throughout the life of the product should be the driving motivation for management. He also promoted the concepts of quality of management, of the company, and of the human being. A related quality practice he encouraged in Japan was the use of *quality circles.* A quality circle usually would consist of 5 to 10 volunteers from the same work area working toward improving the whole enterprise, including human relations, worker happiness, job satisfaction, and the tapping into and activation of human potential. Members of a quality circle are to have mastered basic tools such as Pareto charts, histograms, cause-and-effect diagrams, Shewhart control charts, and so on. They should engage in self and mutual development and find ways to implement the improvements they come up with. If a quality circle can't implement an improvement without help, they may pressure management to help them with implementation.

Joseph M. Juran (1904–) is recognized as a quality management thinker, writer, teacher and consultant. He was born in Romania, and in 1912 came with his family to join his father in the United States. The family was poor, and by working many odd jobs, he was able to earn a bachelor's degree in electrical engineering. He went to work for Western Electric in 1924. Juran's job put him in the inspection department at the Hawthorne Works in Chicago. In 1926 a team of statistical quality control notables, including Walter Shewhart, came to Hawthorn from Bell Laboratories to install statistical procedures the team had developed. Juran was among those trained by the team. He rose to be head of electrical engineering at Western Electric's corporate headquarters. In December of 1941, Juran took a leave of absence to become an administrator in the Lend-Lease Administration, which oversaw the shipping of war-effort materials to friendly nations. At the end of the war he left Western Electric to consult, write, lecture, and promote quality management. He was influential in broadening quality from its statistical side, which led to the concept of total quality management. To Juran, quality meant two things—meeting customer needs, and freedom from trouble. In 1951 Juran published his now famous *Quality Control Handbook.* In 1954 the Union of Japanese Scientists and Engineers invited him to Japan to deliver a series of lectures on managing quality. Deming's visit and Juran's subsequent visit had a tremendous impact on the quality of Japanese goods and industry.

Genichi Taguchi (1924–) contributed significantly to the survival and to the great successes of Japanese industry after World War II. His initial training was as a textile engineer, and he later earned a doctorate in science. His major contribution was to help revolutionize Japanese manufacturing through cost savings. He worked in the Imperial Japanese Navy's Navigation Institute in World War II. After the war he worked for the Ministry of Public Health and Welfare and for the Institute of Statistical Mathematics, Ministry of Education. At the latter, Taguchi learned about experimental designs and orthogonal arrays, which led to his becoming a consultant to a pharmaceuticals company. In 1950 he started working for the Electrical Communications Laboratory of the Nippon Telephone and Telegraph Company, while also consulting with various Japanese industries. He published now famous books on orthogonal arrays and on experimental design. In 1954–1955, he was a visiting professor at the Indian Statistical Institute, where he met R. A. Fisher and Walter Shewhart. Taguchi developed techniques for determining which sources of noise have the greatest impact on variation and cost. He devised a quadratic loss function model for the loss society experiences once a product is shipped. The loss is a function of the distance between the process mean and target value. The loss includes various company costs (e.g., rework and scrap, maintenance costs, warranty claims, even loss of market share) and costs to the customer and society. By Taguchi's model, overall loss is reduced when variation about the target is reduced. Taguchi emphasized the use of particular experimental designs to help identify which sources of noise have the greatest impact on cost and helped companies to manufacture products and set up processes that were robust—that is, would perform even under less-than-ideal conditions. Taguchi worked with industrial statisticians at Bell Laboratories, and from 1964 until 1982, was a professor at Aoyama Gakuin University in Tokyo.

Philip B. Crosby (1926–2001) began his quality career working on an assembly line for Crosley in 1952, after serving in the Navy in the Korean War. In 1957 Crosby began working for Martin-Marietta, and then for ITT beginning in 1965. He was a quality manager at Martin-Marietta, where he created his *zero defects* concept, and was a corporate vice president for ITT. In 1979 he founded Philip Crosby Associates, Inc., a consulting company whose aim was to help management *do it right the first time* by instituting a culture that would prevent defects from occurring, as opposed to inspecting for defects. In 1979 he published *Quality is Free* and in 1984 *Quality Without Tears.* Crosby's approach to quality improvement at a company involved the company's having a core of quality specialists, training of all staff in the tools of quality improvement, inculcating the idea of preventive management in all areas, and developing a process model that would make clear for suppliers and customers the definition of requirements.

There are many sources for finding out more about the history of the quality movement and the lives, contributions, teachings, and philosophies of the movers and shapers in the field of quality. Some are: Besterfield (2003), Crawford, Bodine, and Hoglund (1993), Crosby (1975, 1984, 1996), Deming (1986), Goetsch and Davis (2003), Gitlow and Gitlow (1987), Grant and Leavenworth (1996), Ishikawa (1985, 1986), Juran (1988) and Juran and Gryna (1988), Neave (1990), Pyzdek (1991), Roy (2001), Scherkenbach (1991), Shewhart (1931), Shores (1990), and Walton (1990). The reader may also want to visit some Web sites—for example, ASQ—About History (2003), Phil's Page (2003), Juran—Our Founder (2003), Quality Gurus (2003), and Resources—Leaders (2003).

PROBLEMS

Identify one or more of Deming's 14 points that are violated in the following real incidents. Sometimes a setting may be modified slightly in order to protect someone's privacy. There may be multiple answers.

1. A husband and wife were driving into a town, and, having missed lunch, were extremely hungry. Getting a hamburger sounded wonderful as they approached a succession of fast-food restaurants. They parked, walked into Fast Food Restaurant (FFR) 1, and got at the end of one of three long lines. There were plenty of servers behind the counter, but they were milling around, without an obvious game plan. After several minutes, it was not clear that there was a process for filling orders or for taking orders. Sensing that service would take forever, they got back in the car and drove to FFR 2. There was one line and plenty of staff. But only one of them was taking and filling orders, and the line was really long. They left their car there and walked next door to FFR 3. They were the only customers at the time and walked right up to the counter. Employees could be heard talking in the back. The couple yelled out some *Hellos,* but no one seemed very interested in coming out to wait on them. So they walked next door to FFR 4, a fried chicken place. One server was taking the order of the man ahead of them. All were served promptly and were soon eating.
2. This conversation is relayed by a faculty member we refer to as John Smith, who teaches at a college. Because parking is difficult to find

near the campus, most employees and faculty pay for a parking sticker to park in a campus lot. Each sticker is for a particular lot, lot assignments are based on seniority, and the cost of a sticker decreases as the distance from campus to lot increases. Smith has a sticker for "A Lot," one of the highest priced lots. Premium lots are marked: *24 hour decal required—towing enforced.* It has become common knowledge that lots are not policed after 6 P.M. and haven't been for years. Evening students with stickers for other lots and nonstudents have begun parking wherever it is most convenient. In the last two months, faculty who teach during the afternoon and evening and who go out for dinner risk returning at night to a full lot. It is permissible to park in an equivalent or less costly lot if your assigned lot is full. The conversation between Smith and the officer answering the telephone at the public safety office went as follows.

S: Hi. This is John Smith in the Chemistry department. I just want to let you know where I parked and why, so I don't get a ticket. I have a sticker for A Lot, just got back from dinner, and A Lot is full. I drove to B Lot, and it's full, too. So is C Lot. I have a class to teach in four minutes, so I drove from C Lot back to B Lot and parked in a handicap slot. All of them are empty. I know I'm not supposed to park there, but I have to teach.

P: You can't park in a handicap slot.

S: I know I'm not supposed to. I tried to find another lot, but class is about to start.

P: I can't let you park in a handicap slot.

S: I know I'm not supposed to be there, but everybody knows you guys aren't ticketing at night, so people without stickers have begun parking where they're not supposed to. I'm asking you to please not give me a ticket. I'm in the parking space next to Dunlap Street.

P: We ticket at night. You'll have to find another lot.

S: What does having an A sticker entitle me to? I'm willing to play musical chairs up to a point, but I have to get to class.

P: I can't let you park in a handicap slot. You'll need to find another lot.

S: I'm not reaching you, am I?

P: Would you like to talk to a supervisor?

S: Please. Thanks.

A similar conversation ensued with the supervisor, who said that the next shift was about to go on duty, that they actually would start ticketing in about 15 minutes, and that he could not give permission to park in a handicap slot. The supervisor put Smith on the phone with his supervisor, who made the executive decision to take the car description and license number, and promised that a ticket would not be given.

3. The first model of M-16 rifles used by American soldiers who fought in Vietnam developed a reputation for jamming after cooling down after a firefight—they were supersensitive to dirt. Some soldiers

complained that the new weapon was not tested enough prior to the war and must have been made by the low bidder.

4. A mathematics department offered two introductory statistics courses, Course A for math majors and Course B, which had a very modest prerequisite and which was taken primarily by social science and business students. The Business Department requested that the Math Department design and offer a statistics course specifically for business majors, one that required a stronger mathematics background than the prerequisite for Course B. If such a course—call it Course C—were created, there would be enough business and social science majors to assure that many sections of Course C would be offered each semester. In an effort to hold down the number of different kinds of introductory statistics courses, the Math Department ceased offering Course A, offering instead a Course C that was to be taken by both math majors and business majors. A faculty member who frequently taught Course A warned that the mathematical maturity difference in the two kinds of students in Course C would create problems that would affect the quality of the course. After several semesters, it became clear to all who taught Course C that the mismatch in math backgrounds of the two kinds of students was creating terrible teaching and grading issues. The department resumed offering Course A for math majors, while also offering Courses B and C.
5. Suppose the first versions of a model of a four-burner gas range had burners and knobs laid out as depicted in Figure 6.1 and that all future versions of the model do also.
6. Joe currently is a floor supervisor for part A of a product that consists of parts A and B. Initially, Joe started in customer service, rose to its top spot, and later was shifted to quality manager. Joe is highly respected for his accomplishments in all three areas. Management sees in Joe a potential future plant manager. In an effort to pave the way for Joe to understand the big picture, and since there have been quality problems recently in making part B,

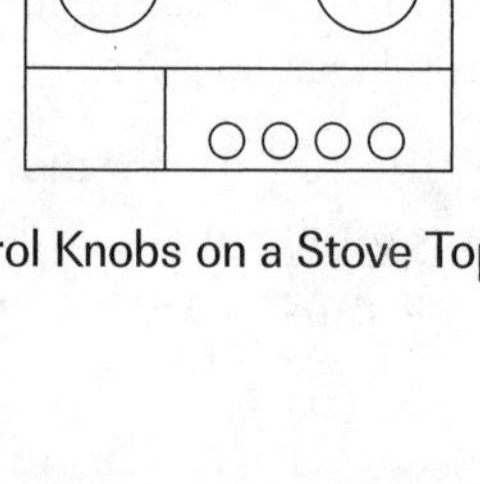

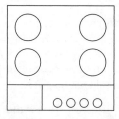

FIGURE 6.1 Burners and Control Knobs on a Stove Top.

Joe is made floor supervisor for part B. Understanding the subcomponents of part B requires understanding physics, and in particular, optics. Joe has never had a physics course. Part B operators determine that Joe doesn't know enough to lead them, and resent that one of their own was not given the supervisor job. The problems with part B persist, operator morale is not high, and Joe gets a poor annual review for his first year on the new job. Joe takes a job working for a competitor as quality manager.

7. When Fred, the quality manager, came to visit line operators or shipping operators, it was fairly common for him to bring along and hold up for viewing some faulty widgets that had been returned by a customer. "Look at what one of you let through," he would say. "Whoever did so has to be more careful." Customers used the widget as a subcomponent in their product, and a defective widget would cause product damage and generate a customer claim. While such an admonishment might be well-intended, it doesn't change the system that, in it current state, lets occasional defects reach the customer. Workers don't intentionally fail to cull out faulty widgets, and operating procedures don't change as a result of a harangue. Many operators found these visits demoralizing.

8. The FBI had projected that in 1990 there would be a 10% increase in violent crime nationwide. In 15 of the 19 rural states that reported crime figures, violent crime had increased an average of just 8.4%. Nevertheless, 6 of those 15 experienced increases above the projected 10% average increase. Senate Judiciary Chairman Joseph R. Biden Jr. (D-Del.), said that this and other data showed that rural criminal violence is "growing at an astonishing pace" and that rural America was suffering a "plague of violent crime . . . " The other data included a comparison of the percentage increase of murders in Montana from 1989 to 1990 to the percentage increase in Los Angeles (see "How Not to Think—an Example" in Section 1.1). The Senator introduced a bill that called for spending more than $100 million to combat the plague (see "Senate Panel Cites 'plague' of Rural Criminal Violence," 1991). Data from the article are used with permission of The Associated Press.

9. In a particular academic department, the most important factor in whether a faculty member gets a high annual evaluation is the number of papers the person has published during the year. Counting papers is an easy way to measure, but discourages work on projects that can't be done in a year, even if the completed project would have a tremendous impact on the subject discipline, on teaching, or on programs at the institution.

10. This is a conversation that occurred as I was walking with an upper-level manager known for her intimidating tactics with subordinates.

    Q: What can we do to help make it easier for our people to want to buy into this change?

    ULM: In my experience, it has always been the case that if you put a gun to their heads, their hearts will follow.

    Actually, I misquote. Putting a gun to their heads is more tasteful than what she really said. Her terror tactics demoralized many subordinates. Fortunately, there was enough discernable negative karma about her that top management forced her out.

11. An instructor who taught freshman-level mathematics had become tired of seeing many students be content to get partial credit on certain skills when, with a little extra effort, they could get full credit. One example was finding solutions to an equation such as $|3x - 5| = 7$. Despite his teaching them how to find both solutions, too many students would simply ignore the absolute values, give an answer of 4, and hope for half credit. When no credit at all was the result, some students would even complain. The instructor tried a teaching experiment the next semester by adding a supplemental twist to the usual rules for assigning grades based on semester average. This particular math problem was established as one of four kinds that students had to get completely right at least once during the semester if they wanted a C or better in the course. Students had multiple opportunities to work each type of problem during the semester, but if a student could not get full credit at least once on each of the four problem types, the student could do no better than a D for the course, even if the student's semester average would otherwise have warranted an A, B, or C. Students scrambled to get full credit and the class did better on the final exam than any he had ever taught. However, his student evaluations were lower than any he had ever received. Students had been forced to exceed their desired work level. Because the department head took student evaluations into account when doing annual reviews, the instructor became less willing to experiment with approaches to teaching.

12. The dean of the school of education at a university in the Midwest had created the Teacher Education Council, a committee composed of representatives from a number of groups—education faculty, noneducation faculty who train future teachers in the various subject matter areas, K–12 teachers who supervise practice teachers, K–12 teachers, and former graduates who are teachers. The existence of such a council was mandated by an accreditation agency to facilitate a healthy information exchange between groups. The school's dean, who chaired the council, used the TEC to promote school of education

activities, but turned a deaf ear to legitimate complaints from K–12 teachers about preparation of practice teachers, because making changes would have meant shaking up the system.

13. It is systemic at colleges and universities for scientists and mathematicians to complain that the teacher preparation curriculum contains too many education courses, not enough science/mathematics content courses, and that education people get to make all the rules. It is also systemic for education faculty to argue that the education content component of teacher preparation is just as important as the other components and to resist attempts to add more subject area courses. A reason for the resistance is that education courses can't be cut without risking loss of accreditation. Also, simply adding more science or mathematics courses to a degree program lengthens the time to complete the degree and makes it less marketable.

    At one particular university, the College of Education and the College of Arts and Sciences developed a unique master of education degree program in science and mathematics for teachers. There were some required education courses, but much of the curriculum consisted of science and/or math content courses taught by scientists or mathematicians who wanted to work with teachers. Curriculum design and graduation requirements in the degree program reflected exceptional compromise and teamwork among faculty from seven departments. The program was governed by a steering committee with equal representation from each college. Mathematics and science departments endorsed the program because their faculty did most of the teaching, and power was equally shared. Local K–12 teachers, and even out-of-state teachers, were attracted to the unique program. With a critical shortage of K–12 science and math teachers nationwide, program faculty and students viewed the program as a jewel.

    Periodically, all teacher education programs at the university were reviewed to ensure that they met the standards of an agency that accredits teacher preparation programs. One reviewer for the agency criticized the program for its shared governance. He wrote that it was unclear whether there was a need for a science and math program, that a two-school structure was awkward, and that the program needed to be governed by the College of Education, like all the other education degree programs at the university.

14. An academic department at a four-year college had a weekly colloquium series. Topics had been on advanced research given by faculty in or visiting the department. The faculty member in charge of the colloquium series decided to have a few weekly topics that would be accessible to undergraduate majors as well as faculty. One faculty member complained about this. Time was precious and she

didn't want to attend any presentations that could be understood by undergraduates.

15. Under a recently adopted Missouri state law, police had to record the age, sex, and minority group of motorists they pulled over, an effort to prevent racial profiling. A south St. Louis police sergeant said the law made police more timid and conscious of being labeled racist. Arrests in south St. Louis plunged 21 percent in the nine months since enforcement began. (See "Police Shootings in St. Louis Have Officers on Edge and on the Defensive," 2002.) Used with permission of the Associated Press.
16. In a particular academic department, every paper submitted for publication had to include the name of the department head as a co-author. The department head became known as prolific, while faculty members in the department, with their contribution lessened, resented having to share recognition with him.
17. A small company had no computer tech person on staff. The one computer programmer at the company found herself continually bombarded by requests to do computer tech jobs for others in the workplace. As one example, a management decision was made to switch from one word-processing software package to another. The switch was made on each employee's computer, but no training was implemented to help people learn how to use the new software. With the constant interruptions by staff needing help, the programmer was always down to the wire on completing her programming assignments and would like to have spent more time on some of them. She felt hamstrung in making contributions to the company's system.
18. At the two-year college where Professor X works, department chairs evaluate their faculty annually in teaching, professional development, and service. At the end of each course, a student opinion questionnaire is given. A major indicator used by a chair in evaluating someone's teaching is how students respond to question 12, which asks students to express their opinion of the professor's overall quality of instruction. Students circle an integer from 1 (well below average) to 6 (outstanding). The annual evaluation of an instructor who taught three sophomore and three freshman English courses contained the following: . . . *Your hourly tests in the freshman-level classes each contained three or fewer essay questions, which is atypical for our 100 level courses. Students may have some discomfort with that, which could be reflected in your student ratings. You might look at the student comments to see whether they express some reaction to the shorter-than-usual tests. Your tests were very good, though, and were appropriate for this course. Your mean ratings on Course-Instructor #12 ranged from 4.2 to 5.8 this year, with an overall mean about 5.0, near the mean for all faculty in our department. Your lowest ratings (4.2 and 4.9) were in your two noncredit developmental English courses.*

6.2 SOME PROBLEM-SOLVING TOOLS

Even the best of systems will have occasional problems. When a problem occurs and the cause is not obvious, it is typical for a quality circle or for an appropriate team of individuals to brainstorm in an effort to find the cause. Problems at a chemical plant, for example, could call for a team that includes chemists, who will know if any proposed change in the chemistry of the product will be acceptable; a quality manager, who understands spec limits and customer requirements; operators, who are experts on doing equipment adjustment; a computer programmer, who understands how data is transferred from product specific measuring equipment to statistical software and spreadsheets; engineers, who understand equipment parameters and who will let everyone know whether a proposed change will blow the plant up; and the vice president in charge of production, who knows a lot about everything that goes on at the plant.

This section is devoted to some of the standard tools used in problem solving. We point out that the histogram, discussed in Section 2.2, is one such tool. We now introduce another commonly used tool and an example of how *not* to use it.

The Pareto Chart

The data given in this illustration are in the spirit of data from a real survey that was taken as the result of a real incident that occurred long enough ago to pose difficulty in locating the newspaper article that contained the actual data and a description of the horrific crime. However, the reader should get the idea.

EXAMPLE 1

Two children committed a crime, one that raised the issue of whether they should be tried as adults. *How could kids their age do something like that?* people asked. The question, *Where do you think children should learn values?* was the key question on a survey conducted in reaction to the crime. Respondents were allowed to choose one of: neighbors, friends, church, school, family, relatives, and community. Let us assume that the survey responses came from a representative cross section of the citizens in the region. The responses were along the lines of the summary shown in Figure 6.2, which is called a **Pareto chart.** This Pareto chart sorts the categories in the order of descending frequency. In industry, Pareto charts are commonly used to identify, based on historical data, which root cause is the most frequent for an item not meeting specs, which cause is the second most frequent, which is the third most frequent, and so on. Alternatively, causes can be arranged in order of decreasing cost to the company instead of decreasing frequency.

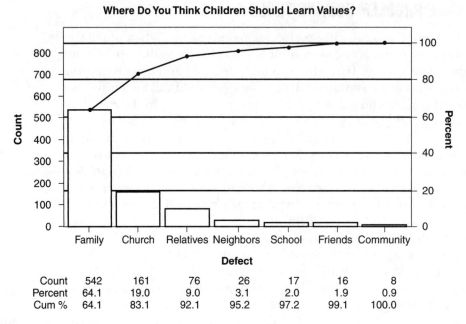

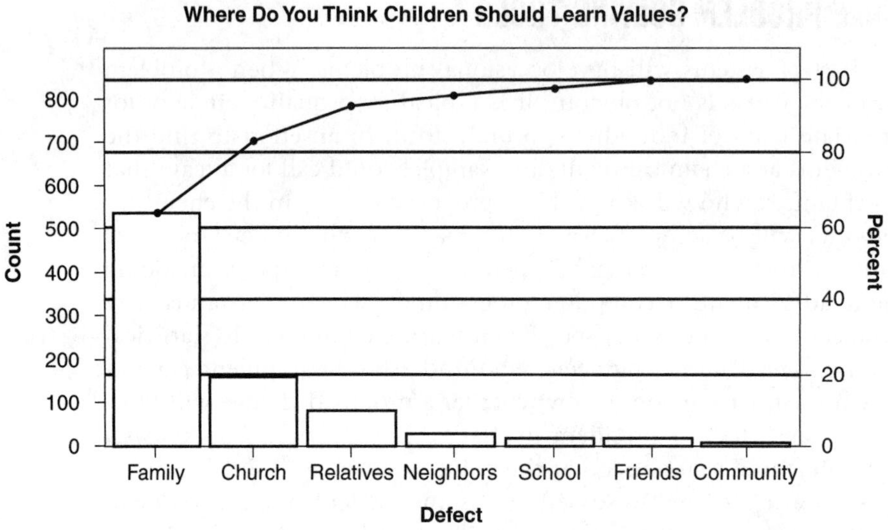

Count	542	161	76	26	17	16	8
Percent	64.1	19.0	9.0	3.1	2.0	1.9	0.9
Cum %	64.1	83.1	92.1	95.2	97.2	99.1	100.0

FIGURE 6.2 Pareto Chart Summarizing a Survey on Where Children Learn Values.

Figure 6.2 shows that *family* was the first choice, comprising 64.1% of the respondents. Next came *church*, with 19.0%, and so on. The polygonal line shows cumulative percent. The first two categories, for example, accounted for 83.1% of all respondents, while 99.1% of the choices were for something other than *community*.

Community businesses, clubs, and other organizations concluded that since *community* had come in dead last, they needed to do more, and requested seed money from city and county governments and from other sources. The money would be used by community organizations to develop value-imparting programs for children. This kind of thinking is backwards. Effort and/or money should not be going to address the cause expected to have the least impact on the problem. The main idea of a Pareto chart is to identify the source or sources that contribute most to defects or cost so that energy and money can be focused on eliminating the most prevalent or costly source of defects, then the second most prevalent or costly, and so on.

For readers with access to MINITAB, we describe how to produce the Pareto chart shown in Figure 6.2. All the counts 8, 16, 17, 161, and so on—order irrelevant—should be entered in one column and the names of the corresponding categories entered into a second column, in the same order that counts were entered. From the menu, choose **Stat,** then **Quality Tools,** then **Pareto Chart.** A box will appear. Click the option **Chart defects table,** then click on the **Frequencies in:** box and identify the column containing the frequencies. Next do the same in the **Labels in:** box to identify the column containing category names. To avoid combining categories with small counts into a single category called "other," enter, say, 99.99 in the next box, to make the surrounding sentence read "**Combine defects after the first 99.99% into one.**" To enter a title for the chart, click the **Options** button and do as indicated. Then click **OK.**

The Cause-and-Effect Diagram

A systematic and hierarchical way to display all the causes of a given effect is in a cause-and-effect diagram (AKA fishbone diagram or Ishikawa diagram). Creating a cause-and-effect diagram can be helpful when brainstorming about which of all possible causes might be the source of the current problem. The diagram's hierarchical structure, similar to a tree or fishbone, allows a team to eliminate or isolate causes by general type. Specific potential causes of like kind are viewed as individual branches off one of the main branches. Figure 6.3 illustrates a cause-and-effect diagram that might be used by a family to determine what spending to adjust in order to have more available income at the end of next year.

Cause-and-effect diagrams can be computer generated. For those with access to MINITAB, six columns need to be keyed in to produce Figure 6.3.

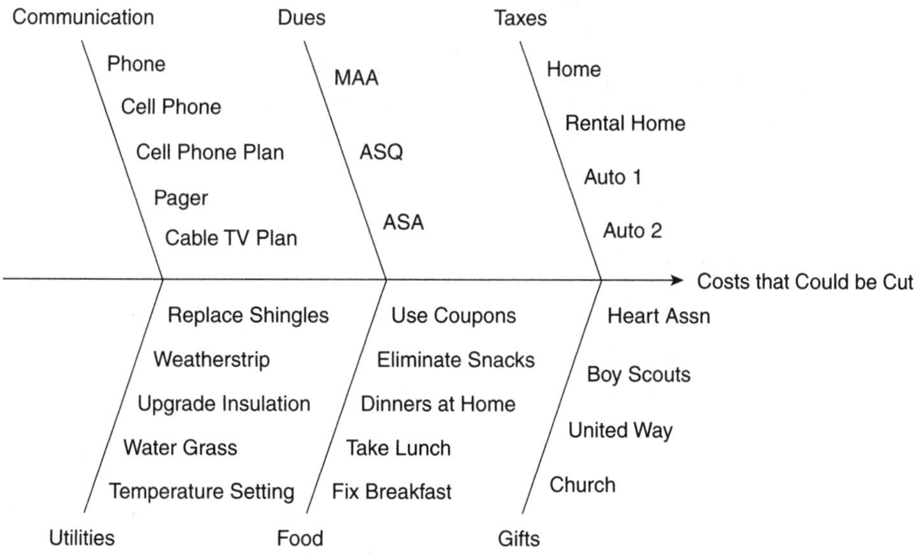

FIGURE 6.3 Cause-and-Effect Diagram for Expenses.

One column, devoted to communication costs, needs to contain five rows, each row containing one of the causes of communication costs—*phone, cell phone, cell phone plan, pager, cable TV plan.* Another column will contain the specific causes of utilities costs, and so on. After specific causes have been entered into the six columns, the user should choose **Stat,** then **Quality Tools,** then **Cause-and-Effect.** A box will appear. In the **Label** column, enter a label name for each of the main branches (taxes, gifts, etc.), one label name per box. Next to each label name, in the **Causes** column, identify the column containing the specific causes for that label. The diagram can be given a title as well as a description of the effect (*Costs That Could Be Cut* is the description used in Figure 6.3). Lastly, click **OK.**

Scatter Diagrams

The **scatter diagram** is a graphical tool that is used to help determine how or whether two variables might be related. One measurement for each of two different characteristics need to taken from each item.

EXAMPLE 2

Suppose two quality characteristics of a brass rod are its length, X, and its average diameter, Y. Suppose the first rod in a sample of size 15 has a length $X_1 = 22.392$ cm and average diameter $Y_1 = 1.124$ cm, the second rod has length $X_2 = 22.539$ cm and average diameter $Y_2 = 1.090$ cm, and so on. Each rod generates a point (X, Y) that can be plotted in an X-Y coordinate system. The scatter diagram (or scatterplot) would consist of the 15 plotted points (22.392,1.124), (22.539 , 1.090), ... Such a scatter diagram is shown in Figure 6.4. This particular point-cloud is approximately horizontally aligned—that is, as X increases, Y does not tend to increase or decrease in any systematic way. One would say here that length and average diameter do not appear to be correlated (or that they appear to be uncorrelated).

EXAMPLE 3

Recall from Section 3.1 the company that made spools of thread and took periodic samples in order to measure the minimum tension needed on the free thread end to make the thread pull freely from the spool without snagging. The company wanted to know whether how the loose thread contacted the spool would affect the minimum tension measurement. The company examined four ways loose thread could hang down: from the top (where the loose end contacts the spool and the thread winds downward), from the middle (where the thread winds downward), from the middle (where the thread winds upward), and from the bottom (where the thread

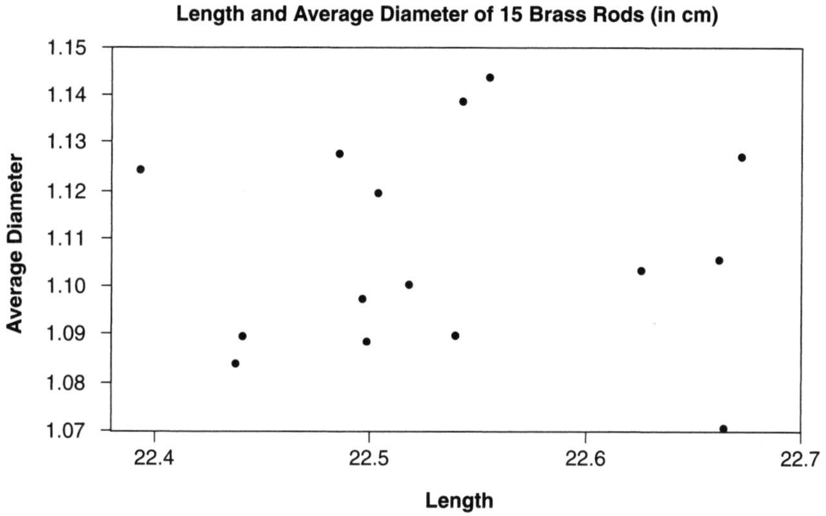

FIGURE 6.4 Scatter Diagram of Average Diameter Versus Length.

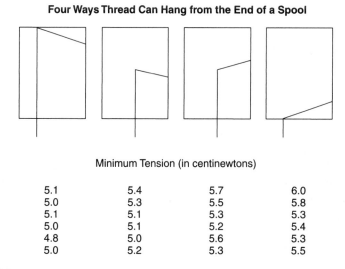

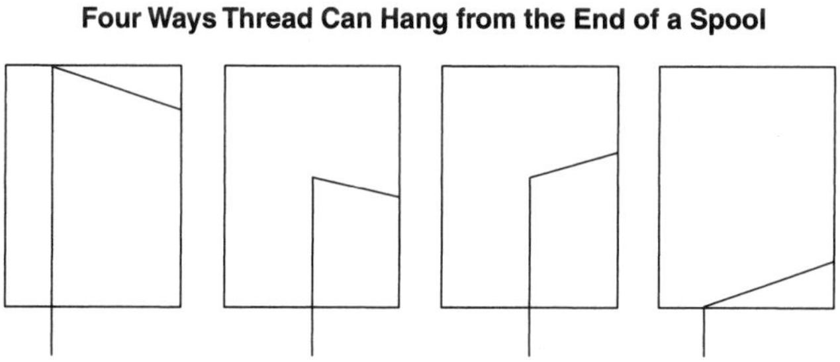

Minimum Tension (in centinewtons)			
5.1	5.4	5.7	6.0
5.0	5.3	5.5	5.8
5.1	5.1	5.3	5.3
5.0	5.1	5.2	5.4
4.8	5.0	5.6	5.3
5.0	5.2	5.3	5.5

FIGURE 6.5 Minimum Tension Needed to Pull Thread from a Spool.

winds upward). Figure 6.5 illustrates. To measure the minimum tension, gradually increasing downward pull was applied to the loose end of the thread until the point was reached when the thread fell freely from the spool. The suspicion was that minimum tension would increase in the left-to-right order shown in Figure 6.5. For a description of how all 24 measurements were obtained from one spool, the reader is referred back to Section 3.1.

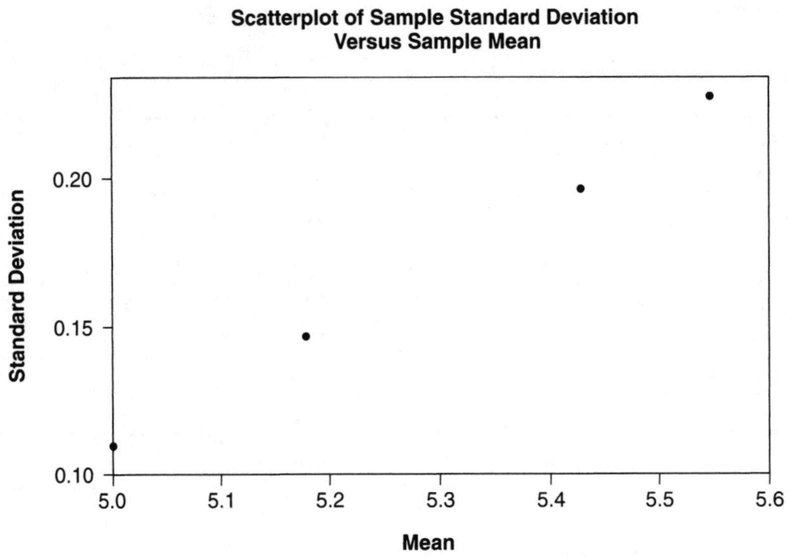

FIGURE 6.6 Positively Correlated Sample Means and Standard Deviations.

Descriptive statistics for these data were:

$\overline{X}$	5.00	5.18	5.43	5.55
S	.110	.147	.229	.288

Prior to this experiment, employees who did the minimum tension tests had paid no attention to how the loose thread contacted the spool. Management had been aware of a positive correlation between $\overline{X}$ and S—that is, the larger the value of one of the variables, the larger the value of the other variable tends to be. So the scatter diagram shown in Figure 6.6 did not come as a complete surprise. But the scatterplot did show something new—that positive correlation between minimum tension and standard deviation is the case, even on the same spool. After the experiment, in order to incorporate this knowledge into the measuring procedure, management decided to select one of the four thread positions as the one to use when taking measurements. The one selected was the *hang from the top* position. This is the method with the smallest standard deviation, and, therefore, is the method that tends to produce the most precise tension measurements. Figure 6.6 shows that the points come very close to falling on a straight line with positive slope. The strength of relationship between tension and standard deviation is quite strong. It is more commonly the case that two variables are correlated in some way but the relationship is not this strong.

TABLE 6.1 Weight Loss Data.

Avg. weight	110.2	115.4	120.8	125.8	126.2	131.8	135.4	140.6
Headstand1 loss	20	48	2	78	31	28	42	83
Headstand2 loss	127	15	29	54	46	39	58	52
Avg. headstand loss	73.5	31.5	15.5	66	38.5	33.5	50	67.5
Standing loss	29	48	74	30	76	50	25	54
Avg. weight	143.6	153.6	164.6	167.4	173.8	174.4	183.2	184.2
Headstand1 loss	75	30	134	60	80	103	18	3
Headstand2 loss	41	93	34	144	119	112	51	117
Avg. headstand loss	58	61.5	84	102	99.5	107.5	34.5	60
Standing loss	−5	48	38	−4	45	91	48	67
Avg. weight	193.6	195.6	216.0	222.8	248.8			
Headstand1 loss	119	29	28	151	0			
Headstand2 loss	80	83	32	106	160			
Avg. headstand loss	99.5	56	30	128.5	80			
Standing loss	93	20	24	82	102			

EXAMPLE 4

Between the mid-1980s and the late 1990s, amateur wrestlers who had been running in order to lose enough weight to qualify for a lower weight class sometimes engaged in a practice that was puzzling to those not connected to the sport. If the weigh-in period was about to end and the wrestler still needed to lose a fraction of an ounce, he might stand on his head for a minute or so, then return to his feet, step back on the scale, and often qualify for the lower weight class. Headstands are now banned. Table 6.1 shows part of a larger data set, printed with the permission of Mick Norton, a former high school and college wrestling official. Twenty-one high school wrestlers jogged/exercised for at least 15 minutes, and were then weighed. A period of 1 minute 45 seconds was then allowed to elapse and the wrestler was then reweighed. Wrestlers would tend to sweat off weight while idle during this period because they had been exercising. On two replications (replicates were done on different days), the wrestler did a headstand during the 1 minute 45 second period. On a third replicate, the wrestler stood upright during the time period. The average weight of the wrestler (based on weights on five different days) and the weight lost in paper clips (one paper clip weighs .00320 pounds) during the 1 minute 45 second period are shown in Table 6.1.

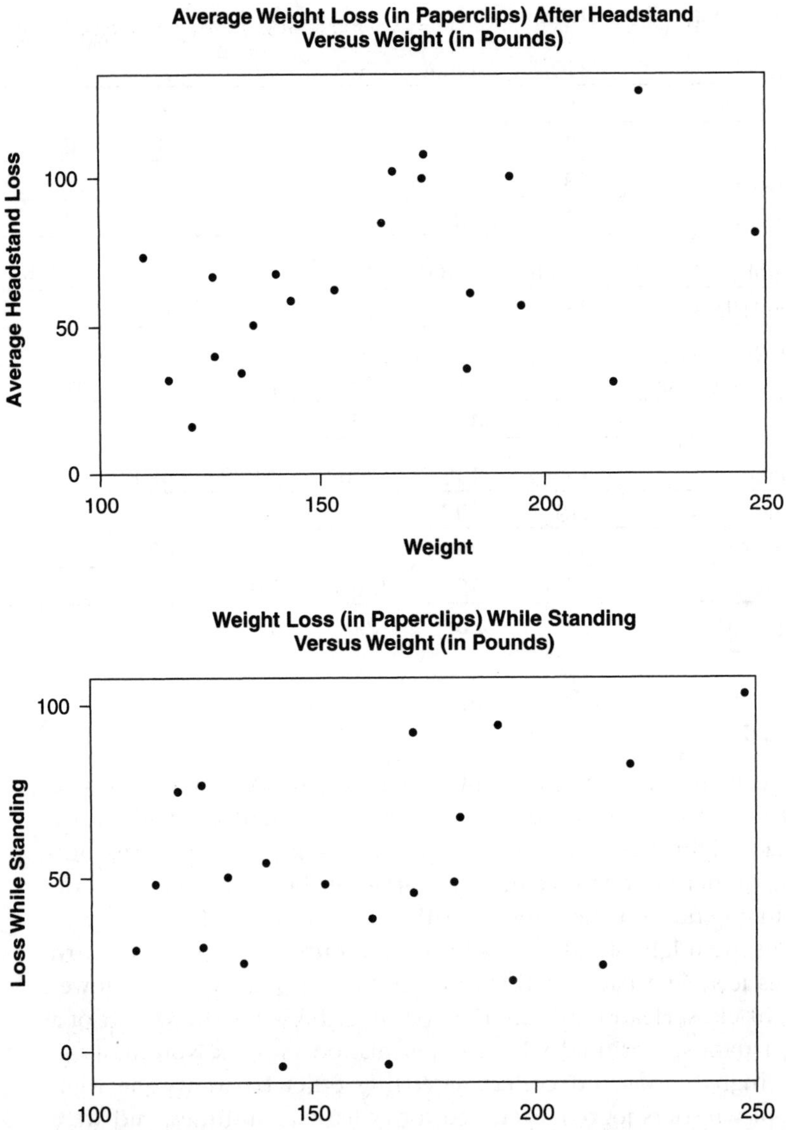

FIGURE 6.7 Scatterplots for Weight Loss During a 1-Minute 45-Second Time Period Following Exercise.

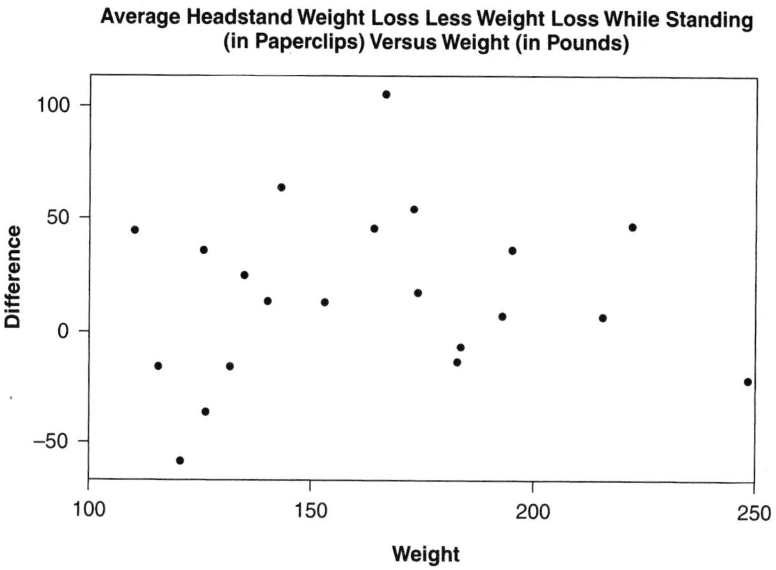

FIGURE 6.7 Continued.

Figure 6.7a is a scatterplot of average headstand weight loss versus wrestler weight. Although the points do not come close to falling on a curve or straight line, the cloud of points does have a definite *Southwest-to-Northeast* orientation. While there are examples of heavier wrestlers who, after exercising, lose less weight during the headstand than lighter wrestlers do, generally, heavier wrestlers *tend* to lose more weight during the headstand. That is, during a headstand, weight loss and weight are positively correlated.

Figure 6.7b is a scatterplot of weight lost while standing versus wrestler weight. The plot is clear here as well. While standing idle after exercise, weight loss and weight are positively correlated.

Figure 6.7c is a scatterplot of average headstand weight loss, less weight loss while standing, versus weight. Fourteen of the 21 wrestlers lost more weight (had positive differences) while doing a headstand, and 7 lost more weight while standing idle. The difference in weight lost by the two methods does not appear to be strongly correlated with weight.

However, the issue as to whether these 14 positive differences and 7 negative differences can make the case that doing a headstand *tends* to help a wrestler lose weight can be addressed by a statistical method called a t-test, a procedure that falls in the realm of hypothesis testing, which is beyond the scope of this book. We can say, however, that the answer is *yes*. An estimate of the average amount of weight lost by doing a headstand is .05 pounds. It might be helpful to point out an explanation. Body liquids such as blood and spinal fluid drain into the head during a headstand, and then drain out after the wrestler returns to an upright position. Thus there is a small amount of accelerating liquid mass that does not contribute to body weight. The apparent weight loss due to the headstand is temporary, lasting until the excess of fluids in the head is completely drained out.

For those with access to MINITAB, we discuss how to obtain a scatter diagram. Refer to Figure 6.7b. The values of X (weight) need to be entered in one column, say C1. The corresponding values of Y (weight lost while standing) must be entered in another column, say C2. Names for these two columns (*weight* and *loss while standing* were used here) should be entered in the cells at the top of the two columns so that the axes will be labeled. Once the data and variable names have been entered in the two columns, the user should choose **Graph,** then **Scatterplot.** A box will appear. Choose the option labeled **Simple** and click **OK.** In the first row under **Y variables,** enter C2, which identifies the column containing the *y*-coordinates. Then, in that place in the window that asks for the column with the *x*-coordinates, enter C1. To give the scatter diagram a title, click the **Labels** button, and enter the desired title. Click **OK** to confirm the title, then **OK** to obtain the scatter diagram.

Flowcharts

A **flowchart** is a graphical representation of the sequence of steps involved in a process or in part of a process. Being able to see the flowchart for a process can be helpful to a team that is brainstorming to determine the source of a problem occurring within the process. An engineer may use a specific geometric shape to represent a step of a particular kind. For example, a diamond usually is used for a step that involves making a decision, whereas a rectangle is a commonly used shape for identifying a physical action. However, that the flow of steps is correct is the most critical thing. A process for filling orders for window glass is shown in Figure 6.8.

Force Field Analysis

Sometimes it is helpful to consider all of the positive and negative forces that bear on making a particular decision or on a given problem. A force field analysis is an organized, pictorial way of showing these forces. Consider a company that plans to implement a zone chart. Whenever the zone chart shows an out-of-control signal, a chemical recipe will be altered to recenter

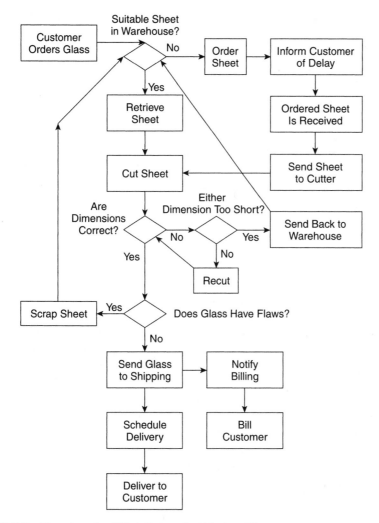

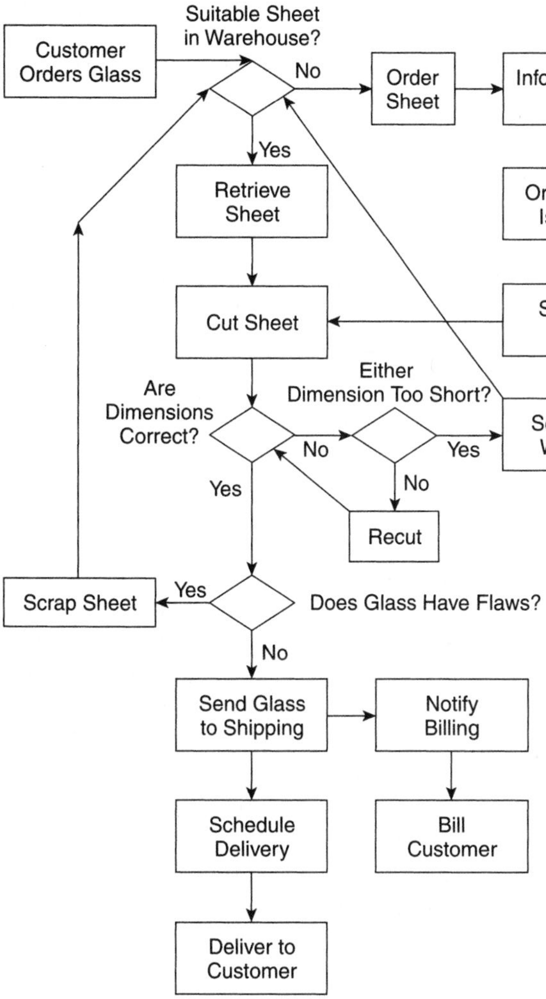

FIGURE 6.8 Flowchart for Filling Orders for Window Glass.

the process. Figure 6.9 shows a force field analysis that might be used to help a group decide whether to have the operators who control the recipe update the zone chart by hand and recognize out-of-control signals themselves or to have the zone chart automated and prompt the operators to change the recipe when necessary.

Working Toward a Decision

Whoever facilitates a group in making a decision must understand decision-making dynamics and politics. While it may be reasonable to assume that everyone in the group wants what is good for the company or enterprise, some individuals may also be influenced in part by the perception of loss

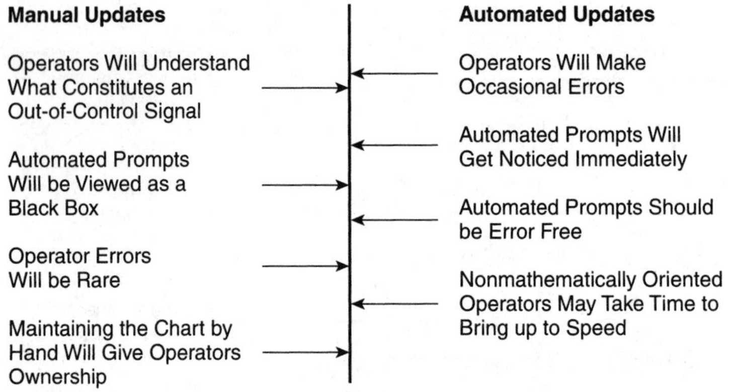

FIGURE 6.9 Force Field Analysis for Choosing Between a Manual or Automated Zone Chart.

of power, by the fear of disagreeing with a superior, by reluctance to make a choice that means having to learn something new, by personal empire building, by ego, and so on. There also may be coalitions within the group. A loud, domineering coalition or individual can be demoralizing to a group. A good facilitator needs to be able to recognize and deal with compromising motivations, draw ideas from reluctant contributors, and be tactful.

A variety of strategies for generating possible options and solutions from a group may be employed. There is brainstorming, a good group activity that requires someone to record the suggested ideas. Ideas that come from imagining an idealized process is another. Benchmarking—examining how other respected companies or work groups have dealt with the issue—is another. Examining data is still another. Quality measurements before and after a problem occurred, or data of a kind that would be projected to occur if a considered option is taken, including the bottom line, are examples of data that could be examined. Asking *why*, possibly repeatedly, can be helpful (Why is it taking customers two days longer to get their orders? Oh? Why did we insert that new step in the process? It did? Why did that happen? Oh? Why should we assume that it would happen again? Why is it worth adding two days to everybody's order for such a problem? etc.).

Once potential solutions are arrived at, choosing one of them by consensus is desirable if it is possible. People generally feel good about consensus decisions and about working with the same group of individuals in the future. Voting, unfortunately, involves winners and losers. Of course, consensus may not prevail and voting might be necessary. If it comes down to two choices, voting by *majority wins* is easy and readily understood. We now examine several voting schemes that may be used when one option must be chosen from more than two.

Write a 1 beneath your first choice, a 2 beneath your second choice, a 3 beneath your third choice, and so on.

Option 1	Option 2	Option 3	. . .	Option *k*
_____	_____	_____		_____

FIGURE 6.10 Example of a Preference Ballot.

Choosing One Option from More Than Two

Suppose there are *k* options. We consider several commonly used *preference* voting methods. In *preference voting* each voter gets to identify his or her preference for all *k* options, in rank order. That is, each voter gets to identify which option is his or her first choice, which is second choice, and so on, down to the *k*th choice. A voter indicates a preference order on a ballot similar to that shown in Figure 6.10.

Let's suppose that after discussing all perceived pros and cons, the issue of raising the quality of items produced during a period of high demand has been narrowed down to four options:

a. Replace all drill presses with new ones from Vendor A.
b. Replace all drill presses with new ones from Vendor B.
c. Increase frequency of maintenance on existing presses and buy three additional presses from Vendor A.
d. Increase frequency of maintenance on existing presses and buy four additional presses from Vendor Z.

Further suppose that there are seven people in the group, each of whom gets to cast a preference ballot, and that the results of the voting are as given in Figure 6.11.

We now discuss several commonly used vote-counting methods. Since different methods may produce different winners, it is important to adopt the vote-counting method prior to voting.

Preference Order				
First Choice	C	A	A	C
Second Choice	B	C	C	A
Third Choice	A	B	D	D
Fourth Choice	D	D	B	B
Number of ballots	2	3	1	1

FIGURE 6.11 Preference Ranking of Four Options by Seven Voters.

The Borda Count Method

Under the Borda Count Method, a voter's first choice receives k points, his or her second choice receives $k - 1$ points, third choice receives $k - 2$ points, and so on, down to 1 point for last choice. The winning option is the one that receives the most total points. From the seven ballots shown in Figure 6.11, we see that the total points received by each option are

Option A	$4(4) + 1(3) + 2(2) + 0(1) = 23$
Option B	$0(4) + 2(3) + 3(2) + 2(1) = 14$
Option C	$3(4) + 4(3) + 0(2) + 0(1) = 24$
Option D	$0(4) + 0(3) + 2(2) + 5(1) = 9$

So based on the Borda count method, Option C wins. Three new drill presses will be purchased from Vendor A and maintenance will be increased on the existing drill presses. An arithmetic check may be done by noting that the total points received by all four options is 70, which is consistent with the fact that each of seven voters assigns a total of $1 + 2 + 3 + 4 = 10$ points.

Plurality-with-Elimination

A different way to choose a winner from k options that would be quite understandable, but quite impractical, could go this way. Each voter casts one vote for his or her first choice. If there is a choice that receives a majority of votes, voting is over and that choice wins. If not, the option receiving the fewest votes is eliminated and a new vote is taken based on the remaining $k - 1$ options. If there is an option that receives a majority of votes, voting is over and that option wins. If not, the option receiving the fewest votes is eliminated and a new vote is taken based on the remaining $k - 2$ options. This process continues until only two options are left, and a final vote determines the winner. Such a procedure could involve $k - 1$ votes and is therefore impractical.

If a preference vote that allows voters to rank all the options is taken first, one can similarly eliminate candidates in succession based on just the one preference vote. It works this way. Whichever option receives the fewest first place votes is eliminated, and the preference ballots are examined to determine each voter's preference ranking of the remaining options. This step is repeated until a winner is determined.

We illustrate how this works based on the preference vote in Figure 6.11. There is a tie between options B and D for which option receives the fewest first place votes. One of these must be eliminated. As a tie breaker, we eliminate the one with the fewest second place votes, Option D. With Option D eliminated from each ballot, the seven preference ballots appear as in Figure 6.12.

Preference Order				
First Choice	C	A	A	C
Second Choice	B	C	C	A
Third Choice	A	B	B	B
Number of ballots	2	3	1	1

FIGURE 6.12 Preference Ballots with Option D Eliminated.

Preference Order				
First Choice	C	A	A	C
Second Choice	A	C	C	A
Number of ballots	2	3	1	1

FIGURE 6.13 Preference Ballots with Option B Eliminated.

Figure 6.12 shows that Option B receives the fewest first place votes of the remaining three options. So Option B is eliminated from each ballot, and the seven preference ballots on the remaining two options appear in Figure 6.13. Note: It would have been mathematically equivalent at the beginning of this example to have observed that neither B nor D had any first place votes, conclude that whichever of B or D was eliminated first, the other would be eliminated next, and proceed directly to Figure 6.13.

With just A and C left, eliminating the one with the fewest first place votes becomes majority wins. Option A wins, 4 to 3. So the company will replace all drill presses with new ones obtained from Vendor A.

Pairwise Comparison

Using the pairwise comparison method, each possible pair of options is pitted in a head-to-head competition as if no other options exist, with the winner of each comparison receiving one point. If two options tie in head-to-head competition, the two options each receive $\frac{1}{2}$ point. The winner is the option having the most total points after all such comparisons are made.

Since there are four options in our illustration, there will be six pairwise comparisons: A/B, A/C, A/D, B/C, B/D, and C/D. We first consider A/B. After comparing voter preferences between A and B in Figure 6.14, we see that five voters prefer A to B and two prefer B to A. So A receives the point from this comparison. The results of all six pairwise comparisons appear in Table 6.2. Option A received the most points and becomes the winner. Based on comparison voting, the company will replace all drill presses with new ones obtained from Vendor A.

For more examples of preference voting, see Pirnot (2001).

Preference Order				
First Choice	C	**A**	**A**	C
Second Choice	**B**	C	C	**A**
Third Choice	**A**	**B**	D	D
Fourth Choice	D	D	**B**	**B**
Number of ballots	2	3	1	1

FIGURE 6.14 Five Voters Prefer A to B While Only Two Prefer B to A.

TABLE 6.2 Results of Comparison Voting—A Wins.

A vs. B	A wins 5 to 2. A gets 1 point.
A vs. C	A wins 4 to 3. A gets 1 point.
A vs. D	A wins 7 to 0. A gets 1 point.
B vs. C	C wins 7 to 0. C gets 1 point.
B vs. D	B wins 5 to 2. B gets 1 point.
C vs. D	C wins 7 to 0. C gets 1 point.

6.3 SOME MOVEMENTS IN QUALITY

A widely accepted truth about quality is that doing those things that make your product competitive today won't be good enough to make your product competitive tomorrow. Another is that one constant in life is change. Successful companies recognize this and have adopted a focus on continuous improvement. Similarly, approaches to quality have changed, with many principles that were once viewed as gospel in the quality arena having been improved. One example is that the concept of routine inspection has been replaced by something better—the concept of making a product so consistent that inspection is not needed. Quality visionaries, some of whom are mentioned in Section 6.1, have changed paradigms and fostered movements in how to do business and how to approach quality. In this section, we examine two such movements, Total Quality Management (TQM) and Six Sigma.

Total Quality Management is management that aims to maximize the competitiveness of an enterprise by means of continuous improvement in the quality of all of its processes, including, but not limited to, its products, services, people, and the mindset established by its workplace culture.

The words *total quality* in the definition of TQM mean that the goal of management is to maximize everything about the enterprise—how it does business, how it treats its customers, how it treats its internal customers, how it interacts with suppliers, and everything else. The definition of TQM may vary slightly from book to book, but the definition generally encompasses the following attitudes/objectives.

- Top-to-bottom commitment by management to quality in the long-term
- Focus on the customer, internal as well as external
- Obsession on quality
- Focus on continuous improvement
- Scientific approach to decision making and problem solving
- Teamwork
- Employee empowerment and involvement
- Focus on supplier quality accompanied by long-term commitments with suppliers
- Education of the entire workforce
- A workplace mindset on quality

Point a.—Top-to-bottom commitment by management to quality in the long-term. Employees more willingly commit their own energies to any effort when they know management is serious about it. Deming, highly sought after for his quality training seminars, also was known to communicate to top management: *If you don't attend along with the employees, I'm not coming,* or words to that effect. It is not sufficient for management simply to establish a quality department and make that department responsible for inculcating quality throughout the enterprise. Management leadership is viewed as essential in establishing total quality. Generally, this is taken to mean that a company must establish a statement of purpose that encompasses quality and long-term quality goals. Quality teams should be established that include all levels of employees, including management.

Point b.—Focus on the customer, internal as well as external. One facet of this idea is to please, even thrill, the customer who buys the product or service. Happy customers bring repeat business and will generate new customers when they tell people about their positive experiences. Customer feedback must be sought, not just to assess the level of satisfaction with the product, but also to seek input that will help the company's research/marketing unit anticipate new markets and meet future customer needs. Treating internal customers well also pays dividends. Employees tend to have more positive job attitudes and company processes tend to work more smoothly when internal customers are well treated. The Golden Rule is a good guide for how to treat all customers. Another good guide is to take the most ethical of possible actions in dealing with customers. An official code of ethics, one that is backed up by management, for how to treat customers and fellow employees contributes to positive workplace attitudes in employees and to trust and long-term relationships with customers.

Point c.—Obsession on quality. The idea is to have quality be defined by the internal or external customer. Once it is known what the customer wants, the objective is to design a product that thrills the customer by meeting the definition, and then some. Simply meeting the need is not enough. At every opportunity, the question to ask is, *How can we make this better?*

Point d.—Focus on continuous improvement. Recall that we spent a great deal of time discussing the merits of increasing the consistency of processes by reducing process standard deviation. This is just one example of the philosophy of continuous improvement. The idea is to continually seek to improve all aspects of the process—for example, cutting scrap, reducing customer waiting time, lowering error rates in shipping and billing, reducing near hits and accident rates, increasing uptime, increasing customer satisfaction, creating an atmosphere in which customer service employees can make common sense decisions, and so on.

Point e.—Scientific approach to decision making and problem solving. Making changes based on intuition or ad hoc approaches creates needless risk. If feasible, it is best to use data-driven approaches when making decisions, monitoring important performance variables, monitoring/controlling important quality variables, setting goals, and solving problems. This means using control charts, SPC, and problem-solving tools such as Pareto charts and cause-and-effect diagrams.

Point f.—Teamwork. Teamwork within and across the units of an organization contributes to the greater good by creating the opportunity for management and workers to feed off one another's good ideas. A workforce that buys into the idea of teamwork is more likely to recognize that doing what is right for the organization may sometimes not be what is best for a person's work unit, and do the right thing nevertheless. After establishing the purpose and membership of a team, members should be trained in teamwork. Recommendations made by the team should be adopted when possible.

Point g.—Employee empowerment and involvement. This is the antithesis of management by fear. Empowered employees take initiative when a situation does not fit the mold, and they make suggestions and speak up when they think improvements can be made. A manager who empowers employees is not giving up power, but rather is increasing the number of good decisions made within a work unit. Employees can be demoralized when they feel blindsided by management decisions that affect them. Workers should be involved from the outset on project teams that are charged to develop new initiatives.

Point h.—Focus on supplier quality accompanied by long-term commitments with suppliers. It is in the interest of both a business and its suppliers to develop long-term relationships. There is more to be gained in product quality by working with a trusted supplier who is involved in product development and who can look forward to long-term business than there is working with a supplier who simply bids low based on product specifications.

Point i.—Education of the entire workforce. To optimize worker contributions, workers need to be trained. All employees should be trained in SPC, TQM, the use of decision-making tools, the knowledge base of their particular responsibilities, and in acting ethically. A trained workforce will contribute more to teams and generate more ideas. Employees with different duties, managers included, logically have different work-related training needs, and it is management's responsibility to ensure that training is afforded.

Point j.—A workplace mindset. People should come to work looking to do their jobs, thinking of ways to do things better, weighing the issues confronting the teams they are on, and obsessing on quality. Such a mindset does not come by chance. Top management has to create a culture in which quality is a way of thinking. Emphasis on quality in a company's mission statement, in its code of ethics, in its reward system, and in its leadership will help establish to the workforce that quality thinking is an expectation in its value system.

Six Sigma

The reader may have noticed similarities between TQM, Deming's 14 Points, and philosophies of quality visionaries described earlier in the chapter. This also will be the case when the Six Sigma movement is described. Debbie Phillips-Donaldson, the editor of *Quality Progress*, observes (see Phillips-Donaldson, 2002) in an editorial message:

> For example, critics say Six Sigma is really nothing new. This is true. Reputable Six Sigma experts often acknowledge the methodology doesn't offer new tools or techniques. Rather, it packages existing, proven tools and concepts in a different way that some companies and people find helpful. . . .
>
> . . . for some organizations, Six Sigma has taken hold and produced outstanding results. Bottom line, it has grabbed the attention and buy-in of the people in the big corner offices. In my view, any quality concept that can do that deserves at least some notice.

It is in this spirit that we briefly discuss the Six Sigma movement. The very words *Six Sigma* indicate that the movement is about improving quality. A so-called *3-sigma process* is a process for which the process mean is three standard deviations from its spec limit(s). We saw for such a process that even the slightest shift of the process mean toward a spec limit results in a substantial increase in the proportion of nonconforming items. We also saw that if the process has upper and lower spec limits, the nonconforming rate of a 3-sigma process that is in control and centered midway between spec limits is 2.7 per thousand, or 2,700 per million, based on normal distributions. The nonconforming rate of a 4-sigma process is better—63.4 per million. That of a 5-sigma process is better still, and so on.

Process Quality Level	Defect Rate
3-sigma	2,700.0 per million
4-sigma	63.4 per million
5-sigma	0.57 per million
6-sigma	0.0020 per million

The Six Sigma movement gets its name from the goal of reducing the process standard deviation to the point that the process mean is Six Sigma from spec limits, a goal introduced by Motorola CEO Robert Galvin in the 1980s. As a practical matter, becoming a Six Sigma company is more of a goal for how employees think and for how quality is approached than a goal for a defect rate, since for most companies, actually reducing the defect rate to the Six Sigma defect rate is unrealistic.

An individual can be certified by a Six Sigma training/consulting firm or by the American Society for Quality (ASQ) as a Six Sigma Black Belt. This means being trained in and/or passing an exam on a Six Sigma Black Belt body of knowledge, which results in a recognition by an employer or potential employer who is seeking people who can lead quality and process improvements. There is variety in the training that leads to someone being certified as a Six Sigma Black Belt, resulting from what company does the training and what the trainee's employer might want its employee to learn. Nevertheless, that an exam—for example, ASQ's—would exist indicates some effort toward standardization in the evolution of the Six Sigma movement. For some personal perspectives about taking the ASQ's Six Sigma Black Belt exam, the reader is referred to Cochrane and Gupta (2002).

Some typical body of knowledge topics are

- Leadership in areas such as managing projects, managing teams, understanding team dynamics and performance, instituting change, and developing goals and objectives of new enterprises
- Managing business processes, including such things as determining project scope, establishing metrics, stating problems, and establishing documentation
- Designing robust processes—that is, processes resistant to failure
- Using probability and statistical tools and methods, process capability, SPC, hypothesis testing, and experimental design
- Using improvement tools and methods such as those used in lean manufacturing

How top management adopts a Six Sigma approach into its company strategy and what the Six Sigma Black Belts do in their organizations depends on the company. At a manufacturing company that makes electrical components such as integrated circuits, say, a Black Belt might divide time between using statistics, teaming with product design staff to anticipate and keep problems from being designed into products during the development stages, designing experiments that help the company produce items that meet customer requirements, and training other employees in Six Sigma methodology.

At another company, a Black Belt might have time divided between working with top management on strategic planning, facilitating teams that benchmark other companies in order to improve processes and/or compare the company to the best competition, teaching employees how to use quality tools, and translating customer response data into process and product improvements and into the creation of new products or services. It also would be common to have Black Belts involved with others in helping existing processes work better—for example, increasing uptime, increasing initial-contact problem resolution, reducing defects and scrap, and so on. A good read on what Black Belts do in different kinds of companies may be found in Faltin, Farrell, Friend, and Friend (2002).

PROBLEMS

In Problems 1–6, discuss one or more quality management principles that are violated.

1. Among drivers seeking license tags, renewing licenses, and so on, South Carolina's Division of Motor Vehicles (DMV) had become notorious for long lines and poor customer service. In the summer of 2002, the DMV instituted Project Phoenix, a computer upgrade that was supposed to alleviate some of the problems. But the new system itself was plagued with problems. One glitch incorrectly changed auto registration information (e.g., produced wrong mailing addresses) for some drivers. Correction took weeks. As a result, instead of painlessly receiving renewed auto registration certificates by mail after having paid personal property taxes, many drivers had to visit the DMV to renew their registration in person because the new system had caused them to receive their personal property tax bills late. Further, DMV office employees refused to waive the late fee that arises when taxes have been paid late. (See "DMV," *The Post and Courier*, 2003.)
2. Company A employs computer programmers who write programs and set up databases for the computer systems of client companies. One frequent client is Company B. A programmer from Company A realizes that the use of the program he is writing for Company B will be to scam money from Company B's customers. The scam is technically legal, but unethical. Company A, which has financial problems, needs clients badly.
3. Company X hires a consultant to teach a three-day short course on SPC. All shift employees and several mid-level managers are required to attend.
4. The manager of a hardware store waited on a customer who had a leak inside his toilet tank. The manager realized the problem could be fixed with a 20-cent neoprene washer, but convinced the customer that he needed to replace the entire mechanism inside the tank. As a result, the customer bought a $95 item.

5. A couple enjoyed treating themselves to breakfast at a nice restaurant every Sunday morning. On one particular Sunday, they decided to split a seafood omelet because they remembered that the husband had ordered it once on an earlier visit and the omelet was huge. This menu items comes with grits and toast. The couple informed the waitress, ordered coffee for each and an additional order of toast. The waitress brought an extra plate. After eating, the husband went to pay the cashier, found the bill to be high, and asked for a cost breakdown. The cashier noted there was a $2 charge for an extra plate. She also said she was sorry, adding that she never notes an extra plate on a bill when she waits tables, but since the waitress noted it in this case, she had to include it. The couple did not leave a tip.
6. A rule that stems from the accumulated observations of an uncanny relationship between consequences and causes in business and industry is *the* 80/20 *Rule*. According to this rule, one may anticipate that in any enterprise 20% of the inventory accounts for 80% of the sales, 20% of the potential causes of problems account for 80% of the trouble, and so on. Just as one can look at a Pareto chart (see Section 6.2) to identify which problem causes have the most impact and should be addressed first, the 80/20 Rule suggests that 80% of the trouble can be potentially removed by addressing the right 20% of all root causes. That one should focus on solving the *vital few* problem causes is a concept promoted by William Juran (for a biographical sketch, see Section 6.1) and sometimes referred to as the *Pareto Principle*. Suppose that at a particular airport, security policy prohibits passenger profiling for purposes of preboarding luggage inspection and wanding of passengers, instead requiring inspection and wanding of randomly chosen passengers.
7. Every individual makes mistakes on the job despite the best of intentions. No doubt, surgeons are particularly vigilant about not leaving sponges, clamps, retractors, or other items inside the patient before closing. Yet even here, mistakes happen. A study spanning 16 years of insurance records in Massachusetts revealed that surgical equipment had been left in 54 patients out of roughly 800,000 surgeries. (See "Study Says . . . ," *The Post and Courier*, 2003. Data used with permission of The Associated Press.) This defect rate for overlooked surgical equipment comes closest to which of the following, from the point of view of the Six Sigma movement: a) 3-sigma process; b) 4-sigma process; c) 5-sigma process; d) 6-sigma process?

Bibliography

American Society for Quality. 2003. *ASQ—About History*. Retrieved November 23, 2003 from <*http://www.asq.org/join/about/history/*>.

Besterfield, D. H. et al. 2003. *Total quality management* (3rd ed.). Upper Saddle River, NJ: Prentice Hall.

Bhote, K. 1991. *World class quality—Using design of experiments to make it happen*. New York: AMACOM.

Cochrane, D., and Gupta, P. ASQ's black belt certification—A personal experience. *Quality Progress* (May, 2002): 33–38.

Crawford, D. K., Bodine, R. J., and Hoglund, R. G. 1993. *The school for quality learning, managing the school and classroom the Deming way*. Washington, DC: The Bookshelf.

Crosby, P. 1975. *Quality is free*. New York: McGraw-Hill.

Crosby, P. 1984. *Quality without tears*. New York. McGraw-Hill.

Crosby, P. 1996. *Quality is still free*. New York. McGraw-Hill.

Deming, W. Edwards. 1986. *Out of the crisis*. Cambridge, MA: MIT Center for Advanced Engineering Study.

DMV's $40 million system still downloading headaches. *The Post and Courier*, Charleston, SC, January 9, 2003.

Faltin, F., Farrell, D., Friend, A., and Friend, T. 2002. A day in the life of a six sigma black belt. *Amstat News*, September, pp. 45–53.

Gitlow, H. S. and Gitlow, S. J. 1987. *The Deming guide to quality and competitive position*, Englewood Cliffs, NJ: Prentice Hall.

Goetsch, D. L., and Davis, S. B. 2003. *Quality management* (4th ed.). Upper Saddle River, NJ: Prentice Hall.

Grant, E. L., and Leavenworth, R. S. 1996. *Statistical quality control* (7th ed.). Boston, MA: McGraw-Hill.

Heart attacks in Montana town drop after indoor smoking ban. *The Post and Courier*, Charleston, SC, April 2, 2003.

Hoyer, R. W., and Ellis, W. C. 1996. A graphical exploration of SPC. *Quality Progress*, June, pp. 57–64.

Hoyer, R. W., and Hoyer, B. B. Y. 2001. What is quality? *Quality Progress*, July, pp. 53–62.

Ishikawa, K. 1985. *What is total quality control? The Japanese way*. Englewood Cliffs, NJ: Prentice Hall.

Ishikawa, K. 1986. *Guide to quality control*. Tokyo: Asian Productivity Organization.

Jaehn, A. 1989. All-purpose chart can make SPC easy. *Quality Progress*, *22*(2), p. 112.

Juran, J. 1988. *Planning for quality*. New York, NY: The Free Press (MacMillan).

Juran, J., and Gryna, F. 1988. *Juran's quality control handbook*. New York, NY: McGraw-Hill.

Juran Institute. 2003. *Juran—Our Founder*. Retrieved November 23, 2003 from <*http://www.juran.com/lower_2.cfm?article_id=21*>.

Kemp, K. W. 1962. The use of cumulative sums for sampling inspection schemes. *Applied Statistics* (Royal Statistical Society), Vol. XI, p. 23.

Minitab Inc. 2003. *Meet Minitalo*, Release 14 for Windows.

Murder rate increased 30% in tri-county. *The Post and Courier*, Charleston, SC, January 16, 2000.

Neave, H. R. 1990. *The Deming dimension*. Knoxville, TN: SPC Press.

Norton, M. 1990. The statistician as an expert witness. Proceedings of the Twentieth Annual Meeting of the Southeast Region of the Decision Sciences Institute; co-editors Charles H. Smith and Robert L. Andrews, Virginia Commonwealth Univ., pp. 288–290.

Norton, M. 2000. Designing an out-of-control signal when distribution location can change frequently with the arrival of new chemical shipments. *Quality Engineering*, Vol. 13, No. 4, pp. 601–605. Used with permission of Marcel Dekker, Inc. N.Y.

Norton, M. 2001–2002. Detecting overcontrol for a continuous variable from a graph diagnostic. *Quality Engineering*, Vol. 14, No. 1, pp. 9–12. Used with permission of Marcel Dekker, Inc. N.Y.

Norton, M., Seaman, S., and Sprankle, M. 1996. Measuring book availability: A monthly sampling method. *College and Undergraduate Libraries*, Vol. 20, No. 2, pp. 101–115.

Ozone-TOMS. 2003. NASA. Retrieved November 24, 2003 from <*http://toms.gsfc.nasa.gov/teacher/ozone_overhead.html?101,59*>.

Perrins, C. M., Lebreton, J. D., and Hirons, G. H. M. Eds. 1991. *Bird population studies* (p. 531). Oxford, UK: Oxford University Press.

Phillips-Donaldson, D. Six sigma: A false god? *Quality Progress*, January, 2002, p. 6.

Phil's Page. 2003. Philip Crosby Associates II. Retrieved November 23, 2003 from <*http://www.philipcrosby.com/philspage/philsbio.htm*>.

Pirnot, T. L. 2001. *Mathematics all around*. Boston, MA: Addison Wesley.

Police shootings in St. Louis have officers on edge and on the defensive. *The Post and Courier*, Charleston, SC, December 13, 2002.

Pyzdek, T., (1991). *What every manager should know about quality*. New York: Marcel Dekker.

Quality Gurus. 2003. Simple Systems International. Retrieved November 23, 2003 from <*http://www.simplesystemsintl.com/quality_gurus/*>.

Resources—Leaders. 2003. Skymark Corporation. Retrieved November 23, 2003 from <*http://www.skymark.com/resources/leaders/biomain.asp*>.

Roy, R. K. 2001. *Design of experiments using the Taguchi approach: 16 steps to product and process*. New York: John Wiley & Sons.

Scherkenbach, W. W. 1991. *The Deming route to quality and productivity—Road maps and roadblocks*. Milwaukee, WI: ASQ Quality Press.

Senate panel cites "plague" of rural criminal violence. *The News and Courier*, Charleston, SC, June 19, 1991.

Shewhart, W. A. 1931. *The control of quality of manufactured product*. New York: D. Van Nostrand.

Shores, A. R. 1990. *A TQM approach to achieving manufacturing excellence*. Milwaukee, WI: ASQ Quality Press.

Smith, G. M. 1991. *Statistical process control and quality improvement*. Upper Saddle River, NJ: Prentice Hall.

Study says surgeons leave tools in 1,500 patients each year. *The Post and Courier*, Charleston, SC, January 16, 2003.

Walton, M. 1990. *Deming management at work*. New York: G. P. Putnam's Sons.

Zacks, S. 1992. *Introduction to reliability analysis*. New York, Springer-Verlag.

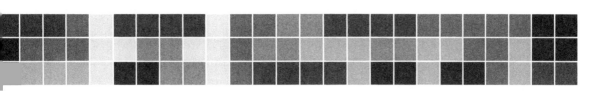

Glossary

assignable cause variation Variation attributable to a source present when a process is not working as designed (also known as *special cause variation*).

capability See process capability.

cause-and-effect diagram A diagram that displays all understood causes of an effect, organized by cause categories (also known as *fishbone diagram* or *Ishikawa diagram*).

combination An unordered arrangement of a collection of objects.

common cause variation Variation that is inherent in a process.

conditional probability The probability of an event occurring under the condition that some second event has already taken place.

continuous variable A variable that has no gaps in its measurement range.

control chart A graph of quality characteristic measurements versus time, designed as a tool for detecting when the underlying distribution has changed.

control limits Values on a control chart beyond which a plotted point is considered to be an out-of-control signal.

in control Term used to describe a process when the quality characteristic being measured has a distribution that is not changing over time.

discrete variable A variable that has gaps in its measurement range.

Empirical Rule For normally or approximately normally distributed variables, roughly 68% of the values fall within one standard deviation of the mean, roughly 95% within two standard deviations of the mean, and roughly 99.7% of the values fall within three standard deviations of the mean.

failure rate The rate at which a given kind of item fails during a prescribed time period, in units of failures/unit time or failures/cycle.

fishbone diagram *See* cause-and-effect diagram.

flowchart A graphical representation of the sequence of steps involved in an activity.

grand mean Mean of sample means ($\overline{\overline{X}}$).

historical mean Value of the process mean, based on recent process data during a sustained period of process stability.

historical standard deviation (historical sigma) Value of the process standard deviation, based on recent process data during a sustained period of process stability.

Ishikawa diagram *See* cause-and-effect diagram.

lower control limit (LCL) That value on a control chart below which a plotted point is considered to be an out-of-control signal.

lower specification limit (LSL) The value below which the characteristic being measured does not meet specifications.

LSL *See* Lower Specification Limit.

mean Average value of a sample or a population.

mutually exclusive events Events that cannot occur at the same time.

natural tolerance of a process The length of the interval in which almost all values of a measured quality characteristic fall (6 times the process standard deviation).

nonconforming Not meeting product specifications.

out-of-control (out of statistical control) A term used to describe a process working at less than the quality level of which it is capable.

out-of-control signal On a control chart, a point or combination of successive points that indicate there may be a process problem.

out-of-spec Not meeting product specifications.

overcontrol Making frequent process adjustments based on other than statistical signals (e.g., noise).

parallel system of components A system that fails only when every component fails.

Pareto chart Chart used in revealing those causes that have the most impact and those that have the least.

permutation Ordered arrangement of a collection of objects.

pooled standard deviation An estimate of the process standard deviation obtained by taking the square root of the mean of a collection of sample variances (S_p).

population The collection of all possible items of a kind, or the collection of measurements of a characteristic obtained from every possible item that could be generated by a process before the process changes.

probability A value between 0 and 1 that measures how likely it is that some event will occur.

probability distribution A table or formula that identifies every possible outcome of an experiment or value of a random variable and the probability of each.

probability limit (exact probability limit) Control limit determined so that the probability that a plotted point falls beyond it is a specified value when the process is in control.

process capability State a process is said to be in when it is centered at least 3σ from relevant specification limits.

process capability index A measure of how well a process is doing, or could be doing, at being centered at least 3σ from relevant specification limits.

random sample A small portion of the population taken in such a way that every member of the population has the same chance of being in the sample.

range A measure of the amount of variation in a sample, given by the distance between the largest and smallest sample values.

reliability The probability that a product or system will not fail during a given time period.

sample A small portion of the population.

sample size The number of items in a sample.

scatter diagram A graph of plotted points used to help determine how or whether two variables might be related.

series system of components A system that fails when any single component in the system fails.

SPC *See* statistical process control.

special cause variation Variation attributable to a source present when a process is not working as designed (also known as *assignable cause variation*).

standard deviation A measure of the amount of variation in a sample or population.

statistical process control (SPC) The use of statistical methods to control/improve a process.

target (or target value) The intended value of a quality characteristic.

total quality management Management that aims to maximize the competitiveness of an enterprise by means of continuous improvement in the quality of all of its processes, including, but not limited to, its products, services, people, and the mindset established by its workplace culture.

upper control limit (UCL) That value on a control chart above which a plotted point is considered to be an out-of-control signal.

upper specification limit (USL) The value above which the characteristic being measured does not meet specifications.

USL *See* upper specification limit.

variance The square of the standard deviation; a measure of the amount of variation in a sample or population.

Appendix

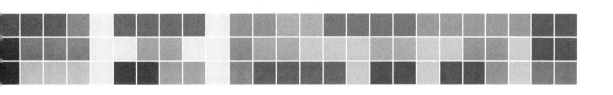

TABLE I Estimating σ

$$\sigma \approx \frac{\overline{R}}{d_2} \qquad \sigma \approx \frac{\overline{S}}{c_4} \qquad \sigma \approx S_p = \sqrt{\frac{\sum_{i=1}^{k} S_i^2}{k}}$$

n	d_2	c_4	n	d_2	c_4
			21	3.778	.9876
2	1.128	.7979	22	3.819	.9882
3	1.693	.8862	23	3.858	.9887
4	2.059	.9213	24	3.895	.9892
5	2.326	.9400	25	3.931	.9896
6	2.534	.9515	30	4.086	.9914
7	2.704	.9594	35	4.213	.9927
8	2.847	.9650	40	4.322	.9936
9	2.970	.9693	45	4.415	.9943
10	3.078	.9727	50	4.498	.9949
11	3.173	.9754	55	4.572	.9954
12	3.258	.9776	60	4.639	.9958
13	3.336	.9794	65	4.699	.9961
14	3.407	.9810	70	4.755	.9964
15	3.472	.9823	75	4.806	.9966
16	3.532	.9835	80	4.854	.9968
17	3.588	.9845	85	4.898	.9970
18	3.640	.9854	90	4.939	.9972
19	3.689	.9862	95	4.978	.9973
20	3.735	.9869	100	5.015	.9975

TABLE II Control Limits for an $\overline{X}$-Chart

Center Line: $\overline{\overline{X}}$ or Target

$$\text{Control Limits} = \text{Center Line} \pm 3\frac{\sigma}{\sqrt{n}} \text{ (if } \sigma \text{ or an estimate of } \sigma \text{ is known)}$$

or

$$\text{Control Limits} = \overline{\overline{X}} \pm A_2\overline{R} \text{ or Control Limits} = \overline{\overline{X}} \pm A_3\overline{S}$$

n	A_2
2	1.88
3	1.02
4	.73
5	.58
6	.48
7	.42
8	.37
9	.34
10	.31
11	.29
12	.27
13	.25
14	.24
15	.22
16	.21
17	.20
18	.19
19	.19
20	.18

n	A_3
2	2.66
3	1.95
4	1.63
5	1.43
6	1.29
7	1.18
8	1.10
9	1.03
10	.98
11	.93
12	.89
13	.85
14	.82
15	.79
16	.76
17	.74
18	.72
19	.70
20	.68

n	A_3
21	.66
22	.65
23	.63
24	.62
25	.61
30	.55
35	.51
40	.48
45	.45
50	.43
55	.41
60	.39
65	.37
70	.36
75	.35
80	.34
85	.33
90	.32
95	.31
100	.30

TABLE III Control Limits for an S-Chart

Center Line: $\overline{S}$ or Center Line: $c_4\sigma$

$LCL_S = B_3\overline{S}$, $UCL_s = B_4\overline{S}$ $LCL_S = B_5\sigma$, $UCL_s = B_6\sigma$

n	B_3	B_4	B_5	B_6
2	0	3.27	0	2.61
3	0	2.57	0	2.28
4	0	2.27	0	2.09
5	0	2.09	0	1.96
6	.03	1.97	.03	1.87
7	.12	1.88	.11	1.81
8	.19	1.81	.18	1.75
9	.24	1.76	.23	1.71
10	.28	1.72	.28	1.67
11	.32	1.68	.31	1.64
12	.35	1.65	.35	1.61
13	.38	1.62	.37	1.59
14	.41	1.59	.40	1.56
15	.43	1.57	.42	1.54
16	.45	1.55	.44	1.53
17	.47	1.53	.46	1.51
18	.48	1.52	.48	1.50
19	.50	1.50	.49	1.48
20	.51	1.49	.50	1.47

n	B_3	B_4	B_5	B_6
21	.52	1.48	.52	1.46
22	.53	1.47	.53	1.45
23	.54	1.46	.54	1.44
24	.55	1.45	.55	1.43
25	.56	1.44	.56	1.42
30	.60	1.40	.60	1.38
35	.63	1.37	.63	1.36
40	.66	1.34	.66	1.33
45	.68	1.32	.68	1.31
50	.70	1.30	.69	1.30
55	.71	1.29	.71	1.28
60	.72	1.28	.72	1.27
65	.73	1.27	.73	1.26
70	.74	1.26	.74	1.25
75	.75	1.25	.75	1.24
80	.76	1.24	.76	1.24
85	.77	1.23	.77	1.23
90	.77	1.23	.77	1.22
95	.78	1.23	.78	1.22
100	.79	1.21	.78	1.21

TABLE IV Control Limits for an R-Chart

Center Line: $\overline{R}$ or Center Line: $d_2\sigma$

$LCL_R = D_3\overline{R}$, $UCL_R = D_4\overline{R}$ $LCL_R = D_1\sigma$, $UCL_R = D_2\sigma$

n	D_1	D_2	D_3	D_4
2	0	3.69	0	3.27
3	0	4.36	0	2.57
4	0	4.70	0	2.28
5	0	4.92	0	2.11
6	0	5.08	0	2.00
7	.20	5.20	.08	1.92
8	.39	5.31	.14	1.86
9	.55	5.39	.18	1.82
10	.69	5.47	.22	1.78
11	.81	5.53	.26	1.74
12	.92	5.59	.28	1.72
13	1.03	5.65	.31	1.69
14	1.12	5.69	.33	1.67
15	1.21	5.74	.35	1.65
16	1.28	5.78	.36	1.64
17	1.36	5.82	.38	1.62
18	1.43	5.85	.39	1.61
19	1.49	5.89	.40	1.60
20	1.55	5.92	.41	1.59

TABLE V Cumulative Binomial Probabilities

$$P(X \le k) = \sum_{r=0}^{k} \binom{n}{r} p^r (1-p)^{n-r}$$

n	k	.01	.02	.03	.04	.05	.06	.07	.08	.09	.10
2	0	0.980	0.960	0.941	0.922	0.903	0.884	0.865	0.846	0.828	0.810
	1	1.000	1.000	0.999	0.998	0.997	0.996	0.995	0.994	0.992	0.990
	2	1.000	1.000	1.000	1.000	1.000	1.000	1.000	1.000	1.000	1.000

n	k	.11	.12	.13	.14	.15	.16	.17	.18	.19	.20
2	0	0.792	0.774	0.757	0.740	0.722	0.706	0.689	0.672	0.656	0.640
	1	0.988	0.986	0.983	0.980	0.977	0.974	0.971	0.968	0.964	0.960
	2	1.000	1.000	1.000	1.000	1.000	1.000	1.000	1.000	1.000	1.000

n	k	.21	.22	.23	.24	.25	.30	.35	.40	.45	.50
2	0	0.624	0.608	0.593	0.578	0.562	0.490	0.422	0.360	0.302	0.250
	1	0.956	0.952	0.947	0.942	0.937	0.910	0.878	0.840	0.797	0.750
	2	1.000	1.000	1.000	1.000	1.000	1.000	1.000	1.000	1.000	1.000

n	k	.01	.02	.03	.04	.05	.06	.07	.08	.09	.10
3	0	0.970	0.941	0.913	0.885	0.857	0.831	0.804	0.779	0.754	0.729
	1	1.000	0.999	0.997	0.995	0.993	0.990	0.986	0.982	0.977	0.972
	2	1.000	1.000	1.000	1.000	1.000	1.000	1.000	0.999	0.999	0.999
	3	1.000	1.000	1.000	1.000	1.000	1.000	1.000	1.000	1.000	1.000

n	k	.11	.12	.13	.14	.15	.16	.17	.18	.19	.20
3	0	0.705	0.681	0.659	0.636	0.614	0.593	0.572	0.551	0.531	0.512
	1	0.966	0.960	0.954	0.947	0.939	0.931	0.923	0.914	0.905	0.896
	2	0.999	0.998	0.998	0.997	0.997	0.996	0.995	0.994	0.993	0.992
	3	1.000	1.000	1.000	1.000	1.000	1.000	1.000	1.000	1.000	1.000

n	k	.21	.22	.23	.24	.25	.30	.35	.40	.45	.50
3	0	0.493	0.475	0.457	0.439	0.422	0.343	0.275	0.216	0.166	0.125
	1	0.886	0.876	0.866	0.855	0.844	0.784	0.718	0.648	0.575	0.500
	2	0.991	0.989	0.988	0.986	0.984	0.973	0.957	0.936	0.909	0.875
	3	1.000	1.000	1.000	1.000	1.000	1.000	1.000	1.000	1.000	1.000

n	k	.01	.02	.03	.04	.05	.06	.07	.08	.09	.10
4	0	0.961	0.922	0.885	0.849	0.815	0.781	0.748	0.716	0.686	0.656
	1	0.999	0.998	0.995	0.991	0.986	0.980	0.973	0.966	0.957	0.948
	2	1.000	1.000	1.000	1.000	1.000	0.999	0.999	0.998	0.997	0.996
	3	1.000	1.000	1.000	1.000	1.000	1.000	1.000	1.000	1.000	1.000

n	k	.11	.12	.13	.14	.15	.16	.17	.18	.19	.20
4	0	0.627	0.600	0.573	0.547	0.522	0.498	0.475	0.452	0.430	0.410
	1	0.938	0.927	0.915	0.903	0.890	0.877	0.863	0.849	0.834	0.819
	2	0.995	0.994	0.992	0.990	0.988	0.986	0.983	0.980	0.976	0.973
	3	1.000	1.000	1.000	1.000	0.999	0.999	0.999	0.999	0.999	0.998
	4	1.000	1.000	1.000	1.000	1.000	1.000	1.000	1.000	1.000	1.000

n	k	.21	.22	.23	.24	.25	.30	.35	.40	.45	.50
4	0	0.390	0.370	0.352	0.334	0.316	0.240	0.179	0.130	0.092	0.063
	1	0.804	0.788	0.772	0.755	0.738	0.652	0.563	0.475	0.391	0.313
	2	0.969	0.964	0.960	0.955	0.949	0.916	0.874	0.821	0.759	0.688
	3	0.998	0.998	0.997	0.997	0.996	0.992	0.985	0.974	0.959	0.938
	4	1.000	1.000	1.000	1.000	1.000	1.000	1.000	1.000	1.000	1.000

n	k	.01	.02	.03	.04	.05	.06	.07	.08	.09	.10
5	0	0.951	0.904	0.859	0.815	0.774	0.734	0.696	0.659	0.624	0.590
	1	0.999	0.996	0.992	0.985	0.977	0.968	0.958	0.946	0.933	0.919
	2	1.000	1.000	1.000	0.999	0.999	0.998	0.997	0.995	0.994	0.991
	3	1.000	1.000	1.000	1.000	1.000	1.000	1.000	1.000	1.000	1.000

(continued)

n	k	.11	.12	.13	.14	.15	.16	.17	.18	.19	.20
5	0	0.558	0.528	0.498	0.470	0.444	0.418	0.394	0.371	0.349	0.328
	1	0.903	0.888	0.871	0.853	0.835	0.817	0.797	0.778	0.758	0.737
	2	0.989	0.986	0.982	0.978	0.973	0.968	0.963	0.956	0.949	0.942
	3	0.999	0.999	0.999	0.998	0.998	0.997	0.996	0.996	0.994	0.993
	4	1.000	1.000	1.000	1.000	1.000	1.000	1.000	1.000	1.000	1.000

n	k	.21	.22	.23	.24	.25	.30	.35	.40	.45	.50
5	0	0.308	0.289	0.271	0.254	0.237	0.168	0.116	0.078	0.050	0.031
	1	0.717	0.696	0.675	0.654	0.633	0.528	0.428	0.337	0.256	0.188
	2	0.934	0.926	0.916	0.907	0.896	0.837	0.765	0.683	0.593	0.500
	3	0.992	0.990	0.989	0.987	0.984	0.969	0.946	0.913	0.869	0.813
	4	1.000	0.999	0.999	0.999	0.999	0.998	0.995	0.990	0.982	0.969
	5	1.000	1.000	1.000	1.000	1.000	1.000	1.000	1.000	1.000	1.000

n	k	.01	.02	.03	.04	.05	.06	.07	.08	.09	.10
6	0	0.941	0.886	0.833	0.783	0.735	0.690	0.647	0.606	0.568	0.531
	1	0.999	0.994	0.988	0.978	0.967	0.954	0.939	0.923	0.905	0.886
	2	1.000	1.000	0.999	0.999	0.998	0.996	0.994	0.991	0.988	0.984
	3	1.000	1.000	1.000	1.000	1.000	1.000	1.000	0.999	0.999	0.999
	4	1.000	1.000	1.000	1.000	1.000	1.000	1.000	1.000	1.000	1.000

n	k	.11	.12	.13	.14	.15	.16	.17	.18	.19	.20
6	0	0.497	0.464	0.434	0.405	0.377	0.351	0.327	0.304	0.282	0.262
	1	0.866	0.844	0.822	0.800	0.776	0.753	0.729	0.704	0.680	0.655
	2	0.979	0.974	0.968	0.961	0.953	0.944	0.934	0.924	0.913	0.901
	3	0.998	0.997	0.997	0.995	0.994	0.993	0.991	0.988	0.986	0.983
	4	1.000	1.000	1.000	1.000	1.000	0.999	0.999	0.999	0.999	0.998
	5	1.000	1.000	1.000	1.000	1.000	1.000	1.000	1.000	1.000	1.000

n	k	.21	.22	.23	.24	.25	.30	.35	.40	.45	.50
6	0	0.243	0.225	0.208	0.193	0.178	0.118	0.075	0.047	0.028	0.016
	1	0.631	0.606	0.582	0.558	0.534	0.420	0.319	0.233	0.164	0.109
	2	0.888	0.875	0.861	0.846	0.831	0.744	0.647	0.544	0.442	0.344
	3	0.980	0.976	0.972	0.967	0.962	0.930	0.883	0.821	0.745	0.656
	4	0.998	0.997	0.997	0.996	0.995	0.989	0.978	0.959	0.931	0.891
	5	1.000	1.000	1.000	1.000	1.000	0.999	0.998	0.996	0.992	0.984
	6	1.000	1.000	1.000	1.000	1.000	1.000	1.000	1.000	1.000	1.000

n	k	.01	.02	.03	.04	.05	.06	.07	.08	.09	.10
7	0	0.932	0.868	0.808	0.751	0.698	0.648	0.602	0.558	0.517	0.478
	1	0.998	0.992	0.983	0.971	0.956	0.938	0.919	0.897	0.875	0.850
	2	1.000	1.000	0.999	0.998	0.996	0.994	0.990	0.986	0.981	0.974
	3	1.000	1.000	1.000	1.000	1.000	1.000	0.999	0.999	0.998	0.997
	4	1.000	1.000	1.000	1.000	1.000	1.000	1.000	1.000	1.000	1.000

n	k	.11	.12	.13	.14	.15	.16	.17	.18	.19	.20
7	0	0.442	0.409	0.377	0.348	0.321	0.295	0.271	0.249	0.229	0.210
	1	0.825	0.799	0.772	0.744	0.717	0.689	0.660	0.632	0.604	0.577
	2	0.967	0.958	0.949	0.938	0.926	0.913	0.899	0.885	0.869	0.852
	3	0.996	0.995	0.993	0.991	0.988	0.985	0.981	0.977	0.972	0.967
	4	1.000	1.000	0.999	0.999	0.999	0.998	0.998	0.997	0.996	0.995
	5	1.000	1.000	1.000	1.000	1.000	1.000	1.000	1.000	1.000	1.000

n	k	.21	.22	.23	.24	.25	.30	.35	.40	.45	.50
7	0	0.192	0.176	0.160	0.146	0.133	0.082	0.049	0.028	0.015	0.008
	1	0.549	0.522	0.496	0.470	0.445	0.329	0.234	0.159	0.102	0.063
	2	0.834	0.816	0.797	0.777	0.756	0.647	0.532	0.420	0.316	0.227
	3	0.961	0.954	0.946	0.938	0.929	0.874	0.800	0.710	0.608	0.500
	4	0.994	0.993	0.991	0.989	0.987	0.971	0.944	0.904	0.847	0.773
	5	1.000	0.999	0.999	0.999	0.999	0.996	0.991	0.981	0.964	0.938
	6	1.000	1.000	1.000	1.000	1.000	1.000	0.999	0.998	0.996	0.992
	7	1.000	1.000	1.000	1.000	1.000	1.000	1.000	1.000	1.000	1.000

(continued)

n	k	.01	.02	.03	.04	.05	.06	.07	.08	.09	.10
8	0	0.923	0.851	0.784	0.721	0.663	0.610	0.560	0.513	0.470	0.430
	1	0.997	0.990	0.978	0.962	0.943	0.921	0.897	0.870	0.842	0.813
	2	1.000	1.000	0.999	0.997	0.994	0.990	0.985	0.979	0.971	0.962
	3	1.000	1.000	1.000	1.000	1.000	0.999	0.999	0.998	0.997	0.995
	4	1.000	1.000	1.000	1.000	1.000	1.000	1.000	1.000	1.000	1.000

n	k	.11	.12	.13	.14	.15	.16	.17	.18	.19	.20
8	0	0.394	0.360	0.328	0.299	0.272	0.248	0.225	0.204	0.185	0.168
	1	0.783	0.752	0.721	0.689	0.657	0.626	0.594	0.563	0.533	0.503
	2	0.951	0.939	0.926	0.911	0.895	0.877	0.859	0.839	0.819	0.797
	3	0.993	0.990	0.987	0.983	0.979	0.973	0.967	0.960	0.952	0.944
	4	0.999	0.999	0.999	0.998	0.997	0.996	0.995	0.993	0.992	0.990
	5	1.000	1.000	1.000	1.000	1.000	1.000	1.000	0.999	0.999	0.999
	6	1.000	1.000	1.000	1.000	1.000	1.000	1.000	1.000	1.000	1.000

n	k	.21	.22	.23	.24	.25	.30	.35	.40	.45	.50
8	0	0.152	0.137	0.124	0.111	0.100	0.058	0.032	0.017	0.008	0.004
	1	0.474	0.446	0.419	0.392	0.367	0.255	0.169	0.106	0.063	0.035
	2	0.775	0.751	0.728	0.703	0.679	0.552	0.428	0.315	0.220	0.145
	3	0.934	0.924	0.912	0.900	0.886	0.806	0.706	0.594	0.477	0.363
	4	0.987	0.984	0.981	0.977	0.973	0.942	0.894	0.826	0.740	0.637
	5	0.998	0.998	0.997	0.997	0.996	0.989	0.975	0.950	0.912	0.855
	6	1.000	1.000	1.000	1.000	1.000	0.999	0.996	0.991	0.982	0.965
	7	1.000	1.000	1.000	1.000	1.000	1.000	1.000	0.999	0.998	0.996
	8	1.000	1.000	1.000	1.000	1.000	1.000	1.000	1.000	1.000	1.000

n	k	.01	.02	.03	.04	.05	.06	.07	.08	.09	.10
9	0	0.914	0.834	0.760	0.693	0.630	0.573	0.520	0.472	0.428	0.387
	1	0.997	0.987	0.972	0.952	0.929	0.902	0.873	0.842	0.809	0.775
	2	1.000	0.999	0.998	0.996	0.992	0.986	0.979	0.970	0.960	0.947
	3	1.000	1.000	1.000	1.000	0.999	0.999	0.998	0.996	0.994	0.992
	4	1.000	1.000	1.000	1.000	1.000	1.000	1.000	1.000	0.999	0.999
	5	1.000	1.000	1.000	1.000	1.000	1.000	1.000	1.000	1.000	1.000

n	k	.11	.12	.13	.14	.15	.16	.17	.18	.19	.20
9	0	0.350	0.316	0.286	0.257	0.232	0.208	0.187	0.168	0.150	0.134
	1	0.740	0.705	0.670	0.634	0.599	0.565	0.532	0.499	0.467	0.436
	2	0.933	0.917	0.899	0.880	0.859	0.837	0.814	0.790	0.764	0.738
	3	0.988	0.984	0.979	0.973	0.966	0.958	0.949	0.938	0.927	0.914
	4	0.999	0.998	0.997	0.996	0.994	0.993	0.990	0.988	0.984	0.980
	5	1.000	1.000	1.000	1.000	0.999	0.999	0.999	0.998	0.998	0.997
	6	1.000	1.000	1.000	1.000	1.000	1.000	1.000	1.000	1.000	1.000

n	k	.21	.22	.23	.24	.25	.30	.35	.40	.45	.50
9	0	0.120	0.107	0.095	0.085	0.075	0.040	0.021	0.010	0.005	0.002
	1	0.407	0.378	0.351	0.325	0.300	0.196	0.121	0.071	0.039	0.020
	2	0.711	0.684	0.657	0.629	0.601	0.463	0.337	0.232	0.150	0.090
	3	0.901	0.886	0.870	0.852	0.834	0.730	0.609	0.483	0.361	0.254
	4	0.976	0.971	0.965	0.958	0.951	0.901	0.828	0.733	0.621	0.500
	5	0.996	0.995	0.994	0.992	0.990	0.975	0.946	0.901	0.834	0.746
	6	1.000	0.999	0.999	0.999	0.999	0.996	0.989	0.975	0.950	0.910
	7	1.000	1.000	1.000	1.000	1.000	1.000	0.999	0.996	0.991	0.980
	8	1.000	1.000	1.000	1.000	1.000	1.000	1.000	1.000	0.999	0.998
	9	1.000	1.000	1.000	1.000	1.000	1.000	1.000	1.000	1.000	1.000

n	k	.01	.02	.03	.04	.05	.06	.07	.08	.09	.10
10	0	0.904	0.817	0.737	0.665	0.599	0.539	0.484	0.434	0.389	0.349
	1	0.996	0.984	0.965	0.942	0.914	0.882	0.848	0.812	0.775	0.736
	2	1.000	0.999	0.997	0.994	0.988	0.981	0.972	0.960	0.946	0.930
	3	1.000	1.000	1.000	1.000	0.999	0.998	0.996	0.994	0.991	0.987
	4	1.000	1.000	1.000	1.000	1.000	1.000	1.000	0.999	0.999	0.998
	5	1.000	1.000	1.000	1.000	1.000	1.000	1.000	1.000	1.000	1.000

(continued)

n	k	.11	.12	.13	.14	.15	.16	.17	.18	.19	.20
10	0	0.312	0.279	0.248	0.221	0.197	0.175	0.155	0.137	0.122	0.107
	1	0.697	0.658	0.620	0.582	0.544	0.508	0.473	0.439	0.407	0.376
	2	0.912	0.891	0.869	0.845	0.820	0.794	0.766	0.737	0.708	0.678
	3	0.982	0.976	0.969	0.960	0.950	0.939	0.926	0.912	0.896	0.879
	4	0.997	0.996	0.995	0.993	0.990	0.987	0.983	0.979	0.973	0.967
	5	1.000	1.000	0.999	0.999	0.999	0.998	0.997	0.996	0.995	0.994
	6	1.000	1.000	1.000	1.000	1.000	1.000	1.000	1.000	0.999	0.999
	7	1.000	1.000	1.000	1.000	1.000	1.000	1.000	1.000	1.000	1.000

n	k	.21	.22	.23	.24	.25	.30	.35	.40	.45	.50
10	0	0.095	0.083	0.073	0.064	0.056	0.028	0.013	0.006	0.003	0.001
	1	0.346	0.318	0.292	0.267	0.244	0.149	0.086	0.046	0.023	0.011
	2	0.647	0.617	0.586	0.556	0.526	0.383	0.262	0.167	0.100	0.055
	3	0.861	0.841	0.821	0.799	0.776	0.650	0.514	0.382	0.266	0.172
	4	0.960	0.952	0.943	0.933	0.922	0.850	0.751	0.633	0.504	0.377
	5	0.992	0.990	0.987	0.984	0.980	0.953	0.905	0.834	0.738	0.623
	6	0.999	0.998	0.998	0.997	0.996	0.989	0.974	0.945	0.898	0.828
	7	1.000	1.000	1.000	1.000	1.000	0.998	0.995	0.988	0.973	0.945
	8	1.000	1.000	1.000	1.000	1.000	1.000	0.999	0.998	0.995	0.989
	9	1.000	1.000	1.000	1.000	1.000	1.000	1.000	1.000	1.000	0.999
	10	1.000	1.000	1.000	1.000	1.000	1.000	1.000	1.000	1.000	1.000

n	k	.01	.02	.03	.04	.05	.06	.07	.08	.09	.10
11	0	0.895	0.801	0.715	0.638	0.569	0.506	0.450	0.400	0.354	0.314
	1	0.995	0.980	0.959	0.931	0.898	0.862	0.823	0.782	0.740	0.697
	2	1.000	0.999	0.996	0.992	0.985	0.975	0.963	0.948	0.931	0.910
	3	1.000	1.000	1.000	0.999	0.998	0.997	0.995	0.991	0.987	0.981
	4	1.000	1.000	1.000	1.000	1.000	1.000	0.999	0.999	0.998	0.997
	5	1.000	1.000	1.000	1.000	1.000	1.000	1.000	1.000	1.000	1.000

n	k	.11	.12	.13	.14	.15	.16	.17	.18	.19	.20
11	0	0.278	0.245	0.216	0.190	0.167	0.147	0.129	0.113	0.098	0.086
	1	0.655	0.613	0.571	0.531	0.492	0.455	0.419	0.385	0.353	0.322
	2	0.888	0.863	0.837	0.809	0.779	0.748	0.716	0.684	0.651	0.617
	3	0.974	0.966	0.956	0.944	0.931	0.915	0.899	0.880	0.860	0.839
	4	0.996	0.994	0.991	0.988	0.984	0.979	0.973	0.967	0.959	0.950
	5	1.000	0.999	0.999	0.998	0.997	0.996	0.995	0.993	0.991	0.988
	6	1.000	1.000	1.000	1.000	1.000	1.000	0.999	0.999	0.999	0.998
	7	1.000	1.000	1.000	1.000	1.000	1.000	1.000	1.000	1.000	1.000

n	k	.21	.22	.23	.24	.25	.30	.35	.40	.45	.50
11	0	0.075	0.065	0.056	0.049	0.042	0.020	0.009	0.004	0.001	0.000
	1	0.294	0.267	0.242	0.219	0.197	0.113	0.061	0.030	0.014	0.006
	2	0.584	0.551	0.519	0.487	0.455	0.313	0.200	0.119	0.065	0.033
	3	0.816	0.792	0.767	0.740	0.713	0.570	0.426	0.296	0.191	0.113
	4	0.939	0.928	0.915	0.901	0.885	0.790	0.668	0.533	0.397	0.274
	5	0.985	0.981	0.977	0.972	0.966	0.922	0.851	0.753	0.633	0.500
	6	0.997	0.996	0.995	0.994	0.992	0.978	0.950	0.901	0.826	0.726
	7	1.000	1.000	0.999	0.999	0.999	0.996	0.988	0.971	0.939	0.887
	8	1.000	1.000	1.000	1.000	1.000	0.999	0.998	0.994	0.985	0.967
	9	1.000	1.000	1.000	1.000	1.000	1.000	1.000	0.999	0.998	0.994
	10	1.000	1.000	1.000	1.000	1.000	1.000	1.000	1.000	1.000	1.000

n	k	.01	.02	.03	.04	.05	.06	.07	.08	.09	.10
12	0	0.886	0.785	0.694	0.613	0.540	0.476	0.419	0.368	0.322	0.282
	1	0.994	0.977	0.951	0.919	0.882	0.840	0.797	0.751	0.705	0.659
	2	1.000	0.998	0.995	0.989	0.980	0.968	0.953	0.935	0.913	0.889
	3	1.000	1.000	1.000	0.999	0.998	0.996	0.992	0.988	0.982	0.974
	4	1.000	1.000	1.000	1.000	1.000	1.000	0.999	0.998	0.997	0.996
	5	1.000	1.000	1.000	1.000	1.000	1.000	1.000	1.000	1.000	0.999
	6	1.000	1.000	1.000	1.000	1.000	1.000	1.000	1.000	1.000	1.000

(continued)

n	k	.11	.12	.13	.14	.15	.16	.17	.18	.19	.20
12	0	0.247	0.216	0.188	0.164	0.142	0.123	0.107	0.092	0.080	0.069
	1	0.613	0.569	0.525	0.483	0.443	0.405	0.370	0.336	0.304	0.275
	2	0.862	0.833	0.802	0.770	0.736	0.701	0.666	0.630	0.594	0.558
	3	0.965	0.954	0.940	0.925	0.908	0.889	0.868	0.845	0.820	0.795
	4	0.993	0.991	0.987	0.982	0.976	0.969	0.961	0.951	0.940	0.927
	5	0.999	0.999	0.998	0.997	0.995	0.994	0.991	0.988	0.985	0.981
	6	1.000	1.000	1.000	1.000	0.999	0.999	0.999	0.998	0.997	0.996
	7	1.000	1.000	1.000	1.000	1.000	1.000	1.000	1.000	1.000	0.999
	8	1.000	1.000	1.000	1.000	1.000	1.000	1.000	1.000	1.000	1.000

n	k	.21	.22	.23	.24	.25	.30	.35	.40	.45	.50
12	0	0.059	0.051	0.043	0.037	0.032	0.014	0.006	0.002	0.001	0.000
	1	0.248	0.222	0.199	0.178	0.158	0.085	0.042	0.020	0.008	0.003
	2	0.523	0.489	0.455	0.422	0.391	0.253	0.151	0.083	0.042	0.019
	3	0.767	0.739	0.710	0.680	0.649	0.493	0.347	0.225	0.134	0.073
	4	0.913	0.898	0.881	0.862	0.842	0.724	0.583	0.438	0.304	0.194
	5	0.976	0.970	0.963	0.955	0.946	0.882	0.787	0.665	0.527	0.387
	6	0.995	0.993	0.991	0.989	0.986	0.961	0.915	0.842	0.739	0.613
	7	0.999	0.999	0.998	0.998	0.997	0.991	0.974	0.943	0.888	0.806
	8	1.000	1.000	1.000	1.000	1.000	0.998	0.994	0.985	0.964	0.927
	9	1.000	1.000	1.000	1.000	1.000	1.000	0.999	0.997	0.992	0.981
	10	1.000	1.000	1.000	1.000	1.000	1.000	1.000	1.000	0.999	0.997
	11	1.000	1.000	1.000	1.000	1.000	1.000	1.000	1.000	1.000	1.000

n	k	.01	.02	.03	.04	.05	.06	.07	.08	.09	.10
15	0	0.860	0.739	0.633	0.542	0.463	0.395	0.337	0.286	0.243	0.206
	1	0.990	0.965	0.927	0.881	0.829	0.774	0.717	0.660	0.604	0.549
	2	1.000	0.997	0.991	0.980	0.964	0.943	0.917	0.887	0.853	0.816
	3	1.000	1.000	0.999	0.998	0.995	0.990	0.982	0.973	0.960	0.944
	4	1.000	1.000	1.000	1.000	0.999	0.999	0.997	0.995	0.992	0.987
	5	1.000	1.000	1.000	1.000	1.000	1.000	1.000	0.999	0.999	0.998
	6	1.000	1.000	1.000	1.000	1.000	1.000	1.000	1.000	1.000	1.000

n	k	.11	.12	.13	.14	.15	.16	.17	.18	.19	.20
15	0	0.174	0.147	0.124	0.104	0.087	0.073	0.061	0.051	0.042	0.035
	1	0.497	0.448	0.401	0.358	0.319	0.282	0.249	0.219	0.192	0.167
	2	0.776	0.735	0.692	0.648	0.604	0.561	0.518	0.477	0.436	0.398
	3	0.926	0.904	0.880	0.852	0.823	0.791	0.757	0.722	0.685	0.648
	4	0.981	0.974	0.964	0.952	0.938	0.922	0.904	0.883	0.861	0.836
	5	0.996	0.994	0.992	0.988	0.983	0.977	0.970	0.961	0.951	0.939
	6	0.999	0.999	0.998	0.998	0.996	0.995	0.993	0.990	0.986	0.982
	7	1.000	1.000	1.000	1.000	0.999	0.999	0.999	0.998	0.997	0.996
	8	1.000	1.000	1.000	1.000	1.000	1.000	1.000	1.000	0.999	0.999
	9	1.000	1.000	1.000	1.000	1.000	1.000	1.000	1.000	1.000	1.000

n	k	.21	.22	.23	.24	.25	.30	.35	.40	.45	.50
15	0	0.029	0.024	0.020	0.016	0.013	0.005	0.002	0.000	0.000	0.000
	1	0.145	0.126	0.109	0.094	0.080	0.035	0.014	0.005	0.002	0.000
	2	0.361	0.327	0.294	0.264	0.236	0.127	0.062	0.027	0.011	0.004
	3	0.610	0.573	0.535	0.498	0.461	0.297	0.173	0.091	0.042	0.018
	4	0.809	0.781	0.750	0.719	0.686	0.515	0.352	0.217	0.120	0.059
	5	0.925	0.910	0.892	0.873	0.852	0.722	0.564	0.403	0.261	0.151
	6	0.977	0.970	0.963	0.954	0.943	0.869	0.755	0.610	0.452	0.304
	7	0.994	0.992	0.990	0.987	0.983	0.950	0.887	0.787	0.654	0.500
	8	0.999	0.998	0.998	0.997	0.996	0.985	0.958	0.905	0.818	0.696
	9	1.000	1.000	1.000	0.999	0.999	0.996	0.988	0.966	0.923	0.849
	10	1.000	1.000	1.000	1.000	1.000	0.999	0.997	0.991	0.975	0.941
	11	1.000	1.000	1.000	1.000	1.000	1.000	1.000	0.998	0.994	0.982
	12	1.000	1.000	1.000	1.000	1.000	1.000	1.000	1.000	0.999	0.996
	13	1.000	1.000	1.000	1.000	1.000	1.000	1.000	1.000	1.000	1.000

n	k	.01	.02	.03	.04	.05	.06	.07	.08	.09	.10
20	0	0.818	0.668	0.544	0.442	0.358	0.290	0.234	0.189	0.152	0.122
	1	0.983	0.940	0.880	0.810	0.736	0.660	0.587	0.517	0.452	0.392
	2	0.999	0.993	0.979	0.956	0.925	0.885	0.839	0.788	0.733	0.677
	3	1.000	0.999	0.997	0.993	0.984	0.971	0.953	0.929	0.901	0.867
	4	1.000	1.000	1.000	0.999	0.997	0.994	0.989	0.982	0.971	0.957
	5	1.000	1.000	1.000	1.000	1.000	0.999	0.998	0.996	0.993	0.989
	6	1.000	1.000	1.000	1.000	1.000	1.000	1.000	0.999	0.999	0.998
	7	1.000	1.000	1.000	1.000	1.000	1.000	1.000	1.000	1.000	1.000

(continued)

n	k	.11	.12	.13	.14	.15	.16	.17	.18	.19	.20
20	0	0.097	0.078	0.062	0.049	0.039	0.031	0.024	0.019	0.015	0.012
	1	0.338	0.289	0.246	0.208	0.176	0.147	0.123	0.102	0.084	0.069
	2	0.620	0.563	0.508	0.455	0.405	0.358	0.315	0.275	0.239	0.206
	3	0.829	0.787	0.743	0.696	0.648	0.599	0.550	0.503	0.456	0.411
	4	0.939	0.917	0.892	0.863	0.830	0.794	0.756	0.715	0.673	0.630
	5	0.982	0.974	0.963	0.949	0.933	0.913	0.890	0.864	0.836	0.804
	6	0.996	0.993	0.990	0.985	0.978	0.970	0.959	0.946	0.931	0.913
	7	0.999	0.999	0.998	0.996	0.994	0.991	0.987	0.982	0.976	0.968
	8	1.000	1.000	1.000	0.999	0.999	0.998	0.997	0.995	0.993	0.990
	9	1.000	1.000	1.000	1.000	1.000	1.000	0.999	0.999	0.998	0.997
	10	1.000	1.000	1.000	1.000	1.000	1.000	1.000	1.000	1.000	0.999
	11	1.000	1.000	1.000	1.000	1.000	1.000	1.000	1.000	1.000	1.000

n	k	.21	.22	.23	.24	.25	.30	.35	.40	.45	.50
20	0	0.009	0.007	0.005	0.004	0.003	0.001	0.000	0.000	0.000	0.000
	1	0.057	0.046	0.037	0.030	0.024	0.008	0.002	0.001	0.000	0.000
	2	0.177	0.151	0.128	0.109	0.091	0.035	0.012	0.004	0.001	0.000
	3	0.369	0.329	0.292	0.257	0.225	0.107	0.044	0.016	0.005	0.001
	4	0.586	0.542	0.499	0.456	0.415	0.238	0.118	0.051	0.019	0.006
	5	0.770	0.734	0.696	0.657	0.617	0.416	0.245	0.126	0.055	0.021
	6	0.893	0.870	0.844	0.816	0.786	0.608	0.417	0.250	0.130	0.058
	7	0.958	0.946	0.933	0.917	0.898	0.772	0.601	0.416	0.252	0.132
	8	0.986	0.981	0.975	0.968	0.959	0.887	0.762	0.596	0.414	0.252
	9	0.996	0.995	0.992	0.990	0.986	0.952	0.878	0.755	0.591	0.412
	10	0.999	0.999	0.998	0.997	0.996	0.983	0.947	0.872	0.751	0.588
	11	1.000	1.000	1.000	0.999	0.999	0.995	0.980	0.943	0.869	0.748
	12	1.000	1.000	1.000	1.000	1.000	0.999	0.994	0.979	0.942	0.868
	13	1.000	1.000	1.000	1.000	1.000	1.000	0.998	0.994	0.979	0.942
	14	1.000	1.000	1.000	1.000	1.000	1.000	1.000	0.998	0.994	0.979
	15	1.000	1.000	1.000	1.000	1.000	1.000	1.000	1.000	0.998	0.994
	16	1.000	1.000	1.000	1.000	1.000	1.000	1.000	1.000	1.000	0.999
	17	1.000	1.000	1.000	1.000	1.000	1.000	1.000	1.000	1.000	1.000

n	k	.01	.02	.03	.04	.05	.06	.07	.08	.09	.10
25	0	0.778	0.603	0.467	0.360	0.277	0.213	0.163	0.124	0.095	0.072
	1	0.974	0.911	0.828	0.736	0.642	0.553	0.470	0.395	0.329	0.271
	2	0.998	0.987	0.962	0.924	0.873	0.813	0.747	0.677	0.606	0.537
	3	1.000	0.999	0.994	0.983	0.966	0.940	0.906	0.865	0.817	0.764
	4	1.000	1.000	0.999	0.997	0.993	0.985	0.973	0.955	0.931	0.902
	5	1.000	1.000	1.000	1.000	0.999	0.997	0.993	0.988	0.979	0.967
	6	1.000	1.000	1.000	1.000	1.000	0.999	0.999	0.997	0.995	0.991
	7	1.000	1.000	1.000	1.000	1.000	1.000	1.000	0.999	0.999	0.998
	8	1.000	1.000	1.000	1.000	1.000	1.000	1.000	1.000	1.000	1.000

n	k	.11	.12	.13	.14	.15	.16	.17	.18	.19	.20
25	0	0.054	0.041	0.031	0.023	0.017	0.013	0.009	0.007	0.005	0.004
	1	0.222	0.180	0.146	0.117	0.093	0.074	0.058	0.045	0.035	0.027
	2	0.471	0.409	0.352	0.300	0.254	0.213	0.177	0.147	0.120	0.098
	3	0.707	0.648	0.588	0.529	0.471	0.416	0.365	0.317	0.273	0.234
	4	0.867	0.827	0.782	0.733	0.682	0.629	0.576	0.523	0.471	0.421
	5	0.950	0.929	0.903	0.873	0.838	0.800	0.758	0.712	0.665	0.617
	6	0.984	0.976	0.964	0.949	0.930	0.908	0.882	0.851	0.817	0.780
	7	0.996	0.993	0.989	0.983	0.975	0.964	0.950	0.934	0.914	0.891
	8	0.999	0.998	0.997	0.995	0.992	0.988	0.982	0.975	0.965	0.953
	9	1.000	1.000	0.999	0.999	0.998	0.997	0.995	0.992	0.988	0.983
	10	1.000	1.000	1.000	1.000	1.000	0.999	0.999	0.998	0.996	0.994
	11	1.000	1.000	1.000	1.000	1.000	1.000	1.000	0.999	0.999	0.998
	12	1.000	1.000	1.000	1.000	1.000	1.000	1.000	1.000	1.000	1.000

(continued)

n	k	.21	.22	.23	.24	.25	.30	.35	.40	.45	.50
25	0	0.003	0.002	0.001	0.001	0.001	0.000	0.000	0.000	0.000	0.000
	1	0.021	0.016	0.012	0.009	0.007	0.002	0.000	0.000	0.000	0.000
	2	0.080	0.064	0.051	0.041	0.032	0.009	0.002	0.000	0.000	0.000
	3	0.199	0.168	0.140	0.117	0.096	0.033	0.010	0.002	0.000	0.000
	4	0.373	0.328	0.287	0.248	0.214	0.090	0.032	0.009	0.002	0.000
	5	0.568	0.518	0.470	0.423	0.378	0.193	0.083	0.029	0.009	0.002
	6	0.740	0.697	0.653	0.607	0.561	0.341	0.173	0.074	0.026	0.007
	7	0.864	0.834	0.801	0.765	0.727	0.512	0.306	0.154	0.064	0.022
	8	0.939	0.921	0.901	0.877	0.851	0.677	0.467	0.274	0.134	0.054
	9	0.976	0.967	0.957	0.944	0.929	0.811	0.630	0.425	0.242	0.115
	10	0.992	0.988	0.984	0.978	0.970	0.902	0.771	0.586	0.384	0.212
	11	0.998	0.996	0.995	0.992	0.989	0.956	0.875	0.732	0.543	0.345
	12	0.999	0.999	0.998	0.998	0.997	0.983	0.940	0.846	0.694	0.500
	13	1.000	1.000	1.000	0.999	0.999	0.994	0.975	0.922	0.817	0.655
	14	1.000	1.000	1.000	1.000	1.000	0.998	0.991	0.966	0.904	0.788
	15	1.000	1.000	1.000	1.000	1.000	1.000	0.997	0.987	0.956	0.885
	16	1.000	1.000	1.000	1.000	1.000	1.000	0.999	0.996	0.983	0.946
	17	1.000	1.000	1.000	1.000	1.000	1.000	1.000	0.999	0.994	0.978
	18	1.000	1.000	1.000	1.000	1.000	1.000	1.000	1.000	0.998	0.993
	19	1.000	1.000	1.000	1.000	1.000	1.000	1.000	1.000	1.000	0.998
	20	1.000	1.000	1.000	1.000	1.000	1.000	1.000	1.000	1.000	1.000

TABLE VI Cumulative Poisson Probabilities

$$P(X \leq k) = \sum_{r=0}^{k} \frac{(\mu_C)^r e^{-\mu_c}}{r!}$$

μ_c or np

k	.1	.2	.3	.4	.5	.6	.7	.8	.9	1.0
0	0.905	0.819	0.741	0.670	0.607	0.549	0.497	0.449	0.407	0.368
1	0.995	0.982	0.963	0.938	0.910	0.878	0.844	0.809	0.772	0.736
2	1.000	0.999	0.996	0.992	0.986	0.977	0.966	0.953	0.937	0.920
3	1.000	1.000	1.000	0.999	0.998	0.997	0.994	0.991	0.987	0.981
4	1.000	1.000	1.000	1.000	1.000	1.000	0.999	0.999	0.998	0.996
5	1.000	1.000	1.000	1.000	1.000	1.000	1.000	1.000	1.000	0.999
6	1.000	1.000	1.000	1.000	1.000	1.000	1.000	1.000	1.000	1.000

k	1.1	1.2	1.3	1.4	1.5	1.6	1.7	1.8	1.9	2.0
0	0.333	0.301	0.273	0.247	0.223	0.202	0.183	0.165	0.150	0.135
1	0.699	0.663	0.627	0.592	0.558	0.525	0.493	0.463	0.434	0.406
2	0.900	0.879	0.857	0.833	0.809	0.783	0.757	0.731	0.704	0.677
3	0.974	0.966	0.957	0.946	0.934	0.921	0.907	0.891	0.875	0.857
4	0.995	0.992	0.989	0.986	0.981	0.976	0.970	0.964	0.956	0.947
5	0.999	0.998	0.998	0.997	0.996	0.994	0.992	0.990	0.987	0.983
6	1.000	1.000	1.000	0.999	0.999	0.999	0.998	0.997	0.997	0.995
7	1.000	1.000	1.000	1.000	1.000	1.000	1.000	0.999	0.999	0.999
8	1.000	1.000	1.000	1.000	1.000	1.000	1.000	1.000	1.000	1.000

k	2.1	2.2	2.3	2.4	2.5	2.6	2.7	2.8	2.9	3.0
0	0.122	0.111	0.100	0.091	0.082	0.074	0.067	0.061	0.055	0.050
1	0.380	0.355	0.331	0.308	0.287	0.267	0.249	0.231	0.215	0.199
2	0.650	0.623	0.596	0.570	0.544	0.518	0.494	0.469	0.446	0.423
3	0.839	0.819	0.799	0.779	0.758	0.736	0.714	0.692	0.670	0.647
4	0.938	0.928	0.916	0.904	0.891	0.877	0.863	0.848	0.832	0.815
5	0.980	0.975	0.970	0.964	0.958	0.951	0.943	0.935	0.926	0.916
6	0.994	0.993	0.991	0.988	0.986	0.983	0.979	0.976	0.971	0.966
7	0.999	0.998	0.997	0.997	0.996	0.995	0.993	0.992	0.990	0.988
8	1.000	1.000	0.999	0.999	0.999	0.999	0.998	0.998	0.997	0.996
9	1.000	1.000	1.000	1.000	1.000	1.000	0.999	0.999	0.999	0.999
10	1.000	1.000	1.000	1.000	1.000	1.000	1.000	1.000	1.000	1.000

(continued)

μ_c or np

k	3.1	3.2	3.3	3.4	3.5	3.6	3.7	3.8	3.9	4.0
0	0.045	0.041	0.037	0.033	0.030	0.027	0.025	0.022	0.020	0.018
1	0.185	0.171	0.159	0.147	0.136	0.126	0.116	0.107	0.099	0.092
2	0.401	0.380	0.359	0.340	0.321	0.303	0.285	0.269	0.253	0.238
3	0.625	0.603	0.580	0.558	0.537	0.515	0.494	0.473	0.453	0.433
4	0.798	0.781	0.763	0.744	0.725	0.706	0.687	0.668	0.648	0.629
5	0.906	0.895	0.883	0.871	0.858	0.844	0.830	0.816	0.801	0.785
6	0.961	0.955	0.949	0.942	0.935	0.927	0.918	0.909	0.899	0.889
7	0.986	0.983	0.980	0.977	0.973	0.969	0.965	0.960	0.955	0.949
8	0.995	0.994	0.993	0.992	0.990	0.988	0.986	0.984	0.981	0.979
9	0.999	0.998	0.998	0.997	0.997	0.996	0.995	0.994	0.993	0.992
10	1.000	1.000	0.999	0.999	0.999	0.999	0.998	0.998	0.998	0.997
11	1.000	1.000	1.000	1.000	1.000	1.000	1.000	0.999	0.999	0.999
12	1.000	1.000	1.000	1.000	1.000	1.000	1.000	1.000	1.000	1.000

k	4.1	4.2	4.3	4.4	4.5	4.6	4.7	4.8	4.9	5.0
0	0.017	0.015	0.014	0.012	0.011	0.010	0.009	0.008	0.007	0.007
1	0.085	0.078	0.072	0.066	0.061	0.056	0.052	0.048	0.044	0.040
2	0.224	0.210	0.197	0.185	0.174	0.163	0.152	0.143	0.133	0.125
3	0.414	0.395	0.377	0.359	0.342	0.326	0.310	0.294	0.279	0.265
4	0.609	0.590	0.570	0.551	0.532	0.513	0.495	0.476	0.458	0.440
5	0.769	0.753	0.737	0.720	0.703	0.686	0.668	0.651	0.634	0.616
6	0.879	0.867	0.856	0.844	0.831	0.818	0.805	0.791	0.777	0.762
7	0.943	0.936	0.929	0.921	0.913	0.905	0.896	0.887	0.877	0.867
8	0.976	0.972	0.968	0.964	0.960	0.955	0.950	0.944	0.938	0.932
9	0.990	0.989	0.987	0.985	0.983	0.980	0.978	0.975	0.972	0.968
10	0.997	0.996	0.995	0.994	0.993	0.992	0.991	0.990	0.988	0.986
11	0.999	0.999	0.998	0.998	0.998	0.997	0.997	0.996	0.995	0.995
12	1.000	1.000	0.999	0.999	0.999	0.999	0.999	0.999	0.998	0.998
13	1.000	1.000	1.000	1.000	1.000	1.000	1.000	1.000	0.999	0.999
14	1.000	1.000	1.000	1.000	1.000	1.000	1.000	1.000	1.000	1.000

μ_c or np

k	5.1	5.2	5.3	5.4	5.5	5.6	5.7	5.8	5.9	6.0
0	0.006	0.006	0.005	0.005	0.004	0.004	0.003	0.003	0.003	0.002
1	0.037	0.034	0.031	0.029	0.027	0.024	0.022	0.021	0.019	0.017
2	0.116	0.109	0.102	0.095	0.088	0.082	0.077	0.072	0.067	0.062
3	0.251	0.238	0.225	0.213	0.202	0.191	0.180	0.170	0.160	0.151
4	0.423	0.406	0.390	0.373	0.358	0.342	0.327	0.313	0.299	0.285
5	0.598	0.581	0.563	0.546	0.529	0.512	0.495	0.478	0.462	0.446
6	0.747	0.732	0.717	0.702	0.686	0.670	0.654	0.638	0.622	0.606
7	0.856	0.845	0.833	0.822	0.809	0.797	0.784	0.771	0.758	0.744
8	0.925	0.918	0.911	0.903	0.894	0.886	0.877	0.867	0.857	0.847
9	0.964	0.960	0.956	0.951	0.946	0.941	0.935	0.929	0.923	0.916
10	0.984	0.982	0.980	0.977	0.975	0.972	0.969	0.965	0.961	0.957
11	0.994	0.993	0.992	0.990	0.989	0.988	0.986	0.984	0.982	0.980
12	0.998	0.997	0.997	0.996	0.996	0.995	0.994	0.993	0.992	0.991
13	0.999	0.999	0.999	0.999	0.998	0.998	0.998	0.997	0.997	0.996
14	1.000	1.000	1.000	1.000	0.999	0.999	0.999	0.999	0.999	0.999
15	1.000	1.000	1.000	1.000	1.000	1.000	1.000	1.000	1.000	0.999
16	1.000	1.000	1.000	1.000	1.000	1.000	1.000	1.000	1.000	1.000

k	6.1	6.2	6.3	6.4	6.5	6.6	6.7	6.8	6.9	7.0
0	0.002	0.002	0.002	0.002	0.002	0.001	0.001	0.001	0.001	0.001
1	0.016	0.015	0.013	0.012	0.011	0.010	0.009	0.009	0.008	0.007
2	0.058	0.054	0.050	0.046	0.043	0.040	0.037	0.034	0.032	0.030
3	0.143	0.134	0.126	0.119	0.112	0.105	0.099	0.093	0.087	0.082
4	0.272	0.259	0.247	0.235	0.224	0.213	0.202	0.192	0.182	0.173
5	0.430	0.414	0.399	0.384	0.369	0.355	0.341	0.327	0.314	0.301
6	0.590	0.574	0.558	0.542	0.527	0.511	0.495	0.480	0.465	0.450
7	0.730	0.716	0.702	0.687	0.673	0.658	0.643	0.628	0.614	0.599
8	0.837	0.826	0.815	0.803	0.792	0.780	0.767	0.755	0.742	0.729
9	0.909	0.902	0.894	0.886	0.877	0.869	0.860	0.850	0.840	0.830
10	0.953	0.949	0.944	0.939	0.933	0.927	0.921	0.915	0.908	0.901
11	0.978	0.975	0.972	0.969	0.966	0.963	0.959	0.955	0.951	0.947
12	0.990	0.989	0.987	0.986	0.984	0.982	0.980	0.978	0.976	0.973
13	0.996	0.995	0.995	0.994	0.993	0.992	0.991	0.990	0.989	0.987
14	0.998	0.998	0.998	0.997	0.997	0.997	0.996	0.996	0.995	0.994
15	0.999	0.999	0.999	0.999	0.999	0.999	0.998	0.998	0.998	0.998

(continued)

					μ_c or np					
16	1.000	1.000	1.000	1.000	1.000	0.999	0.999	0.999	0.999	0.999
17	1.000	1.000	1.000	1.000	1.000	1.000	1.000	1.000	1.000	1.000
k	**7.1**	**7.2**	**7.3**	**7.4**	**7.5**	**7.6**	**7.7**	**7.8**	**7.9**	**8.0**
0	0.001	0.001	0.001	0.001	0.001	0.001	0.000	0.000	0.000	0.000
1	0.007	0.006	0.006	0.005	0.005	0.004	0.004	0.004	0.003	0.003
2	0.027	0.025	0.024	0.022	0.020	0.019	0.017	0.016	0.015	0.014
3	0.077	0.072	0.067	0.063	0.059	0.055	0.052	0.048	0.045	0.042
4	0.164	0.156	0.147	0.140	0.132	0.125	0.118	0.112	0.106	0.100
5	0.288	0.276	0.264	0.253	0.241	0.231	0.220	0.210	0.201	0.191
6	0.435	0.420	0.406	0.392	0.378	0.365	0.351	0.338	0.326	0.313
7	0.584	0.569	0.554	0.539	0.525	0.510	0.496	0.481	0.467	0.453
8	0.716	0.703	0.689	0.676	0.662	0.648	0.634	0.620	0.607	0.593
9	0.820	0.810	0.799	0.788	0.776	0.765	0.753	0.741	0.729	0.717
10	0.894	0.887	0.879	0.871	0.862	0.854	0.845	0.835	0.826	0.816
11	0.942	0.937	0.932	0.926	0.921	0.915	0.909	0.902	0.895	0.888
12	0.970	0.967	0.964	0.961	0.957	0.954	0.950	0.945	0.941	0.936
13	0.986	0.984	0.982	0.980	0.978	0.976	0.974	0.971	0.969	0.966
14	0.994	0.993	0.992	0.991	0.990	0.989	0.987	0.986	0.984	0.983
15	0.997	0.997	0.996	0.996	0.995	0.995	0.994	0.993	0.993	0.992
16	0.999	0.999	0.999	0.998	0.998	0.998	0.997	0.997	0.997	0.996
17	1.000	1.000	0.999	0.999	0.999	0.999	0.999	0.999	0.999	0.998
18	1.000	1.000	1.000	1.000	1.000	1.000	1.000	1.000	0.999	0.999
19	1.000	1.000	1.000	1.000	1.000	1.000	1.000	1.000	1.000	1.000
k	**8.1**	**8.2**	**8.3**	**8.4**	**8.5**	**8.6**	**8.7**	**8.8**	**8.9**	**9.0**
0	0.000	0.000	0.000	0.000	0.000	0.000	0.000	0.000	0.000	0.000
1	0.003	0.003	0.002	0.002	0.002	0.002	0.002	0.001	0.001	0.001
2	0.013	0.012	0.011	0.010	0.009	0.009	0.008	0.007	0.007	0.006
3	0.040	0.037	0.035	0.032	0.030	0.028	0.026	0.024	0.023	0.021
4	0.094	0.089	0.084	0.079	0.074	0.070	0.066	0.062	0.058	0.055
5	0.182	0.174	0.165	0.157	0.150	0.142	0.135	0.128	0.122	0.116
6	0.301	0.290	0.278	0.267	0.256	0.246	0.235	0.226	0.216	0.207
7	0.439	0.425	0.412	0.399	0.386	0.373	0.360	0.348	0.336	0.324
8	0.579	0.565	0.551	0.537	0.523	0.509	0.496	0.482	0.469	0.456
9	0.704	0.692	0.679	0.666	0.653	0.640	0.627	0.614	0.601	0.587
10	0.806	0.796	0.785	0.774	0.763	0.752	0.741	0.729	0.718	0.706

μ_c or np

11	0.881	0.873	0.865	0.857	0.849	0.840	0.831	0.822	0.813	0.803
12	0.931	0.926	0.921	0.915	0.909	0.903	0.897	0.890	0.883	0.876
13	0.963	0.960	0.956	0.952	0.949	0.945	0.940	0.936	0.931	0.926
14	0.981	0.979	0.977	0.975	0.973	0.970	0.967	0.965	0.962	0.959
15	0.991	0.990	0.989	0.987	0.986	0.985	0.983	0.982	0.980	0.978
16	0.996	0.995	0.995	0.994	0.993	0.993	0.992	0.991	0.990	0.989
17	0.998	0.998	0.998	0.997	0.997	0.997	0.996	0.996	0.995	0.995
18	0.999	0.999	0.999	0.999	0.999	0.999	0.998	0.998	0.998	0.998
19	1.000	1.000	1.000	1.000	0.999	0.999	0.999	0.999	0.999	0.999
20	1.000	1.000	1.000	1.000	1.000	1.000	1.000	1.000	1.000	1.000

k	**9.1**	**9.2**	**9.3**	**9.4**	**9.5**	**9.6**	**9.7**	**9.8**	**9.9**	**10.0**
0	0.000	0.000	0.000	0.000	0.000	0.000	0.000	0.000	0.000	0.000
1	0.001	0.001	0.001	0.001	0.001	0.001	0.001	0.001	0.001	0.000
2	0.006	0.005	0.005	0.005	0.004	0.004	0.004	0.003	0.003	0.003
3	0.020	0.018	0.017	0.016	0.015	0.014	0.013	0.012	0.011	0.010
4	0.052	0.049	0.046	0.043	0.040	0.038	0.035	0.033	0.031	0.029
5	0.110	0.104	0.099	0.093	0.089	0.084	0.079	0.075	0.071	0.067
6	0.198	0.189	0.181	0.173	0.165	0.157	0.150	0.143	0.137	0.130
7	0.312	0.301	0.290	0.279	0.269	0.258	0.248	0.239	0.229	0.220
8	0.443	0.430	0.417	0.404	0.392	0.380	0.368	0.356	0.344	0.333
9	0.574	0.561	0.548	0.535	0.522	0.509	0.496	0.483	0.471	0.458
10	0.694	0.682	0.670	0.658	0.645	0.633	0.621	0.608	0.596	0.583
11	0.793	0.783	0.773	0.763	0.752	0.741	0.730	0.719	0.708	0.697
12	0.868	0.861	0.853	0.845	0.836	0.828	0.819	0.810	0.801	0.792
13	0.921	0.916	0.910	0.904	0.898	0.892	0.885	0.879	0.872	0.864
14	0.955	0.952	0.948	0.944	0.940	0.936	0.931	0.927	0.922	0.917
15	0.976	0.974	0.972	0.969	0.967	0.964	0.961	0.958	0.955	0.951
16	0.988	0.987	0.985	0.984	0.982	0.981	0.979	0.977	0.975	0.973
17	0.994	0.993	0.993	0.992	0.991	0.990	0.989	0.988	0.987	0.986
18	0.997	0.997	0.997	0.996	0.996	0.995	0.995	0.994	0.993	0.993
19	0.999	0.999	0.998	0.998	0.998	0.998	0.998	0.997	0.997	0.997
20	0.999	0.999	0.999	0.999	0.999	0.999	0.999	0.999	0.999	0.998
21	1.000	1.000	1.000	1.000	1.000	1.000	1.000	0.999	0.999	0.999
22	1.000	1.000	1.000	1.000	1.000	1.000	1.000	1.000	1.000	1.000

(continued)

μ_c or np

k	10.5	11.0	11.5	12.0	12.5	13.0	13.5	14.0	14.5	15.0
1	0.000	0.000	0.000	0.000	0.000	0.000	0.000	0.000	0.000	0.000
2	0.002	0.001	0.001	0.001	0.000	0.000	0.000	0.000	0.000	0.000
3	0.007	0.005	0.003	0.002	0.002	0.001	0.001	0.000	0.000	0.000
4	0.021	0.015	0.011	0.008	0.005	0.004	0.003	0.002	0.001	0.001
5	0.050	0.038	0.028	0.020	0.015	0.011	0.008	0.006	0.004	0.003
6	0.102	0.079	0.060	0.046	0.035	0.026	0.019	0.014	0.010	0.008
7	0.179	0.143	0.114	0.090	0.070	0.054	0.041	0.032	0.024	0.018
8	0.279	0.232	0.191	0.155	0.125	0.100	0.079	0.062	0.048	0.037
9	0.397	0.341	0.289	0.242	0.201	0.166	0.135	0.109	0.088	0.070
10	0.521	0.460	0.402	0.347	0.297	0.252	0.211	0.176	0.145	0.118
11	0.639	0.579	0.520	0.462	0.406	0.353	0.304	0.260	0.220	0.185
12	0.742	0.689	0.633	0.576	0.519	0.463	0.409	0.358	0.311	0.268
13	0.825	0.781	0.733	0.682	0.628	0.573	0.518	0.464	0.413	0.363
14	0.888	0.854	0.815	0.772	0.725	0.675	0.623	0.570	0.518	0.466
15	0.932	0.907	0.878	0.844	0.806	0.764	0.718	0.669	0.619	0.568
16	0.960	0.944	0.924	0.899	0.869	0.835	0.798	0.756	0.711	0.664
17	0.978	0.968	0.954	0.937	0.916	0.890	0.861	0.827	0.790	0.749
18	0.988	0.982	0.974	0.963	0.948	0.930	0.908	0.883	0.853	0.819
19	0.994	0.991	0.986	0.979	0.969	0.957	0.942	0.923	0.901	0.875
20	0.997	0.995	0.992	0.988	0.983	0.975	0.965	0.952	0.936	0.917
21	0.999	0.998	0.996	0.994	0.991	0.986	0.980	0.971	0.960	0.947
22	0.999	0.999	0.998	0.997	0.995	0.992	0.989	0.983	0.976	0.967
23	1.000	1.000	0.999	0.999	0.998	0.996	0.994	0.991	0.986	0.981
24	1.000	1.000	1.000	0.999	0.999	0.998	0.997	0.995	0.992	0.989
25	1.000	1.000	1.000	1.000	0.999	0.999	0.998	0.997	0.996	0.994
26	1.000	1.000	1.000	1.000	1.000	1.000	0.999	0.999	0.998	0.997
27	1.000	1.000	1.000	1.000	1.000	1.000	1.000	0.999	0.999	0.998
28	1.000	1.000	1.000	1.000	1.000	1.000	1.000	1.000	0.999	0.999
29	1.000	1.000	1.000	1.000	1.000	1.000	1.000	1.000	1.000	1.000

k	16.0	17.0	18.0	19.0	20.0	21.0	22.0	23.0	24.0	25.0
4	0.000	0.000	0.000	0.000	0.000	0.000	0.000	0.000	0.000	0.000
5	0.001	0.001	0.000	0.000	0.000	0.000	0.000	0.000	0.000	0.000
6	0.004	0.002	0.001	0.001	0.000	0.000	0.000	0.000	0.000	0.000
7	0.010	0.005	0.003	0.002	0.001	0.000	0.000	0.000	0.000	0.000

μ_c or np

k	16.0	17.0	18.0	19.0	20.0	21.0	22.0	23.0	24.0	25.0
8	0.022	0.013	0.007	0.004	0.002	0.001	0.001	0.000	0.000	0.000
9	0.043	0.026	0.015	0.009	0.005	0.003	0.002	0.001	0.000	0.000
10	0.077	0.049	0.030	0.018	0.011	0.006	0.004	0.002	0.001	0.001
11	0.127	0.085	0.055	0.035	0.021	0.013	0.008	0.004	0.003	0.001
12	0.193	0.135	0.092	0.061	0.039	0.025	0.015	0.009	0.005	0.003
13	0.275	0.201	0.143	0.098	0.066	0.043	0.028	0.017	0.011	0.006
14	0.368	0.281	0.208	0.150	0.105	0.072	0.048	0.031	0.020	0.012
15	0.467	0.371	0.287	0.215	0.157	0.111	0.077	0.052	0.034	0.022
16	0.566	0.468	0.375	0.292	0.221	0.163	0.117	0.082	0.056	0.038
17	0.659	0.564	0.469	0.378	0.297	0.227	0.169	0.123	0.087	0.060
18	0.742	0.655	0.562	0.469	0.381	0.302	0.232	0.175	0.128	0.092
19	0.812	0.736	0.651	0.561	0.470	0.384	0.306	0.238	0.180	0.134
20	0.868	0.805	0.731	0.647	0.559	0.471	0.387	0.310	0.243	0.185
21	0.911	0.861	0.799	0.725	0.644	0.558	0.472	0.389	0.314	0.247
22	0.942	0.905	0.855	0.793	0.721	0.640	0.556	0.472	0.392	0.318
23	0.963	0.937	0.899	0.849	0.787	0.716	0.637	0.555	0.473	0.394
24	0.978	0.959	0.932	0.893	0.843	0.782	0.712	0.635	0.554	0.473
25	0.987	0.975	0.955	0.927	0.888	0.838	0.777	0.708	0.632	0.553
26	0.993	0.985	0.972	0.951	0.922	0.883	0.832	0.772	0.704	0.629
27	0.996	0.991	0.983	0.969	0.948	0.917	0.877	0.827	0.768	0.700
28	0.998	0.995	0.990	0.980	0.966	0.944	0.913	0.873	0.823	0.763
29	0.999	0.997	0.994	0.988	0.978	0.963	0.940	0.908	0.868	0.818
30	0.999	0.999	0.997	0.993	0.987	0.976	0.959	0.936	0.904	0.863
31	1.000	0.999	0.998	0.996	0.992	0.985	0.973	0.956	0.932	0.900
32	1.000	1.000	0.999	0.998	0.995	0.991	0.983	0.971	0.953	0.929
33	1.000	1.000	1.000	0.999	0.997	0.994	0.989	0.981	0.969	0.950
34	1.000	1.000	1.000	0.999	0.999	0.997	0.994	0.988	0.979	0.966
35	1.000	1.000	1.000	1.000	0.999	0.998	0.996	0.993	0.987	0.978
36	1.000	1.000	1.000	1.000	1.000	0.999	0.998	0.996	0.992	0.985
37	1.000	1.000	1.000	1.000	1.000	0.999	0.999	0.997	0.995	0.991
38	1.000	1.000	1.000	1.000	1.000	1.000	0.999	0.999	0.997	0.994
39	1.000	1.000	1.000	1.000	1.000	1.000	1.000	0.999	0.998	0.997
40	1.000	1.000	1.000	1.000	1.000	1.000	1.000	1.000	0.999	0.998
41	1.000	1.000	1.000	1.000	1.000	1.000	1.000	1.000	0.999	0.999
42	1.000	1.000	1.000	1.000	1.000	1.000	1.000	1.000	1.000	0.999
43	1.000	1.000	1.000	1.000	1.000	1.000	1.000	1.000	1.000	1.000

TABLE VII Standard Normal Cumulative Probabilities

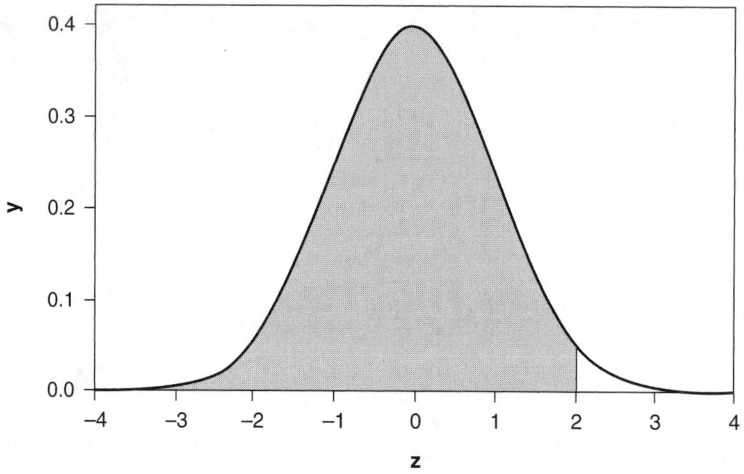

$P(Z \le k)$

k	.00	.01	.02	.03	.04	.05	.06	.07	.08	.09
−4.0	.4(0)32	.4(0)30	.4(0)29	.4(0)28	.4(0)27	.4(0)26	.4(0)25	.4(0)24	.4(0)23	.4(0)22
−3.9	.4(0)48	.4(0)46	.4(0)44	.4(0)42	.4(0)41	.4(0)39	.4(0)37	.4(0)36	.4(0)34	.4(0)33
−3.8	.4(0)72	.4(0)69	.4(0)67	.4(0)64	.4(0)62	.4(0)59	.4(0)57	.4(0)54	.4(0)52	.4(0)50
−3.7	.00011	.00010	.00010	.4(0)96	.4(0)92	.4(0)88	.4(0)85	.4(0)82	.4(0)78	.4(0)75
−3.6	.00016	.00015	.00015	.00014	.00014	.00013	.00013	.00012	.00012	.00011
−3.5	.00023	.00022	.00022	.00021	.00020	.00019	.00019	.00018	.00017	.00017
−3.4	.00034	.00032	.00031	.00030	.00029	.00028	.00027	.00026	.00025	.00024
−3.3	.00048	.00047	.00045	.00043	.00042	.00040	.00039	.00038	.00036	.00035
−3.2	.00069	.00066	.00064	.00062	.00060	.00058	.00056	.00054	.00052	.00050
−3.1	.00097	.00094	.00090	.00087	.00084	.00082	.00079	.00076	.00074	.00071
−3.0	.00135	.00131	.00126	.00122	.00118	.00114	.00111	.00107	.00104	.00100

k	.00	.01	.02	.03	.04	.05	.06	.07	.08	.09
−2.9	.0019	.0018	.0018	.0017	.0016	.0016	.0015	.0015	.0014	.0014
−2.8	.0026	.0025	.0024	.0023	.0023	.0022	.0021	.0021	.0020	.0019
−2.7	.0035	.0034	.0033	.0032	.0031	.0030	.0029	.0028	.0027	.0026
−2.6	.0047	.0045	.0044	.0043	.0041	.0040	.0039	.0038	.0037	.0036
−2.5	.0062	.0060	.0059	.0057	.0055	.0054	.0052	.0051	.0049	.0048
−2.4	.0082	.0080	.0078	.0075	.0073	.0071	.0069	.0068	.0066	.0064
−2.3	.0107	.0104	.0102	.0099	.0096	.0094	.0091	.0089	.0087	.0084

k	**.00**	**.01**	**.02**	**.03**	**.04**	**.05**	**.06**	**.07**	**.08**	**.09**
−2.2	.0139	.0136	.0132	.0129	.0125	.0122	.0119	.0116	.0113	.0110
−2.1	.0179	.0174	.0170	.0166	.0162	.0158	.0154	.0150	.0146	.0143
−2.0	.0228	.0222	.0217	.0212	.0207	.0202	.0197	.0192	.0188	.0183
−1.9	.0287	.0281	.0274	.0268	.0262	.0256	.0250	.0244	.0239	.0233
−1.8	.0359	.0351	.0344	.0336	.0329	.0322	.0314	.0307	.0301	.0294
−1.7	.0446	.0436	.0427	.0418	.0409	.0401	.0392	.0384	.0375	.0367
−1.6	.0548	.0537	.0526	.0516	.0505	.0495	.0485	.0475	.0465	.0455
−1.5	.0668	.0655	.0643	.0630	.0618	.0606	.0594	.0582	.0571	.0559
−1.4	.0808	.0793	.0778	.0764	.0749	.0735	.0721	.0708	.0694	.0681
−1.3	.0968	.0951	.0934	.0918	.0901	.0885	.0869	.0853	.0838	.0823
−1.2	.1151	.1131	.1112	.1093	.1075	.1056	.1038	.1020	.1003	.0985
−1.1	.1357	.1335	.1314	.1292	.1271	.1251	.1230	.1210	.1190	.1170
−1.0	.1587	.1562	.1539	.1515	.1492	.1469	.1446	.1423	.1401	.1379
−.9	.1841	.1814	.1788	.1762	.1736	.1711	.1685	.1660	.1635	.1611
−.8	.2119	.2090	.2061	.2033	.2005	.1977	.1949	.1922	.1894	.1867
−.7	.2420	.2389	.2358	.2327	.2297	.2266	.2236	.2206	.2177	.2148
−.6	.2743	.2709	.2676	.2643	.2611	.2578	.2546	.2514	.2483	.2451
−.5	.3085	.3050	.3015	.2981	.2946	.2912	.2877	.2843	.2810	.2776
−.4	.3446	.3409	.3372	.3336	.3300	.3264	.3228	.3192	.3156	.3121
−.3	.3821	.3783	.3745	.3707	.3669	.3632	.3594	.3557	.3520	.3483
−.2	.4207	.4168	.4129	.4090	.4052	.4013	.3974	.3936	.3897	.3859
−.1	.4602	.4562	.4522	.4483	.4443	.4404	.4364	.4325	.4286	.4247
−.0	.5000	.4960	.4920	.4880	.4840	.4801	.4761	.4721	.4681	.4641
.0	.5000	.5040	.5080	.5120	.5160	.5199	.5239	.5279	.5319	.5359
.1	.5398	.5438	.5478	.5517	.5557	.5596	.5636	.5675	.5714	.5753
.2	.5793	.5832	.5871	.5910	.5948	.5987	.6026	.6064	.6103	.6141
.3	.6179	.6217	.6255	.6293	.6331	.6368	.6406	.6443	.6480	.6517
.4	.6554	.6591	.6628	.6664	.6700	.6736	.6772	.6808	.6844	.6879
.5	.6915	.6950	.6985	.7019	.7054	.7088	.7123	.7157	.7190	.7224
.6	.7257	.7291	.7324	.7357	.7389	.7422	.7454	.7486	.7517	.7549
.7	.7580	.7611	.7642	.7673	.7703	.7734	.7764	.7794	.7823	.7852
.8	.7881	.7910	.7939	.7967	.7995	.8023	.8051	.8078	.8106	.8133
.9	.8159	.8186	.8212	.8238	.8264	.8289	.8315	.8340	.8365	.8389
1.0	.8413	.8438	.8461	.8485	.8508	.8531	.8554	.8577	.8599	.8621
1.1	.8643	.8665	.8686	.8708	.8729	.8749	.8770	.8790	.8810	.8830
1.2	.8849	.8869	.8888	.8907	.8925	.8944	.8962	.8980	.8997	.9015
1.3	.9032	.9049	.9066	.9082	.9099	.9115	.9131	.9147	.9162	.9177

(continued)

1.4	.9192	.9207	.9222	.9236	.9251	.9265	.9279	.9292	.9306	.9319
1.5	.9332	.9345	.9357	.9370	.9382	.9394	.9406	.9418	.9429	.9441
1.6	.9452	.9463	.9474	.9484	.9495	.9505	.9515	.9525	.9535	.9545
1.7	.9554	.9564	.9573	.9582	.9591	.9599	.9608	.9616	.9625	.9633
1.8	.9641	.9649	.9656	.9664	.9671	.9678	.9686	.9693	.9699	.9706
1.9	.9713	.9719	.9726	.9732	.9738	.9744	.9750	.9756	.9761	.9767
2.0	.9772	.9778	.9783	.9788	.9793	.9798	.9803	.9808	.9812	.9817
2.1	.9821	.9826	.9830	.9834	.9838	.9842	.9846	.9850	.9854	.9857
2.2	.9861	.9864	.9868	.9871	.9875	.9878	.9881	.9884	.9887	.9890
2.3	.9893	.9896	.9898	.9901	.9904	.9906	.9909	.9911	.9913	.9916
2.4	.9918	.9920	.9922	.9925	.9927	.9929	.9931	.9932	.9934	.9936
2.5	.9938	.9940	.9941	.9943	.9945	.9946	.9948	.9949	.9951	.9952
2.6	.9953	.9955	.9956	.9957	.9959	.9960	.9961	.9962	.9963	.9964
2.7	.9965	.9966	.9967	.9968	.9969	.9970	.9971	.9972	.9973	.9974
2.8	.9974	.9975	.9976	.9977	.9977	.9978	.9979	.9979	.9980	.9981
2.9	.9981	.9982	.9982	.9983	.9984	.9984	.9985	.9985	.9986	.9986

k	**.00**	**.01**	**.02**	**.03**	**.04**	**.05**	**.06**	**.07**	**.08**	**.09**
3.0	.99865	.99869	.99874	.99878	.99882	.99886	.99889	.99893	.99896	.99900
3.1	.99903	.99906	.99910	.99913	.99916	.99918	.99921	.99924	.99926	.99929
3.2	.99931	.99934	.99936	.99938	.99940	.99942	.99944	.99946	.99948	.99950
3.3	.99952	.99953	.99955	.99957	.99958	.99960	.99961	.99962	.99964	.99965
3.4	.99966	.99968	.99969	.99970	.99971	.99972	.99973	.99974	.99975	.99976
3.5	.99977	.99978	.99978	.99979	.99980	.99981	.99981	.99982	.99983	.99983
3.6	.99984	.99985	.99985	.99986	.99986	.99987	.99987	.99988	.99988	.99989
3.7	.99989	.99990	.4(9)00	.4(9)04	.4(9)08	.4(9)12	.4(9)15	.4(9)18	.4(9)22	.4(9)25
3.8	.4(9)28	.4(9)31	.4(9)33	.4(9)36	.4(9)38	.4(9)41	.4(9)43	.4(9)46	.4(9)48	.4(9)50
3.9	.4(9)52	.4(9)54	.4(9)56	.4(9)58	.4(9)59	.4(9)61	.4(9)63	.4(9)64	.4(9)66	.4(9)67
4.0	.4(9)68	.4(9)70	.4(9)71	.4(9)72	.4(9)73	.4(9)74	.4(9)75	.4(9)76	.4(9)77	.4(9)78

Answers to Odd-Numbered Problems

SECTION 1.1

1. There is an out-of-control signal on day 8—a point below the lower control limit—and another signal on day 15—a point above the upper control limit. See Figure Answers 1.1.
3. (a) false; (b) false; (c) true; (d) true.

SECTION 1.2

1. $\overline{X} = 17.07$ g $S = 3.06$ g $R = 8.8$ g
3. $\overline{X} = 98.7$ veh/hr $S = 3.5$ veh/hr $R = 12$ veh/hr
5. $\overline{X} = 8.0$ $S = 3.0$
7. The bolts would have lengths 15.99, 15.99, 16.01, and 16.01 cm. $S =$ .0115 cm. The three values would be either 15.99, 15.99, and 16.01 cm or 15.99, 16.01, and 16.01 cm. $S =$ .0115 cm. *Note:* The value of S is the same in each case. Algebra may be used to show the interesting result that when half the sample values equal the maximum possible value

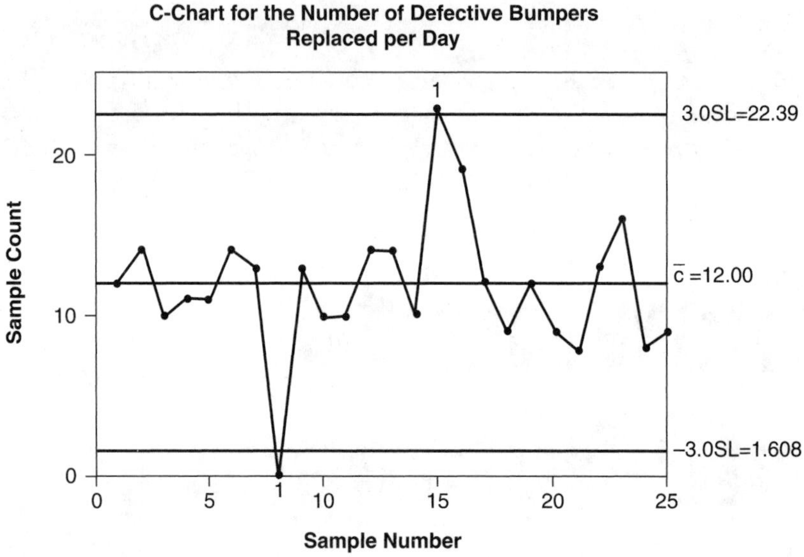

FIGURE ANSWERS 1.1 *C*-Chart for the Number of Defective Bumpers Replaced Per Day.

and the other half equal the minimum possible value, the standard deviation will not change when one of the values is eliminated.

9. (a) continuous; (b) continuous; (c) discrete; (d) discrete; (e) continuous; (f) discrete; (g) continuous.

11. The process reflects lack of control even though there are no points outside the control limits. Resistance is close to on target until sometime near sample 61, after which the resistance appears to drop and stay low. Variation appears to be stable throughout the period—that is, "swings" from one point to the next are of the same order of magnitude before and after the drop in resistance.

SECTION 1.3

1. False. The process could be in control. Roughly three points per thousand will fall outside the control limits when there is no assignable cause. That is why an out-of-control signal is a cue to *investigate* to determine if there is an assignable cause.

3. No, $z = \dfrac{24 - 21}{2.5} = 1.2$

5. Use $\mu \pm 3\sigma = 21 \pm 3 \cdot 2.5$. Then $a = 13.5$ years and $b = 28.5$ years.

7. The z scores are $z = \dfrac{6.6 - 8.1}{.4} = -3.75$ and $z = \dfrac{9.6 - 8.1}{.4} = 3.75$.

The answer is roughly 100%. The actual percentage will be a lot closer to 100% than the 99.7% promised by the Empirical Rule. This really is as precise as we need it to be in this course. *Note:* The actual value is 99.98%.

9. (a) No; (b) No; (c) Yes; (d) No.

SECTION 1.4

1. When $n = 1$, the sample mean is equal to the individual sample value. They would always be equally close to μ.

3. As long as no inputs dominate, the more inputs there are involved in a sum or average, the more like a normal distribution the output distribution will be (the Central Limit Theorem).

5. (a) The limits should lie $3\sigma/\sqrt{n} = 3(2.4)/\sqrt{5} = 3.22$ hours from μ. That is, a = 3 − 3.22 = −.22 hrs, and b = 3 + 3.22 = 6.22 hrs. However, in the context of this problem, where means cannot be negative, it is customary to define the lower limit to be zero, the smallest theoretically possible value of a sample mean. (b) Yes. The 60 means depicted in Figure 1.21b fall between roughly 1.25 and 4.5, well within the bounds a = 0 and b = 6.22. (c) Since the population sampled is not normally distributed (it is mound-shaped but noticeably unsymmetric), three should be viewed as a ballpark estimate.

SECTION 2.1

1. $\overline{\overline{X}} = 8.026$ mm. Without sample 29,

$$\overline{\overline{X}} = \frac{30(8.026) - 8.167}{29} = 8.021 \text{ mm}$$

3.
a. No. The standard deviation plot shows that values within samples are becoming more variable over time—that is, standard deviation is increasing with time. As a result, the sample means are becoming more variable over time, even though the process mean appears to stay the same (approximately 30 kg) throughout.

b. No. The process standard deviation is stable throughout, centered at approximately 1 ohm. However, the process mean has been steadily climbing.

c. The process standard deviation has been stable, centered at roughly 1.8 microns, and the process mean has been stable, centered at roughly 8 microns. The process is in apparent control, so a tentative control chart is justified.

d. Possibly. The process standard deviation is reasonably stable, centered at roughly 1.8%. The process mean declines with time for the initial samples. However, one could argue that for the last 26 samples (at the margin of the rule-of-thumb), the process mean has been stable. Either a target mean or the mean-of-means of the last 26 samples could be used to locate the center line, and all samples could be used, based on methods in the next section, to estimate the process standard deviation σ.

SECTION 2.2

1. b. 8.026 mm, .101 mm, .196 mm, .113 mm
   c. 8.026 mm
   d. .114 mm, .116 mm, .113 mm
   e. 8.026 mm, 7.829 mm, 8.223 mm
   f. 8.026 mm, 7.825 mm, 8.227 mm
   g. 8.026 mm, 7.830 mm, 8.222 mm

3. 2.03 mL, 354.4 mL, 351.91 mL, 356.89 mL.

5. (a) Yes; (b) 7.236 cn, .373 cn; (c) 7.326 cn, 6.444 cn, 8.028 cn.

7. Fifteen samples is not enough to allow for a good decision as to whether the process appears to be in control or to estimate the process mean or standard deviation.

9. 804.975, 6.15.

11. The start-up data may not be typical of the process. If more data show the same pattern, then either the sample size should be increased (unless there is a cost problem) or a data transformation that carries sample data into values from a near normal distribution should be explored. In either case, a statistician should be consulted. See Figure Answers 2.1.

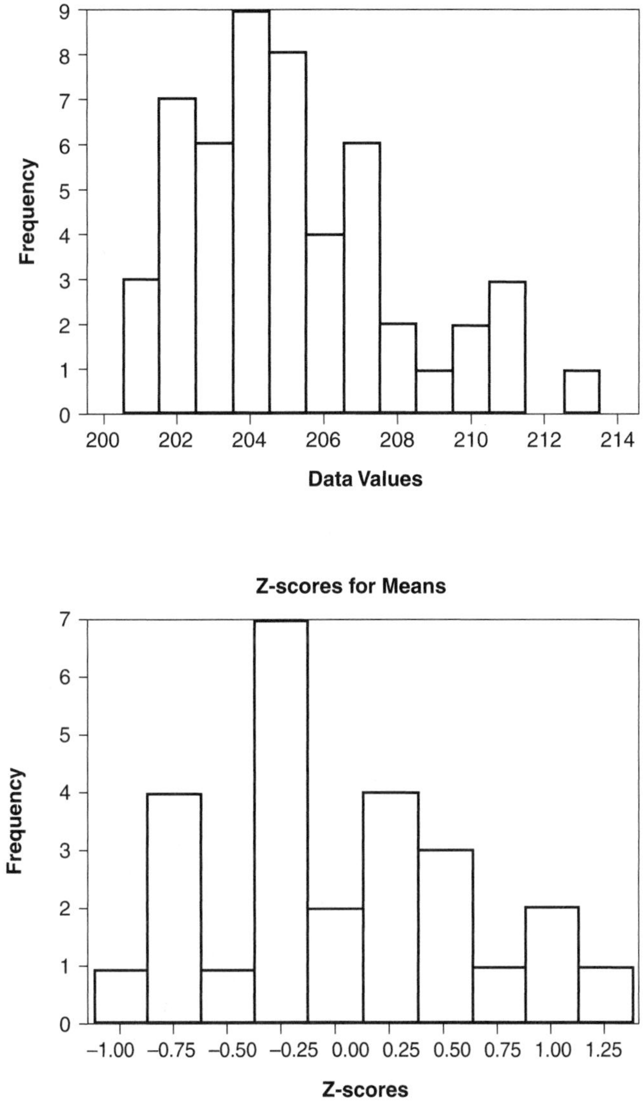

FIGURE ANSWERS 2.1 Start-Up Data Values.

SECTION 2.3

1. (a) .0049 cm; (b) .451 cm, .4425 cm, .4595 cm; (c) .00434 cm, 0 cm, .01117 cm.

3. (a) 1.107 mL; (b) 354.4 mL, 352.74 mL, 356.06 mL; (c) 2.28 mL, 0 mL, 5.198 mL.

5. a. means: .4072 mm, .4002 mm, .4172 mm, .3871 mm. The four standard deviations are all near .012. Some means differ from each other by substantially more than .012.

b. The vast majority of sample standard deviations exceed .012 mm.

c. Each sample typically consists of one value near .4072, one value near .4002, one value near .4172, and one value near .3871, producing a sample mean near .4029 mm. Including in each group one value from each position when between position variation is great is what makes the 3-sigma limits far from the center line.

d. Maintain four sets of control charts, one set for each position. At a given position, a sample could consist of an appropriate number of consecutive measurements.

SECTION 3.1

1. a. Natural tolerance = 6(.20) = 1.20. Routine sampling.

b. Natural tolerance = 6(.32) = 1.92. Frequent sampling.

c. Natural tolerance = 6(.40) = 2.40. 100% inspection.

3. $180.0 + 3\sigma = 180.0 + 3(2.607) = 187.82$ lb.

5. (a) Tiring machine operator.

7. b. Subassembly from temporary supplier put to use and appears in units sampled beginning with sample 18. The process mean appears stable, but variation has been reduced beginning when subassembly parts began coming from the temporary supplier.

SECTION 3.2

1. a. Sample no. 12: 8 in a row above the center line
Sample no. 13: 8 in a row above the center line
Sample no. 20: 4 of 5 in a row below −1.0SL
Sample no. 33: 2 of 3 in a row above 2.0SL

b. Sample no. 14: 8 in a row below the center line
Sample no. 15: 8 in a row below the center line
Sample no. 32: point above 3.0SL

3. Sample no. 10.

5. a. Comparing sample means to spec limits is inappropriate because spec limits are meant to apply to the length of one rod, not to the sample mean.

b. Operators were doing what they were told.

c. USL $- 3\sigma = 182.20 - 3(.017) = 182.149$ cm.

d. No. The z-score of the lower spec limit is

$$z = \frac{182.05 - 182.149}{.017} = -5.82, \text{ or LSL would be nearly}$$

6-sigma below the mean.

SECTION 3.3

1. a.

$$C_p = \frac{USL - LSL}{6\sigma} \approx \frac{12.7 - 3.0}{6(1.5)} = 1.08,\ C_{pL} = \frac{\mu - LSL}{3\sigma} \approx \frac{8.1 - 3.0}{3(1.5)} =$$

$$1.13,\ C_{pU} = \frac{USL - \mu}{3\sigma} \approx \frac{12.7 - 8.1}{3(1.5)} = 1.02, \text{ and}$$

$$C_{pk} = \min\{C_{pL}, C_{pU}\} = \min\{1.13, 1.02\} = 1.02.$$

b. No. C_{pk} would be a maximum if the process mean is

$$\frac{USL + LSL}{2} = \frac{3.0 + 12.7}{2} = 7.85.$$

c. $C_{pk} = \min\{C_{pL}, C_{pU}\} = \min\{1.08, 1.08\} = 1.08 = C_p$.

3. a. C_{pL}.

b. The estimate of the process mean is 3.807 mm, while

$$\sigma \approx \frac{\overline{R}}{d_2} = \frac{.0577}{2.059} = .0280 \text{ mm}.$$

c. $C_{pL} = \frac{\mu - LSL}{3\sigma} = \frac{3.807 - 3.700}{3(.0280)} = 1.274$

5. a. Sampling error—that is, for the samples obtained, it was pure chance that caused the shortage of values in the interval, the kind of variation one gets from different samples from the same process.

7. a. $\overline{\overline{X}} = 3.239$.

b. $\sigma \approx 2.521$.

c. No. The data values appear to come from a distribution that is clearly skewed to the right, very non-normal.

SECTION 4.1

1. When the subgroup size is $n = 1$, all ranges will equal 0. Also, because the divisor under the square root sign is $n - 1$, the sample standard deviation is undefined because division by 0 is undefined.

3. a. $\overline{R} = 4.423$ based on ranges of length 4 is nearly twice as large.

b. $\overline{R}$ based on moving ranges of length 4.

5. a. Yes. Refer to Figure Answers 4.1.

b. $\mu \approx \overline{X} = 19.99$ grams, and $\overline{R} = .555$ grams, so that

$$\sigma \approx \frac{\overline{R}}{d_2} = \frac{.555}{1.128} = .492 \text{ grams.}$$

c. $\mu \pm 3\sigma \approx 20.00 \pm 3(.492) = 20.00 \pm 1.48$, so that $UCL_I = 21.48$ grams and $LCL_I = 18.52$ grams.

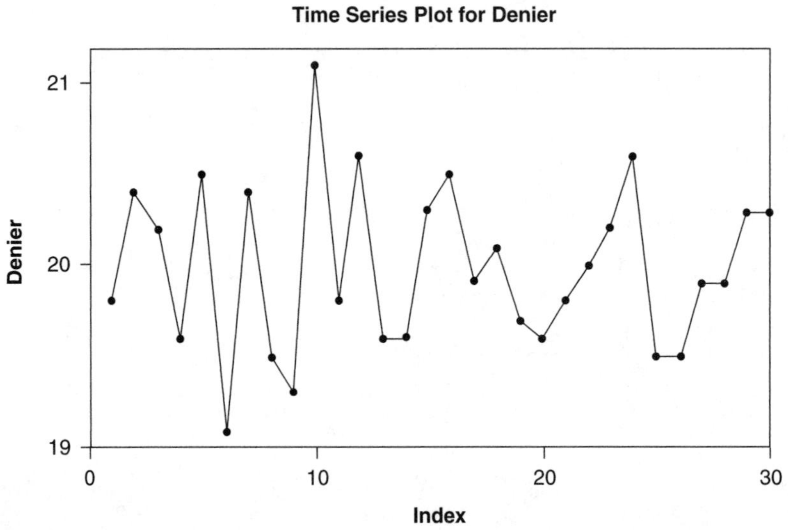

FIGURE ANSWERS 4.1 Time Series Plot for Denier.

SECTION 4.2

1. a. $\bar{p} = \dfrac{535}{8{,}613} = .0621.$

b. The 3-sigma limits are

$$\bar{p} \pm 3\sqrt{\bar{p}(1-\bar{p})/411} = .0621 \pm 3\sqrt{.0621(1-.0621)/411} = .0621 \pm .0357.$$

Hence $UCL_p = .0978$ *and* $LCL_p = .0264$. See Figure Answers 4.2.

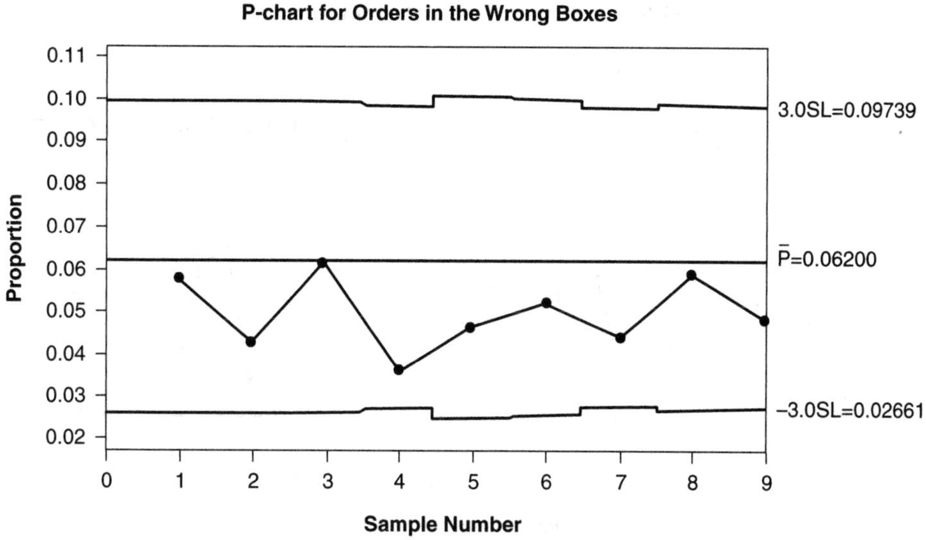

FIGURE ANSWERS 4.2 *p*-Chart for Orders in the Wrong Boxes.

3. a. $\bar{p} = \dfrac{89}{1115} = .0798.$

b. $\bar{p} \pm 3\sqrt{\bar{p}(1-\bar{p})/n} = .0798 \pm 3\sqrt{.0798(.9202)/46} =$

$.0798 \pm .1199$. Hence $UCL_p = .1997$, while LCL_p is either undefined or equated to zero. See Figure Answers 4.3.

5. a. Yes.

b. No. Since $\bar{p} = .0798$, for $n\bar{p} \geq 5$ to hold, one would have to have $n \geq 5/\bar{p} = 5/.0798 = 62.7$. In the five weeks when data were collected, n was always less than 62.7. Taking $n = 50$ as typical, for example, gives $n\bar{p} = 50(.0798) = 3.99$.

7. a. center line value is .024 and 3-sigma limits are given by

$n\bar{p} \pm 3\sqrt{n\bar{p}(1-\bar{p})} = 500(.024) \pm 3\sqrt{500(.024)(.976)} =$
12.0 ± 10.9. So upper and lower 3-sigma limits are 22.9 and 1.1, respectively (or 22.5 and 1.5).

b. There is an out-of-control signal on the fifth day.

9. a. Altogether, there were 125 tools missing out of $20(60) = 1{,}200$ tools, so that $p \cong \bar{p} = \dfrac{125}{1200} = .1042.$

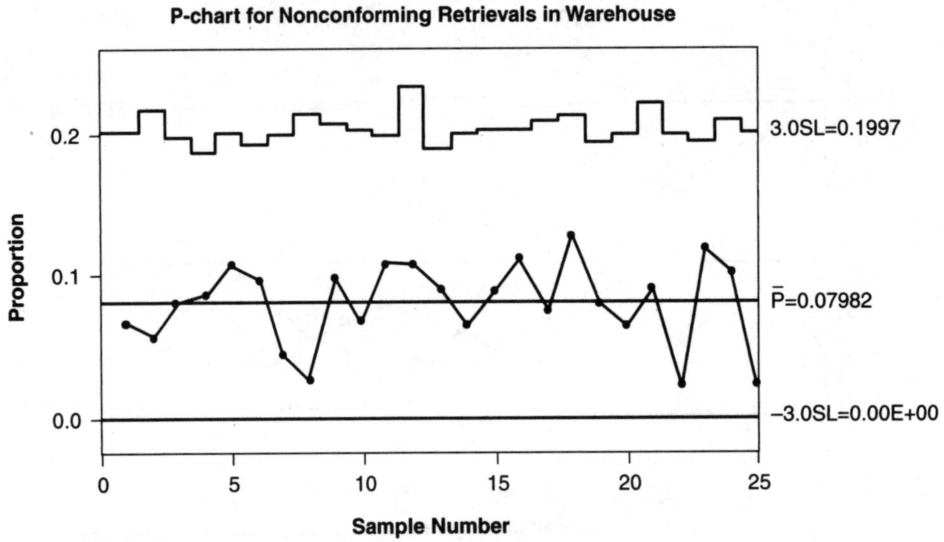

FIGURE ANSWERS 4.3 *p*-Chart for Nonconforming Retrievals in Warehouse.

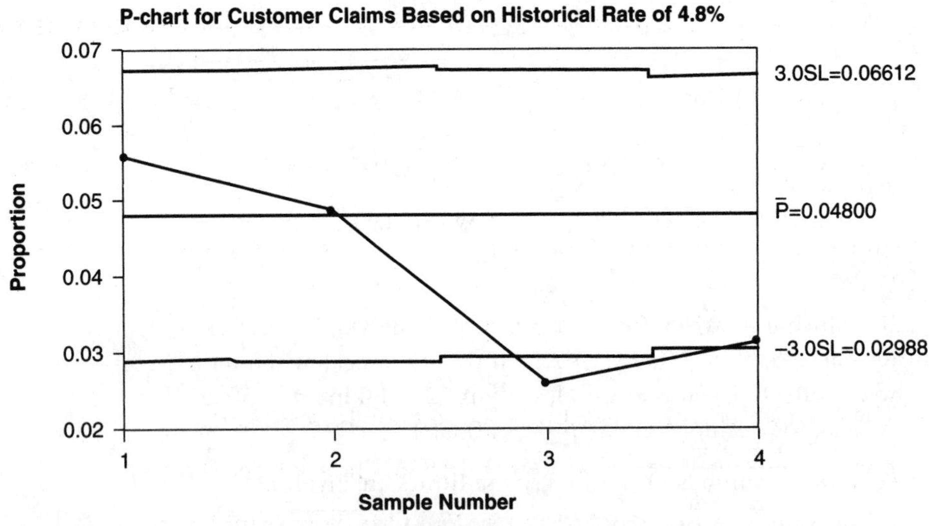

FIGURE ANSWERS 4.4 *p*-Chart for Customer Claims Based on Historical Rate of 4.8%.

b. The center line value is $n\bar{p} = 60(.1042) = 6.252$ and control limits are given by

$$n\bar{p} \pm 3\sqrt{n\bar{p}(1-\bar{p})} = 6.252 \pm 3\sqrt{6.252(.8958)} =$$

6.252 ± 7.100. So UCL $= 13.352$ (or 13.5 for convenience) and LCL is undefined or taken to be 0.

c. See Figure Answers 4.5

d. The process does appear to be in control.

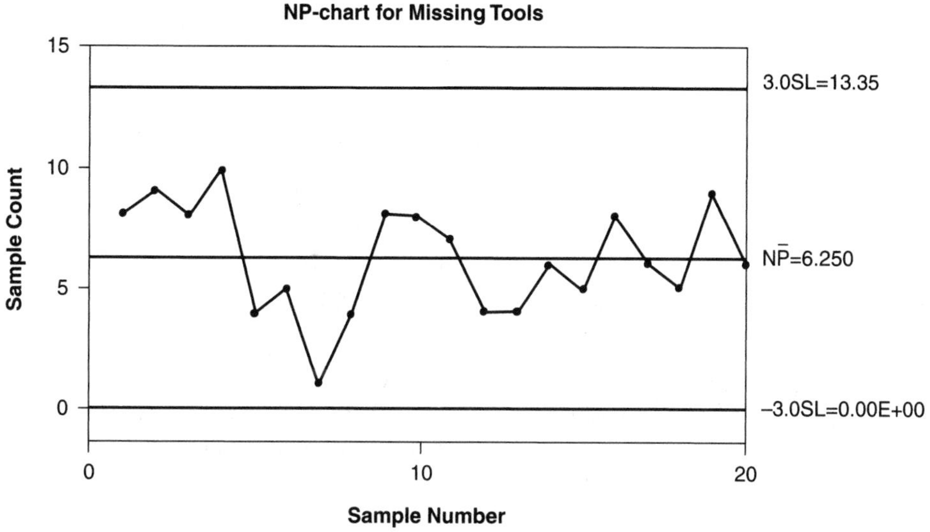

FIGURE ANSWERS 4.5 *np*-Chart for Missing Tools.

SECTION 4.3

1. The 12 missing book counts for 1996 were 1, 3, 0, 0, 2, 0, 0, 2, 1, 1, 1 and 0.

a. The estimate of p, the probability that a book is missing, is $\bar{p} = \dfrac{1 + 3 + 0 + \cdots + 0}{12(50)} = \dfrac{11}{600} = .01833$. The value for the center line is $n\bar{p} = 50(.01833) = .917$ and the 3-sigma limits are

$$n\bar{p} \pm 3\sqrt{n\bar{p}(1 - \bar{p})} = .917 \pm 3\sqrt{.917(1 - .01833)} = .917 \pm 2.846,$$

which leads to 3SL $= 3.763$ (or 3SL $= 3.5$) and to -3SL either being equal to 0 or undefined. For the *c*-chart, $\mu_C \cong \bar{C} = \dfrac{1 + 3 + 0 + \cdots + 0}{12} = .917$, while the 3-sigma limit computations are $\overline{C} \pm 3\sqrt{\overline{C}} = .917 \pm 3\sqrt{.917}$. This gives 3SL = 3.789 (or 3SL = 3.5) and −3SL either equals 0 or is undefined.

b. For the *np*-chart, based on $p = .01833$ and $n = 50$, a relevant portion of the table of cumulative probabilities for a binomial distribution is:

TABLE ANSWERS 4.1

k	$P(X \leq k)$
0	0.39653
1	0.76674
2	0.93609
3	0.98669
4	0.99779
5	0.99970

From this we see that the lower probability limit is undefined or taken to equal 0. The probability of 5 or more missing books equals 1 minus the probability of 4 or fewer, or 1 − .99779 = .00221. So UPL = 4.5.

For the *c*-chart, based on $\mu_C = .917$, a relevant portion of the table of cumulative probabilities for a Poisson distribution is:

TABLE ANSWERS 4.2

k	$P(X \leq k)$
0	0.39972
1	0.76626
2	0.93431
3	0.98568
4	0.99746
5	0.99962

The upper and lower probability limits are the same as for the *np*-chart.

3. a. 3-sigma limits are: $7 \pm 3\sqrt{7} = 7 \pm 7.937$. Hence 3SL = 14.937 (or 14.5) while −3SL = 0 or is undefined.

b. No. The nonconformity count exceeds 3SL on sample 36.

c. The estimate of μ_C should be based on the first 35 samples:

$$\overline{C} = \frac{3 + 6 + 7 + \cdots + 3 + 6}{35} = 6.714.$$

d. This is a judgment call that should depend on whether those involved with the process have reason to believe that the process mean *should* have changed. A conservative approach would be to continue under the assumption that $\mu_C = 7.0$ and see whether future data give further credence to a drop in μ_C.

5. a. When $\mu_C = 23$, $3SL = 23 + 3\sqrt{23} = 37.387$, or for convenience, $3SL = 37.5$. Thirty murders falls roughly midway between $3SL$ and the historical mean, and is not unusual.

b. When $\mu_C = 877$, $3SL = 877 + 3\sqrt{877} = 965.842$, or for convenience, $3SL = 965.5$. A one-year murder count of 877 falls above the upper 3-sigma limit.

c. Los Angeles.

7. a. Yes, with the possible exception of sample 10.

b. $\overline{C} = \dfrac{6 + 3 + \cdots + 3}{27} = 4.296.$

c. The center line value is 4.296, while 3-sigma limits are given by $4.296 \pm 3\sqrt{4.296} = 4.296 \pm 6.218$. Thus $3SL = 10.514$, or 10.5 for convenience. $-3SL$ is 0 or undefined.

d. A relevant portion of the table of cumulative probabilities for a Poisson distribution with mean 4.296 is

TABLE ANSWERS 4.3

k	$P(X \leq k)$
0	0.01362
1	0.07215
2	0.19786
3	0.37787
4	0.57121
5	0.73733
6	0.85627
7	0.92926
8	0.96846
9	0.98717
10	0.99521
11	0.99835

From this we see that there is no k for which the probability of k or fewer flaws is at most .005. So the lower 3-sigma limit is either taken to be 0 or does not exist. The smallest k for which the probability of k or fewer flaws is at least .995 is $k = 10$. So 11 or more flaws constitutes an out-of-control signal and $3SL = 10.5$.

e. See Figure Answers 4.6.

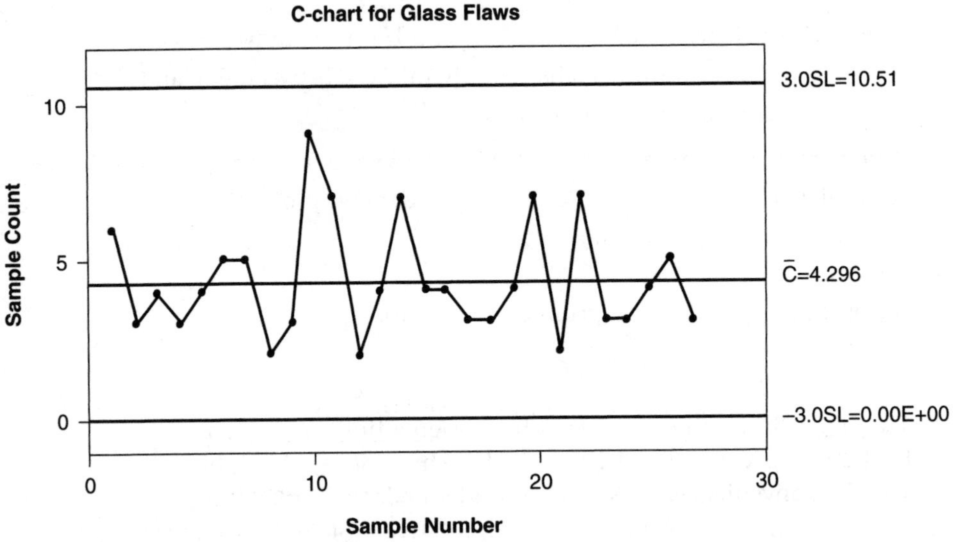

FIGURE ANSWERS 4.6 *c*-Chart for Class Flaws.

f. Based on a historical process mean of 2.5, the process does not reflect control. See Figure Answers 4.7.

9. 5(2.3) = 11.5 flaws/sample.

SECTION 4.4

1. a. $\dfrac{225}{204.44} = 1.101.$

b. $1.101 \pm 3\sqrt{\dfrac{1.101}{6.49}} = 1.101 \pm 1.236$. Hence $-3\text{SL} = 0$ or is undefined and $3\text{SL} = 2.337$.

c. $\dfrac{3}{4.28} = .701.$

d. The chart reflects apparent control (see Figure Answers 4.8).

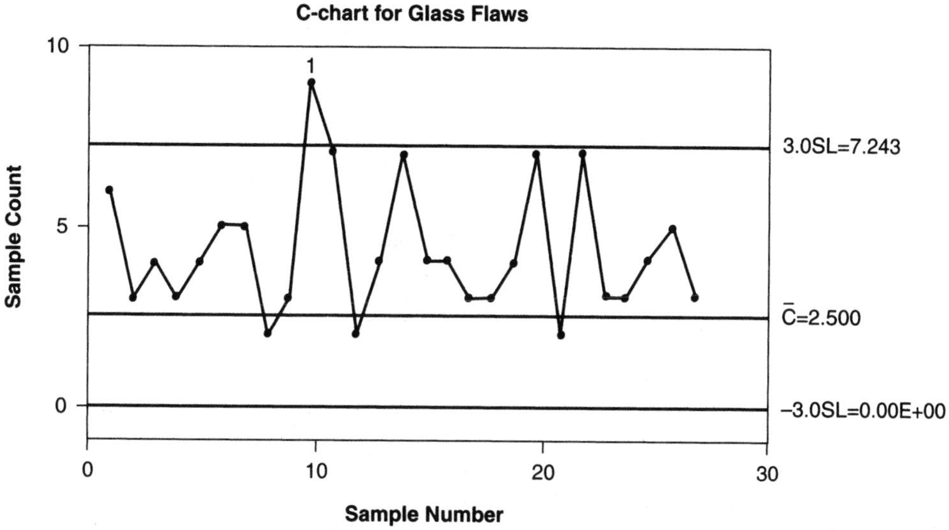

FIGURE ANSWERS 4.7 *c*-Chart for Class Flaws.

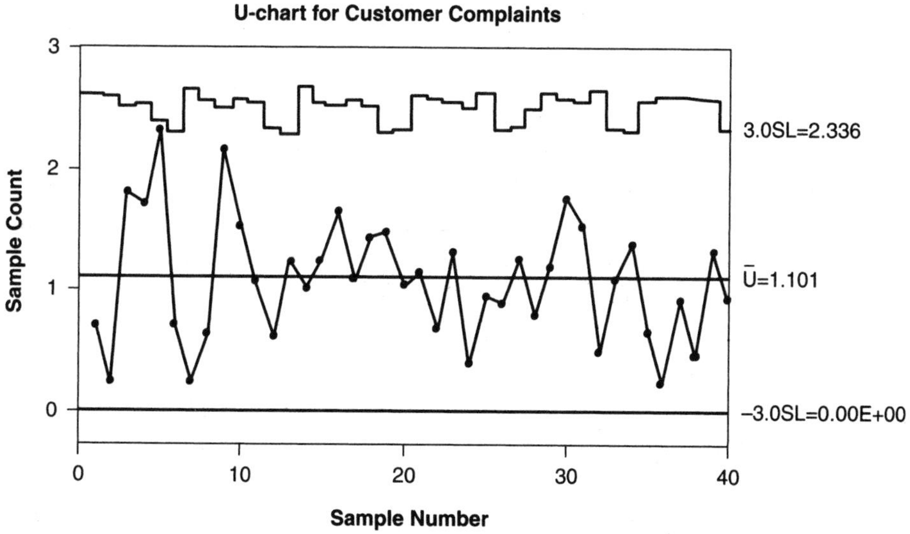

FIGURE ANSWERS 4.8 *u*-Chart for Customer Complaints.

3. a. Center line value: 3.2 accidents/mo; 3-sigma limits:

$$3.2 \pm 3\sqrt{\frac{3.2}{28/30}} = 3.2 \pm 5.55, \text{ or } 3SL = 8.75 \text{ for February and}$$

$-3SL = 0$ or is undefined.

$$3.2 \pm 3\sqrt{\frac{3.2}{30/30}} = 3.2 \pm 5.37, \text{ or } 3SL = 8.57 \text{ for 30-day months}$$

and $-3SL = 0$ or is undefined.

$$3.2 \pm 3\sqrt{\frac{3.2}{31/30}} = 3.2 \pm 5.28, \text{ or } 3SL = 8.48 \text{ for 31-day months}$$

and $-3SL = 0$ or is undefined.

b. See Figure Answers 4.9.
c. The chart reflects control.

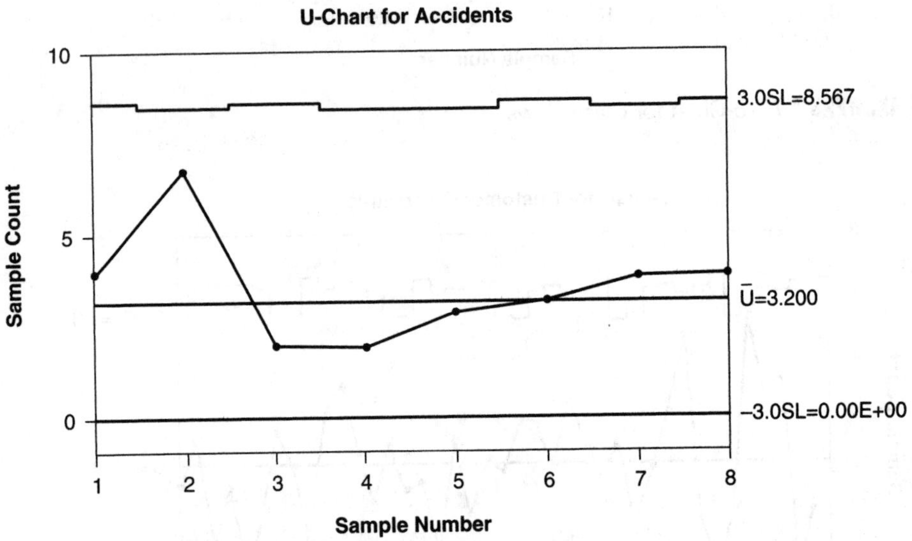

FIGURE ANSWERS 4.9 *U*-Chart for Accidents.

SECTION 4.5

1. Y = 4.25 + .25X; the two are always equal because the points fall on a line.

3. a. Y = 5,385.25 − 2.68333X. See Figure Answers 4.10.

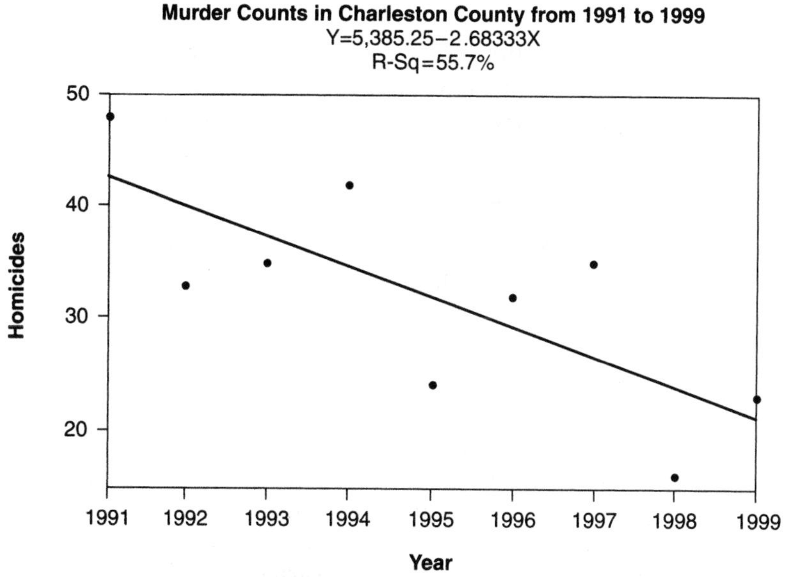

FIGURE ANSWERS 4.10 Murder Counts in Charleston County from 1991 to 1999.

b. Yes. Fallout is random. Whether a point falls above or below the line does not appear to depend on year.

c. During the period from 1991 to 1999, murders decreased at a rate of roughly 2.7 murders per year.

d. 15.9.

e. Substituting 2001 into the equation produces −.19 as the value of Y. One cannot guarantee how far into the future the approximate linear relationship will hold. However, it certainly cannot hold for the years 2007 or beyond because projected murder counts would be negative.

f. No. The focus is on one year's number versus the preceding year's number. What constitutes routine fluctuation up or down from one year to the next has not been taken into account. Murder counts over the nine-year period make it clear that the fluctuation for these two years fits into the larger picture of an overall decline in the murder rate. Taken in isolation, a focus on the 44% rise is misleading and alarmist.

5. a. $S = .0499$

b.

Time	n	S
Before new Chemical D	4	.01893
After new Chemical D	5	.03578
After new Chemical B	3	.07506
After new Chemical C	3	.06028
After new Chemical A	3	.02000
After new Chemical D	2	.00707

$$S_p = \sqrt{\frac{3(.01893)^2 + 4(.03578)^2 + 2(.07506)^2 + 2(.06028)^2 + 2(.02000)^2 + .00707^2}{3 + 4 + 2 + 2 + 2 + 1}} = .0427$$

c. Computing a standard deviation based on combining all values incorporates into the data location shifts associated with the arrival of chemical shipments—that is, the data in part a. involve both between-batch variation and between-shipment variation, but the pooled standard deviation in b. is based on between-batch variation only.

7. 4, 70, $\frac{70}{4} = 17.5$.

9. a. Samples 7 through 11 comprise a two-back-to-back signal, and a CD adjustment should have occurred after batch 11.

b. The concentrations 5.55, 5.53, 5.51, 5.56 and 5.47 comprise the signal. Their mean AB concentration is $\frac{5.55 + 5.53 + 5.51 + 5.56 + 5.47}{5} = 5.524$. Since a 1-kg CD increase results in a decrease in AB of .0221, approximately, then the CD adjustment should be as close as possible to $\frac{5.524 - 5.65}{.0221} = -5.70$ kg. Since AB is to be raised, there should be a 6 kg reduction in the amount of CD in the recipe.

c. Every time a value equal to the target value is to be plotted, a fresh cumulative count is started (see Figure Answers 4.11).

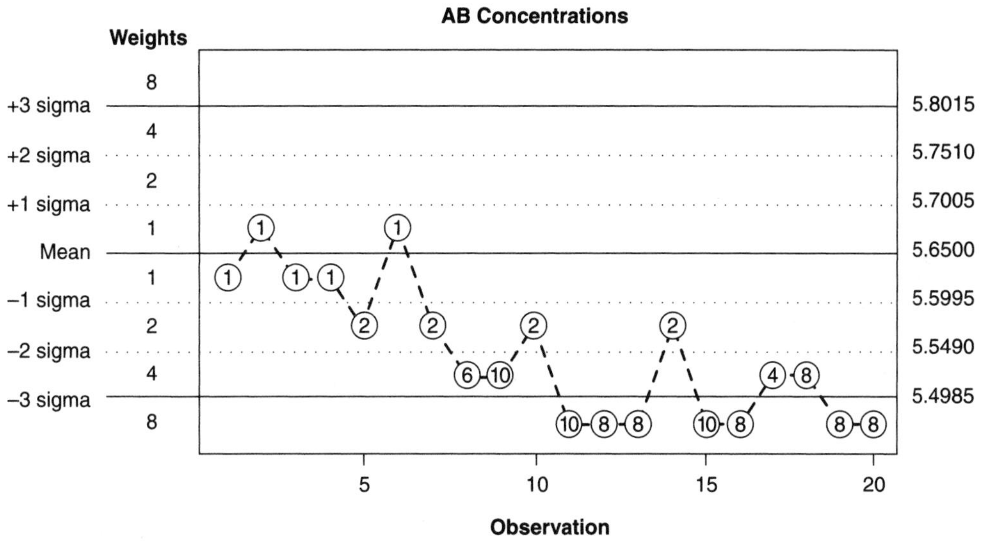

FIGURE ANSWERS 4.11 AB Concentrations.

SECTION 4.6

3. a. 76, 77.6, 78.08, 78.064, 78.651, 79.121.

b. $76 \pm 3\sqrt{\frac{.2}{1-.2}}\ \frac{2}{\sqrt{3}} = 76 \pm 1.155$, which leads to $-3SL =$ 74.845, $3SL = 77.155$

c. All but the first sample.

d. 72.54, 79.464.

e. All but the fourth sample.

5. a. 75.870, 76.299, 77.319, 77.794, 77.285, 76.870, 76.909, 77.086, 77.600, and 77.600.

b. $76 \pm 3\sqrt{\frac{.3}{2-.3}}\ \frac{2}{\sqrt{3}} = 76 \pm 1.455$, which leads to $-3SL =$ 74.545 and $3SL = 77.455$.

c. The fourth sample and the last two samples.

d. For both choices of α, the last two samples produced out-of-control signals. For $\alpha = .3$, the fourth sample produced a signal, but for $\alpha = .2$, the fourth sample barely missed

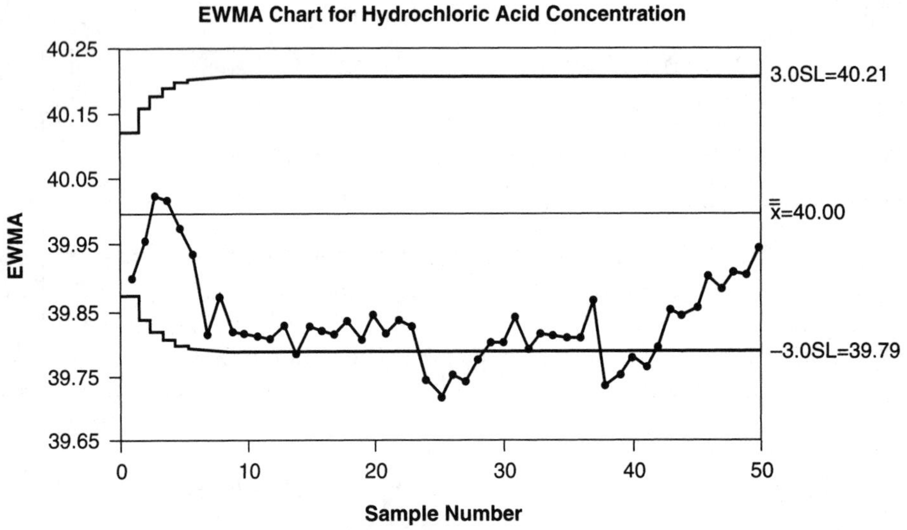

FIGURE ANSWERS 4.12 EWMA Chart for Hydrochloric Acid Concentration.

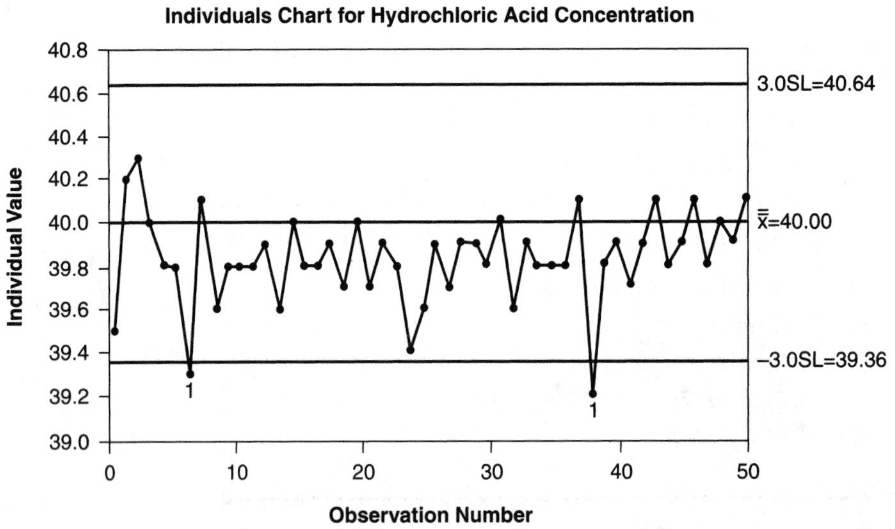

FIGURE ANSWERS 4.13 Individuals Chart for Hydrochloric Acid Concentration.

producing a signal. It is not clear whether $\alpha = .3$ creates a strong advantage or disadvantage.

7. a. No. See Figure Answers 4.12.

b. No. See Figure Answers 4.13.

SECTION 4.7

1. $\mu_r = 52 - 3(.6) = 50.2$, $k^* = (50.2 + 50)/2 = 50.1$, $k = |50.1 - 50|\sqrt{4}/.6 = .333$

3. $\mu_r = 60.5 - 3(.12) = 60.14$, $k^* = (60 + 60.14)/2 = 60.07$, control limits are $\pm 2.8 \times \frac{.12}{\sqrt{3}} = \pm .194$. Then $H(0) = 0$, $H(1) = 0 + (60.075 - 60.070) = .005$, $H(2) = .005 + (60.105 - 60.070) = .040$. In computing $H(3)$, since $.040 + (60.028 - 60.070) < 0$, we set $H(3) = 0$. Similarly, $H(4) = .05$, $H(5) = .107$, $H(6) = .154$, and $H(7) = .202$, which is an out-of-control signal. Then $H(8) = 0 + (60.080 - 60.070) = .010$.

5. $h = 2.75$, $k = .88$

7. a. We are given $\sigma \approx \overline{R}/d_2 = .659/2.059 = .320$ and $k^* = 200.12$. Now, from the prescribed ARLs, $k = .81$. Then $.81 = |200.12 - 200|\sqrt{n}/.320$, so that $\sqrt{n} = 2.16$. Squaring both sides shows that $n = 4.666$, and since n must be an integer, we use $n = 5$ (always round up).

b. Since $n = 5$, then $k = |200.12 - 200|\sqrt{5}/.32 = .84$. This with the prescribed ARL_r gives $ARL_a \approx 770$.

c. $h = 3.1$, $k = .84$.

9. The value of μ_r above the target mean is $\mu_r = 27.5 - 3(.6) = 25.7$, and $k^* = 25.35$. So $k = |25.35 - 25|\sqrt{2}/.6 = .825$. This with the prescribed value of ARL_r gives $h = 2.65$. See Figure Answers 4.14.

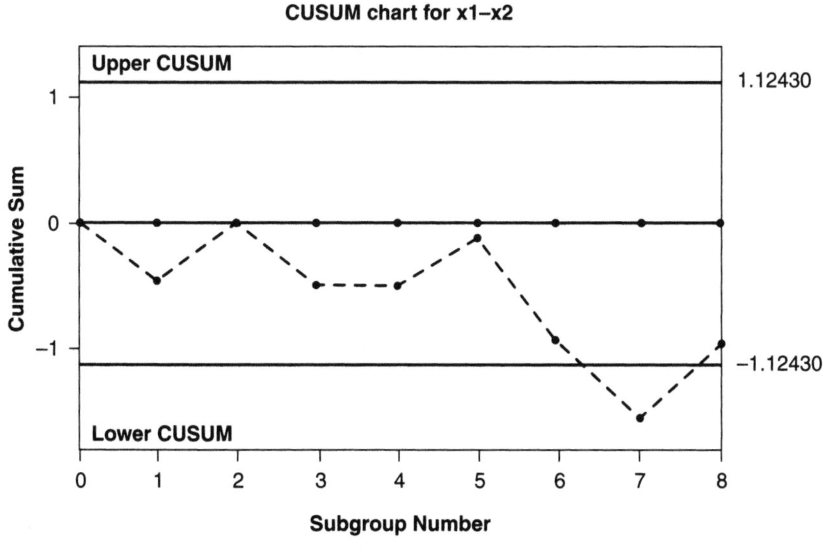

FIGURE ANSWERS 4.14 CUSUM Chart for x1 − x2.

SECTION 4.8

1. a. .00125, −.00450, −.00525, −.00300, .00025 inches.

b. s_p = .004714 inch.

c. $\overline{\overline{X}}_{shortrun}$ = −.00225, $-3SL$ = −.009321, $+3SL$ = .004821.

3. a. $\overline{\overline{X}}_{shortrun}$ = .304, + 2SL = .304 + 2(.051)/2 = .355. The last three points all fall above the center line, and the third and fifth points exceed .355.

b. Not necessarily. With only five samples, the pooled standard deviation is a very imprecise estimate of σ.

SECTION 5.1

1. a. $\frac{18}{352} \approx .05$.

b. $\frac{334}{352} \approx .95$.

c. $\frac{196}{352} \approx .56$.

3. (a) No; (b) Yes; (c) This is P(A′) = 1 − P(A) = 1 − .6 = .4.

5. a.

gallons	.96	.97	.98	.99	1.00	1.01	1.02	1.03	1.04
probability	.04	.08	.12	.16	.20	.16	.12	.08	.04

b. .16 + .12 + .08 + .04 = .40

7. (a) 1/3; (b) 2/5; (c) 1; (d) 1/5; (e) 1/5; (f) 0.

9. 5/8.

11. $P(F) = \frac{1}{5}$ and $P(F|G) = \frac{1}{5}$, so F and G are independent. The events F and E are dependent because $P(F|E) = \frac{1}{3} \neq P(F)$.

13. a. $P(S) = \frac{1}{6}$ and $P(S|T) = \frac{1}{2} \neq \frac{1}{6}$, so S and T are dependent.

b. $P(E) = \frac{1}{6}$ and $P(E|T) = \frac{1}{3} \neq \frac{1}{6}$, so E and T are dependent.

c. $P(S) = \frac{1}{6}$ and $P(S|E) = \frac{1}{6} = P(S)$, so S and E are independent.

15. P(at least two tosses) = 1 − P(one toss is needed) = 1 − P(single toss produces a 4) = 1 − 1/6 = 5/6

17. Let V denote the event customer is very pleased, and K, J, and M denote that the agent is Ken, Jo Ann, and Mike, respectively. Then

$P(K) = .25$, $P(J) = .40$, $P(M) = .35$, and $P(V|K) = .40$, $P(V|J) = .30$ and $P(V|M) = .25$, and

$$P(K|V) = \frac{P(V|K)P(K)}{P(V|K)P(K) + P(V|J)P(J) + P(V|M)P(M)}$$

$$= \frac{.4 \times .25}{(.4 \times .25) + (.3 \times .4) + (.25 \times .35)}$$

$$= \frac{.1}{.1 + .12 + .0875} = .325$$

19. P(at least one point above 6.85) = 1 − P(all 7 points are below 6.85). Suppose the result of the next 7 tosses is ABBABBB, meaning that the first point is above 6.85, the next two are below, the fourth point is above, and the last three points are below. There are $2^7 = 128$ possible outcomes, all equally likely. So

$$\begin{aligned} P(\text{at least one point above } 6.85) &= 1 - P(\text{all 7 points are below } 6.85) \\ &= 1 - P(\text{BBBBBBB}) \\ &= 1 - \frac{1}{128} \\ &= .992 \end{aligned}$$

SECTION 5.2

1. 0, 1, 2, 3, 4.

x	0	1	2	3	4
$p(x)$	.102	.363	.381	.139	.014

$$\mu = 0(.102) + 1(.363) + 2(.381) + 3(.139) + 4(.014) = 1.598$$

3. 0, 1, 2, 3, 4.

x	0	1	2	3	4
$p(x)$	.65884	.29947	.03993	.00174	.00002

For example, $p(2) = P(X = 2) = \binom{4}{2}\binom{48}{3} / \binom{52}{5} = .03993.$

5. Let X be the number of defectives in a sample. We want $P(X \geq 1) \geq .75$, or $P(X = 0) \leq .25$. Substituting test values of n into

$P(X = 0) = \binom{43}{n} / \binom{50}{n}$ shows that the smallest allowable value of n is 9.

7. Let X be the number of points above the center line. $P(X \geq 16) = 1 - P(X \leq 15) = 1 - .994 = .006$.

9. Let X denote the number of Mondays that incur downtime.

a. X has a binomial distribution with parameters $n = 15$ and $p = .35$. From Table V, $P(X \leq 2) = .062$.

b. X has a binomial distribution with parameters $n = 15$ and $p = .06$. Then P(changes are not adopted) $= P(X \geq 3) = 1 - P(X \leq 2) = 1 - .943 = .057$.

11. Let X be the number of patients successfully treated. Then X has a binomial distribution with parameters $n = 25$ and $p = .30$, and

a. $\mu = np = 25 \times .30 = 7.5$, $\sigma = \sqrt{np(1-p)} = \sqrt{25 \times .3 \times .7} = 2.29$.

b. P(treatment is declared superior) $= P(X \geq 15) = 1 - P(X \leq 14) = 1 - .998 = .002$.

13. Let X denote the number of accidents at that intersection on a workday. Treat X as Poisson with a mean of 1.8.

a. $P(X \leq 3) = .891$.

b. $P(X > 5) = 1 - P(X \leq 5) = 1 - .990 = .010$.

c. Let Y be the number of accidents in a workweek. Treat Y as Poisson with a mean of 5 times 1.8, or 9. Then $P(Y \geq 16) = 1 - P(Y \leq 15) = 1 - .978 = .022$.

15. Let X denote the number of customer complaints in a week. X would have a binomial distribution with $n = 3302$ and $p = .001$. Since n is large and p is small, the distribution of X is approximately Poisson with mean $np = 3.3$.

a. From Table VI, to three decimal places, $P(X \geq 12) = 1 - P(X \leq 11) = 1 - 1.000 = .000$.

b. $\sigma = \sqrt{np} = \sqrt{3.3} = 1.817$.

c. $\dfrac{12 - 3.3}{1.817} \approx 4.79$

17. Let X denote the thickness of a plate and Z be the z-score of X.

a. Thicknesses that exceed 2.33 have z-scores that exceed $\dfrac{2.33 - \mu}{\sigma} = \dfrac{2.33 - 2.31}{.0092} = 2.17$. So $P(X > 2.33) = P(Z > 2.17) = 1 - P(Z \leq 2.17) = 1 - .9850 = .0150$.

b. Thicknesses less than 2.27 have z-scores less than
$\frac{2.27 - \mu}{\sigma} = \frac{2.27 - 2.31}{.0092} = -4.35$. So $P(X < -4.35) \approx 0$.

c. .0150 + 0 = .0150.

d. $P(X > 2.33) = P\left(Z > \frac{2.33 - \mu}{\sigma}\right) =$

$P\left(Z > \frac{2.33 - 2.30}{.0092}\right) = P(Z > 3.26) =$

.00056, and

$P(X < 2.27) = P\left(Z < \frac{2.27 - \mu}{\sigma}\right) = P\left(Z < \frac{2.27 - 2.30}{.0092}\right) =$

$P(Z < -3.26) = .00056$. The probability a plate does not meet specs is 2(.00056) = .0011, or 0.11%.

19. Control limits are $UCL = 12.7 + 3\left(\frac{.16}{\sqrt{4}}\right) = 12.94$ and

$LCL = 12.7 - 3\left(\frac{.16}{\sqrt{4}}\right) = 12.46$. After the shift, sample means come

from a normal distribution with mean of 12.7− .1 = 12.6 and

standard deviation $\sigma_{\overline{X}} = \frac{.16}{\sqrt{4}} = .08$.

a. The probability a sample mean yields an out-of-control signal is
w$P(\overline{X} < 12.46) + P(\overline{X} > 12.94) =$

$P\left(Z < \frac{12.46 - 12.6}{.08}\right) + P\left(Z > \frac{12.94 - 12.6}{.08}\right) =$

$P(Z < -1.75) + P(Z > 4.25) = .0401 + .0000 = .0401$.

b. P(at least one out-of-control signal) = 1 − P(no out-of-control signals) = $1 - (1 - .0401)^{10} = .3359$.

c. Let X be the measurement from one product unit.
$P(X < 12.2) + P(X > 13.4) =$

$P\left(Z < \frac{12.2 - 12.6}{.16}\right) + P\left(Z > \frac{13.4 - 12.6}{.16}\right) =$

$P(Z < -2.5) + P(Z > 5) = .0062 + .0000 = .0062$.

SECTION 5.3

1. The diagram for this series system is shown in Figure Answers 5.1.

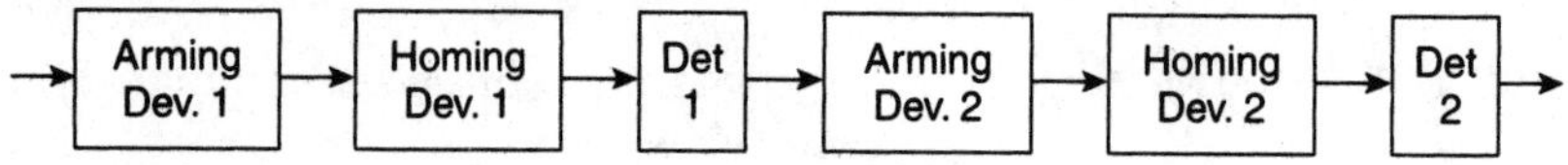

FIGURE ANSWERS 5.1

$$R = (R_A\, R_H\, R_D)\,(R_A\, R_H\, R_D) = .969^2 = .939$$

3. This is a parallel system of inspection.

a. $R = 1 - (1 - .95)(1 - .9) = .995.$

b. $.9999 = 1 - (1 - R_{\text{one-insp}})^2$, so that $R_{\text{one-insp}} = .99.$

5. $R = .96 \times (1 - .04^2) \times .96 = .9201.$

7. $(1 - (1 - R_{\text{component}})^3) \times (1 - (1 - R_{\text{component}})^3) = .99999$

$$(1 - (1 - R_{\text{component}})^3)^2 = .99999$$
$$1 - (1 - R_{\text{component}})^3 = \sqrt{.99999} = .999995$$
$$(1 - R_{\text{component}})^3 = .000005$$
$$(1 - R_{\text{component}}) = .0171$$
$$R_{\text{component}} = .9829.$$

11. (a) .76; (b) .60; (c) .12.

SECTION 6.1

1. Point 2—adopt the new philosophy. Workers are not trained properly.

3. Point 4—don't award business based on price tag alone, and Point 5—improve constantly and forever the system of production and service.

5. Point 5—improve constantly and forever the system of production and service. A customer has to take time to read which knobs control which burners. A company should learn from its mistakes.

7. Point 7—institute leadership. Workers are being admonished for common cause variation. The focus should be on system changes that reduce the frequency of faulty widgets being produced and/or the frequency of their being shipped.

9. This is management by numbers, Point 7. Point 8—drive out fear—also applies because the faculty member, out of fear of the annual evaluation, is afraid to work on extremely worthy research that doesn't fit a short time frame.

11. Point 7—management by numbers. Point 8—drive out fear—also applies because the faculty member, out of fear of the annual evaluation, is afraid to try new ideas in teaching.

13. Point 9—break down barriers. Shared power doesn't equal loss of power.

15. Point 11—eliminate numerical quotas.

17. Point 12—remove barriers that rob people of pride in workmanship.

SECTION 6.3

7. b. The defect rate is

$$\frac{54}{800{,}000} = .0000675 = .0000675 \times \frac{1{,}000{,}000}{1{,}000{,}000} = \frac{67.5}{1{,}000{,}000}.$$ This comes close to the 63.4 per million defect rate of a 4-sigma process.

Index

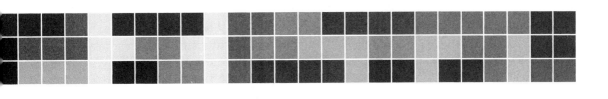